The Writing of Spirit

The Writing of Spirit

SOUL, SYSTEM, AND THE ROOTS OF LANGUAGE SCIENCE

Sarah M. Pourciau

FORDHAM UNIVERSITY PRESS *New York* 2017

THIS BOOK IS MADE POSSIBLE BY A COLLABORATIVE GRANT
FROM THE ANDREW W. MELLON FOUNDATION.

This publication is made possible in part by the Barr Ferree Foundation Fund for Publications, Department of Art and Archaeology, Princeton University.

Fordham University Press has no responsibility for the persistence or accuracy of URLs for external or third-party Internet websites referred to in this publication and does not guarantee that any content on such websites is, or will remain, accurate or appropriate.

Fordham University Press also publishes its books in a variety of electronic formats. Some content that appears in print may not be available in electronic books.

Visit us online at www.fordhampress.com.

Library of Congress Cataloging-in-Publication Data available online at http://catalog.loc.gov.

Printed in the United States of America

19 18 17   5 4 3 2 1

First edition

CONTENTS

CL   Roman Jakobson. *Child Language, Aphasia, and Phonological Universals.* Translated by Allan R. Keiler. The Hague: Mouton, 1968. Original German: Roman Jakobson. *Kindersprache, Aphasie, und Allgemeine Lautgesetze* (1941). In Roman Jakobson, *Selected Writings.* Vol. 1, *Phonological Studies*, 328–401. The Hague: Mouton, 1962.

CLG   Ferdinand de Saussure. *Cours de linguistique générale.* Édition critique. Edited by Rudolf Engler. 2 vols. Wiesbaden, Ger.: Otto Harrassowitz, 1967 and 1974.

LEG   Ferdinand de Saussure. "Légendes et récits d'Europe du Nord: De Sigfrid à Tristan." Selected and edited by Béatrice Turpin. In *Ferdinand de Saussure: Cahiers de l'Herne*, edited by Simon Bouquet, 351–429. Paris: Éditions de l'Herne, 2003.

NRP   Roman Jakobson. "Modern Russian Poetry: Velimir Khlebnikov." Translated by E. J. Brown. In *Major Soviet Writers: Essays in Criticism*, edited by E. J. Brown, 58–82. Oxford: Oxford University Press, 1973. Original Russian: Roman Jakobson. *Novejšaja russkaja poèzija: Nabrosok pervyi: Viktor Xlebnikov* [Newest Russian poetry: First attempt: Viktor Khlebnikov] (1921). In Roman Jakobson, *Selected Writings.* Vol. 5, *On Verse, Its Masters and Explorers*, 299–354. The Hague: Mouton, 1979.

OD   Richard Wagner. *Opera and Drama.* Translated by William Ashton Ellis. In *Richard Wagner's Prose Works.* 8 vols. 1895–99; repr., Omaha: University of Nebraska, 1995. Original German: Richard Wagner. *Oper und Drama.* In

Sämtliche Schriften und Dichtungen, edited by Richard
Sternfeld. Leipzig: Breitkopf & Härtel / C. F. W. Siegel,
1911–16.

PC   Hermann von Helmholtz. "On the Physiological Causes of
Harmony in Music." *Science and Culture: Popular and Phil-
osophical Essays.* Translated by A. J. Ellis. Edited by David
Cahan. Chicago: University of Chicago Press, 1995. Original
German: "Die physiologischen Ursachen der musikalischen
Harmonie" (1857). In *Populäre wissenschaftliche Vorträge.*
Vol. 1, 57–91. Braunschweig: Friedrich Vieweg, 1865.

SP   Roman Jakobson. "Zur Struktur des Phonems." 1939. In
Jakobson, *Selected Writings.* Vol. 1, *Phonological Studies,*
280–10. The Hague: Mouton, 1962.

VA   Roman Jakobson. "On the So-Called Vowel Alliteration
in Germanic Verse." 1963. In Jakobson, *Selected Writings.*
Vol. 5, *On Verse, Its Masters and Explorers,* 189–96. The
Hague: Mouton, 1962.

WGL   Ferdinand de Saussure. *Writings in General Linguistics.*
Translated by Carol Sanders and Matthew Pires. Oxford:
Oxford University Press, 2006. Original French: *Écrits de
linguistique générale,* edited by Simon Bouquet and Rudolf
Engler. Paris: Gallimard, 2002.

WW   Jean Starobinski. *Words upon Words: The Anagrams of
Ferdinand de Saussure.* Translated by Olivia Emmet. New
Haven, CT: Yale University Press, 1979. Original French:
*Les mots sous les mots: Les anagrammes de Ferdinand de
Saussure.* Paris: Éditions Gallimard, 1971.

The Writing of Spirit

Introduction

Systems, today, are generally presumed to have dispensed with their souls. Complex organic beings, according to contemporary biologists, do not originate and develop under the influence of an invisible "life force." World events, according to historians, do not progress toward the realization of goals dictated by Absolute Reason. Individual languages, according to linguists, do not evolve in conformity with the dictates of a shaping spirit or *Sprachgeist*. The premise of an intending Mind—for centuries considered indispensable to any scientific account of self-perpetuating structures such as bodies and dialects—has finally, over the course of the last few centuries, been expunged from the domain of science, and the story of this expulsion belongs, under the banner of a positivist disenchantment, to the heritage of our contemporary age. The consequences for what counts as knowledge are both radical and, by now, cliché: God dies (to science). The subject follows. The passage of time, which no longer moves bodies or eras or languages in the direction of a purpose fulfilled, loses all claim to intrinsic meaning, and thus also to the domain of enduring truth. "History" as an object of investigation ceases to play any necessary role in the quest for present-tense models of the universe. The sciences of nature and culture diverge, institutionally and methodologically, leaving humanists at least nominally responsible for a whole host of suspiciously soul-like phenomena (the German term, after all, is *Geisteswissenschaftler*, or "spirit scientist"), which their natural scientific counterparts have agreed to consider epistemically irrelevant.

The Writing of Spirit revises a crucial aspect of this familiar story about the rise of the "natural sciences" by reinterpreting the historical development of modern system theories within the paradigmatic realm of natural language. The process through which twentieth-century linguists first successfully purged their systems of soul, I argue here, has long been misunderstood precisely because it has never before been conceived primarily *as* a process, and thus also as an ongoing confrontation with its own nineteenth-century preconditions. Much exciting work has been done in recent years, and is currently being done today, on the relevance of a new "organicist" understanding of system for the radical transformation of German thought around 1800, in domains such as life science, literature, and philosophy.[1] Less attention has been paid, in this context, to the domain of language science, despite its exemplary status for the time period in question, and still less to the relationship between the spirit of early nineteenth-century systems and their spiritless twentieth-century successors.[2] *The Writing of Spirit* tackles this essentially unexplored terrain, in an effort to demonstrate that the *way* language spirit disappears from linguistics, when carefully attended to, yields a concept of system that depends nontrivially on the very history it emerges to exclude. The result is not a paradox but rather—in an important addendum to the Foucaultian tradition of archeological antinarrative—an epistemic shift whose very "break" consists in a new approach to historical continuity.

The terms "soul" and "system," linguistically speaking, cite the poles of an oft-rehearsed trajectory. Around 1800, so the standard account goes, German Romantics such as Friedrich Schlegel, Franz Bopp, and Jacob Grimm founded a new discipline called comparative linguistics, which sought to uncover the laws governing the temporal trajectory of language spirit (*Sprachgeist*). After 1900, so the tale continues, European structuralists such as Ferdinand de Saussure and Roman Jakobson shifted the emphasis from the diachronic to the synchronic, replaced the motivated word with the arbitrary sign, and in doing so precipitated a "linguistic turn" that revolutionized the humanistic disciplines. The precise content of this linguistic "revolution" has been interpreted in many different ways—the reception by midcentury French "structuralistes" in fields such as literary analysis and anthropology is only one particularly powerful example of such repurposing, which also helped shape the development of modern psychology, sociology, communication theory, and historiography[3]—but its status as paradigmatic exemplum has remained remarkably constant over the course of the intervening century. The notion of language as a disembodied and despiritualized network of relationally rather than referentially defined signs would appear to correspond emblematically, as few commentators have failed to observe, to the self-understanding of a modernity in which meaning-bestowing absolutes no longer reign.

Existing revisions to this narrative of a structuralist revolution, of which there have been many, tend to proceed by critiquing some aspect of its triumphalist frame. They seek to demonstrate that the structuralist concept of system is (still) haunted by unsystematizable energies, like its ancient Platonic predecessor (Jacques Derrida);[4] or that it imports into linguistics, belatedly and perhaps even somewhat clumsily, a natural scientific shift that began already with Galileo (Jean-Claude Milner);[5] or that it misguidedly ignores the insights of a much earlier, French rationalist notion of structure (Noam Chomsky).[6] *The Writing of Spirit* takes a different approach, since it emphatically does *not* seek to call into question the radical character of the structuralist model. On the contrary: it argues that the linguistic paradigm first envisioned by Saussure, and later realized by Jakobson, unfolds a new system theory of unprecedented conceptual power. This new theory, however, can appear as such—according to the somewhat counterintuitive thesis of the present study—only when interpreted as an *outgrowth* of the very spirit-drenched linguistics it purports to leave entirely behind.

The stakes of this claim require additional clarification. Recent work in the history of linguistics, and particularly in the development of linguistic structuralism, has emphasized Saussure's lifelong engagement with historicist methods and problems in order to complicate the one-sided picture of his thought encapsulated by the student-edited and posthumously published *Cours de linguistique générale*.[7] Much of this revisionist work takes its point of departure, as mine does, from the extraordinary 1996 discovery of Saussurean manuscripts long presumed lost;[8] such treatments challenge the traditional narrative in ways that deserve to be far more widely known, and that have deeply informed my own interpretation. Their questions, however, are not my questions, because the notion of a diachronically informed synchrony does not, in itself, constitute a philosophical puzzle. As Roman Jakobson himself demonstrates repeatedly from the 1920s onward, synchronic and diachronic perspectives can interact fruitfully at nearly every level of linguistic analysis, without in any way challenging conventional contemporary ideas about the relationship between systems and time, so long as the *foundations* of structure remain purely differential.[9] If, however, these foundations themselves could be shown to depend on a particular interpretation of temporal development—if the very definition of what counts as "synchronic" should turn out to consist in a reinterpretation of diachrony—then the whole story of the shift from nineteenth- to twentieth-century models, together with the system theory in which this story culminates, would need to be rethought. *The Writing of Spirit* proposes and performs such a rethinking.

The crux of the argument is a new account of the arbitrary. Saussure's foundational insight into the arbitrary nature of the linguistic sign has nearly always been conceived either as a version of the conventionalist thesis, which asserts that words bear no mimetic relationship to the things they name, or, more radically, as an argument for the priority of relations over relata, which threatens to upend the very notion of a word-thing relationship *per se.* In the first case, the claim is banal to the point of inanity, since few serious thinkers have historically attempted to uphold the "Cratylan" theory of mimetic signs.[10] In the second case, the claim is explosive but incoherent, since the concept of relationship would appear to depend, fundamentally and nonnegotiably, on the concept of "things that relate." *The Writing of Spirit* provides an original solution to this crucial conundrum regarding the real meaning of the structuralist arbitrariness thesis, and in the process offers an equally original interpretation of its nineteenth-century language-scientific roots: from the origins of a uniquely Germanic philology to the quest for an indigenous Germanic poetry to the discovery of an etymological method that embeds both projects in a theory of law-governed time, the spirit of nineteenth-century historical linguistics lives on, I argue, in the spirit-free linguistic structures first envisioned by Saussure.

This spirit lives on, however, only *in the process of evaporating away.* An autonomous system cannot exist, under my interpretation of the structuralist dilemma, without the companion concept of an immanent, unity-bestowing principle such as the one envisioned under the name of a historically developing *Sprachgeist.* Yet such a principle can be considered truly immanent, under my interpretation of the central structuralist innovation, only if its unity-bestowing effect can also be integrated without remainder into the very framework it simultaneously undergirds. Foundational for the Saussurean fantasy of a purely relational mode of meaning, which finds its fullest actualization in the emptied letter-forms of a Jakobsonian phonology, is thus less the nineteenth-century notion of language spirit itself than the nineteenth-century notion of language spirit in the oddly powerful shape of its own parodic undoing: the new structuralist notation seeks to literalize spirit not by subtracting it but by inscribing the ongoing *event* of its subtraction into the very heart of all writable structure.[11] The historical linguists' profoundly temporal, teleological theory of system formation does not simply disappear from language science with the advent of a primarily synchronic approach, nor does it haunt the margins of synchrony like the destabilizing residue of an incompletely accomplished "disenchantment." Rather, it evaporates, and, in evaporating, persists, because the process of evaporation itself acquires the status of systemic ground. The emptying of spirit takes place differentially, according to the formalizable rules of binary opposition, with the result that the structuralist system actually manages to incorporate its own presystemic origins. The relationship of finished forms to formal principle, of ordered pairs to ordering agency, of structure to soul, of writing to spirit, receives here, for the first time in the history of systems theories—according to the breathtakingly ambitious structuralist claim—a rigorously nonparadoxical, system-internal account.

The theory of this dynamic develops in tandem, as the trajectory of the present study will demonstrate at length, with a new theory about the systemic function of poetry, conceived as an eminently "structured" mode of language. Poetic form, according to the logic of language scientists from Jacob Grimm to Roman Jakobson, is essentially language form *re-formed*, which is to say "squared," and the nature of this most concentrated manifestation of order must therefore have much to reveal about the nature of linguistic order more generally. Where, however, the early nineteenth-century thinkers in question see the intensification of language spirit, whose animating energy motivates and sustains the twin "organisms" of idiom and verse, the early twentieth-century structuralists see a redoubling of the spiritual subtraction, whose emptiness subtends all linguistic forms. The concern, here, is still creativity—that traditionally nebulous province of traditional humanist exploration—but the power of poetic expression now derives from evacuation rather than inspiration. Saussure and Jakobson seek to formalize the force of the poetic *Logos* by writing down the rule of its rule-giving power, and they consequently refuse to revel in the unformalizable paradoxes of an ostensibly ecstatic speech.

Both Saussure's reinterpretation of Indo-European verse origins, in the context of his famously enigmatic "anagram studies," and Jakobson's rethinking of a Western poetic future, in the context of his early work on the Russian avant-garde, participate in this project of formalization. Both must therefore be understood, I argue, not as eccentric deviations from their more "mainstream" linguistic endeavors but as profound thought experiments—conducted in polemical dialogue with the nineteenth-century theory of enspirited Germanic letters, or runes, which runs from Jacob Grimm to Richard Wagner—regarding the conditions of a maximally despiritualized language. And both must also be kept wholly separate, with respect to their system-theoretical status, from the many later attempts to deploy the differential principles of phonology *non*linguistically, under the influence of thinkers such as Claude Lévi-Strauss and A. J. Greimas, for the purposes of literary, sociological, or anthropological analysis. Such extralinguistic deployments do not at all succeed at, or, for the most part, even work toward, a truly system-internal account of systemic origins. They thus have a tendency to fall prey to precisely the kind of structural incoherence that so many poststructuralist thinkers later so brilliantly critique. These same poststructuralist critiques, however, overreach when they seek to transpose the blind spots of a 1950s and 1960s French reception history back onto the model of an earlier, purely linguistic poetics. Indeed, from a system-theoretical perspective, Saussure's and Jakobson's poetic visions will turn out to have less in common with the insights of French literary theory than with the revelations of an early twentieth-century "Formalist" mathematics, whose pioneers laid the foundations for the new fields of cybernetics, higher-order logic, and axiomatic set theory: the linguistic hypothesis of a self-reflexive poetic *reemptying* at the outermost limit of language structure closely parallels the modern mathematical emphasis on a self-reflexive principle of functional *recursion* as a means for defining computable sets. The detailed elaboration of such connections lies, of course, beyond the scope of the present study, which seeks only to uncover the particular character of the passage from *Sprachgeist* to modern language science. But the long invisible homology between twentieth-century theories of letter and number, poetry and sets, rhyme and ordered pairs, forms the wider historical backdrop against which the narrative of this linguistic transformation unfolds.

THE WORLD SOUL, EMERGING

The peculiar dynamism of spirit evanescing into structure is the true object of the following pages. In order, however, for an account of this process to make any sense, it must be accompanied by an account of the entity that evanesces. In other words, only an interpretation that takes seriously the specific, epistemic contribution of the specific, nineteenth-century concept of *Sprachgeist* can hope to do justice to the true character of the structuralist subtraction. *The Writing of Spirit* tackles this foundational task by disclosing, in chapter 1, the implicit theory of enspirited systems that paves the way for the establishment of a rule-governed and methodologically unified "science" of Indo-European language-in-time. I seek there to account for the philosophical work performed by the (profoundly nonmetaphorical) figure of language spirit—in the context of a broader, epochal concern for the origins of self-sustaining order—without, on the one hand, subsuming the newly formalized discipline of *Sprachwissenschaft* to a centuries-old tradition of natural scientific notation,[12] or, on the other, identifying its emergence with the origins of a primarily humanist historicism.[13] The distinction of nineteenth-century language science, I argue, lies rather in the way that the peculiarly nineteenth-century nexus of language spirit and system *transcends* our contemporary understanding of the nature/culture divide, yielding a set of tools simultaneously scientific and historical.

My conclusions in this regard thus differ substantially from those of the more established narratives, and particularly from the now-classical account of Michel Foucault. Foucault's 1966 "archeology of the human sciences," *The Order of Things* (*Les mots et les choses*), famously equates the beginnings of what he calls the modern "sciences of man"—exemplified by the early nineteenth-century studies of human life, labor, and language—with the end of a model of knowledge he calls "representation." Foucault means by this, in part, that the possibility of analyzing the world, conceived as a capacity for picturing or mirroring its structure, becomes problematic for late eighteenth-century European thinkers in a way it had not been before. Taking Immanuel Kant's paradigmatic expression of this new problem as a pivot point—Kant's "Copernican revolution," after all, famously transforms the conditions of a scientific world picture into the primary *task* of a critical philosophy—Foucault argues that the historical emergence of the quest for representational foundations

implies the historical emergence of the human subject: Kantian philosophy, and with it, the modern episteme, "discovers" the human as
the site of the sought-after ground of representation, and thus also as
a potential object of scientific investigation.[14]

On the basis of this analysis of an epistemic shift around 1800,
Foucault then goes on to suggest, in the final pages of his encyclope-
dic work, that twentieth-century structuralist theories of language
might already spell the end of "modern man" and his disciplinary
doubles, the human sciences.[15] The structuralist displacement of the
human subject, together with related categories such as "agency"
and "will" from the gravitational center of knowledge systems, performs a perspectival modification he implicitly associates with a second, and markedly un-Kantian, "Copernican revolution." *The Order
of Things,* in other words, presents one of the earliest, most influential, and most elegant versions of the now-familiar "structuralist
breakthrough" story, whereby the unprecedented twentieth-century
discovery of agentless structure—of autonomously emerging, unin-
tentionally rational systems of meaning—relegates the nineteenth century's various human-focused, historical approaches to obsolescence.

This extraordinarily powerful, and in many ways marvelously
insightful, antinarrative of two ruptures fails to do justice, in my
view, to the complexity of the relationship between nineteenth- and
twentieth-century theories of structure, precisely because it purposefully, indeed, *programmatically* fails to do justice to the complexity
of the relationship between nineteenth-century empirical science and
philosophy. The sciences of life, language, and labor give expression
not to an unconsciously refracted version of Kant's transcendental
subject but to a consciously reflected polemic *against* the great boundary drawer's decision to sever mind from world. And the result is a
radical rethinking, not a renunciation, of traditional representational
paradigms. Foucault is certainly correct, therefore, to read the shift
he identifies as post-Kantian. I will argue, however—in line with several important recent works in the history of biology, which transfer
the title of generational "voice" from Kant to his younger contemporary, Friedrich Schelling—that the shift is post-Kantian in the sense
of being *anti*-Kant, and that the scientists in question seek to transcend rather than submit to a (de)limiting equation of rationality with
humanity. The most significant linguists, comparative anatomists,
and economists of the period remain crucially committed, as Kant
does not, to the objective reality of the orders they analyze, and thus

also to the reality of analyzable structure per se, as a feature of the cosmos itself. In doing so, they implicitly align themselves with a fundamentally *representational* theory of knowledge that has its deepest roots in an ancient, Platonic account of concept formation. The task of science is still, here, to model or mirror the inherently rational construction of the universe. The specifically nineteenth-century challenge is thereby to do so without recourse to the Platonic hypothesis of a demiurge (or the Christian one of a creator God). In the wake of Kant but also of Newton, Bacon, Descartes, and Leibniz, with their collective commitment to the ideal of a single, empirically verifiable science, the principle of structure can no longer be assumed to infuse or inspire an otherwise structureless "first matter" from somewhere beyond the realm of nature. How, then, to understand the objective existence of worldly form? And, in particular, how to account for the genesis of self-enclosed, self-regulating, self-producing *systems* such as those of life, money, and language?

The nineteenth-century solution to the challenge of world-immanent, agentless structure—a solution that receives its earliest and most explicit philosophical treatment in the context of Friedrich Schelling's "nature philosophy" but whose reach extends across multiple disciplines, from the beginning of the century to the end—is a theory of History writ *cosmically* large. Where order is presumed to emerge from within nature rather than without, its emergence necessarily takes the shape of a development rather than a punctual beginning. Its truth thus becomes the truth of the history of nature itself, which is to say, of the *telos* of its definitively transhuman time. In the realm of Schellingian nature philosophy, this teleological principle goes by the title of "world soul," and the analytic method required to reveal its trajectory is Schelling's polemical, nature-historical perversion of Kant's "deductive" technique for tracing worldly phenomena back to their roots ("deduction," from Lat. *deducere*, "to lead down"). In the realm of life science, the motor of transformation is called the formative principle, or *Bildungsprinzip*, and the analytic method that allows the scientist to deduce its direction is embryonic dissection. In the realm of language science, the shaping force is the language spirit, or *Sprachgeist*—here reconceived as a microcosmic particularization of Schelling's pantheistic world soul—and the analytic method that gives access to its tendencies is etymology.[16]

All three methods already existed, of course, at least in name, prior to the early nineteenth-century problematization of their objects.

None of them, however, would have appeared familiar, in their respective nineteenth-century forms, to earlier practitioners. Franz Bopp's *On the Conjugation System of the Sanskrit Language in Comparison with that of the Greek, Latin, Persian, and Germanic Languages* (*Über das Conjugationssystem der Sanskritsprache in Vergleichung mit jenem der griechischen, lateinischen, persischen und germanischen Sprache*, 1816), for instance, works to implement Friedrich Schlegel's suggestion for a "comparative anatomy of language" by treating grammatical systems as developing bodies: it uses etymology not to learn the original meanings of individual words but to uncover the underlying mechanism of language-structural "growth." And the second edition of Jacob Grimm's *Germanic Grammar* (*Deutsche Grammatik*), published in 1822, builds on Bopp's approach with a methodological innovation that makes the study of language synonymous with the study of its rule-governed transformations of over time. Grimm's famous formulation of his "sound laws," in a table that quite literally displays the periodic "standing together" (*systemata*) of sequential letter systems, provides the first actual criterion for distinguishing true etymologies from false ones but also language systems from their others.

THE WORLD SOUL, EVAPORATING

The Writing of Spirit traces the consequences of this new thinking of language structure, together with its etymological methodology, through the nineteenth century and into the first half of the twentieth, in the twin language-scientific spheres of linguistics and poetics. Part 1 explores the concept of language system as it develops, against the backdrop of Friedrich Schelling's nature philosophy, in the work of nineteenth-century German linguists such as Jacob Grimm, Franz Bopp, and their successors (chapter 1). It then goes on to investigate the transformations to which Saussure, who was educated in comparative linguistics at the German center of Leipzig, subjects this nature-philosophical model in his unpublished notes (chapter 2). I argue that the Saussurean vision of a purely differential notation takes aim not, as has traditionally been assumed, at some naive, Cratylan investment in the motivated character of the sign but rather at the far more sophisticated model of nineteenth-century language spirit: Saussure, like his predecessors, insists on the scientific necessity of a language-internal dynamism; he accepts this fundamentally nature-philosophical premise, however, only in order to *deflate* the twin notions of interiority and dynamism by uncoupling them from the profundity of any conceivable animating force.

Part 1 then concludes with an extended treatment of the nineteenth-century emergence, and polemical Saussurean reinterpretation, of a specifically Germanic poetics of enspirited letters. The ancient compositional technique of alliteration (*Stabreim*), which gives structure to some of the oldest known poetic works of the Germanic and Scandinavian peoples, is interpreted by nineteenth-century German language scientists as a particularly powerful tool for explicating the structuring energy of the German language. Primal poetry, philologically reconstructed, thus comes to operate as a kind of projection screen for language-scientific fantasies about the writing (down) of language spirit. Saussure's response, I argue, takes the form of a battle with the Germans over the origins of verse, the true character of alliteration, and the etymology of the German word for "letter" (*Stab*). In the context of his famously enigmatic "anagram studies," and in line with his linguistic vision of a fully disenspirited notation, Saussure imagines the roots of Indo-European rhythm as a rule-governed procedure for *emptying* poetry of all mental content, via the counting, rhyming, and cancelling of meaningless letter-signs (chapter 3).

Part 2 moves beyond the boundaries of the comparative linguistic tradition in order to explore the extraordinarily influential historical realization of the two paradigms so daringly imagined by the thinkers in part 1. I begin with Richard Wagner's operatic dramatization of the *Stabreim*, which was inspired by his intense engagement with the language theories of Jacob Grimm; Wagner's poetic project in the *Ring* cycle can be understood, I argue, as an attempt to harness the rhythms of ancient alliterative verse to an all-encompassing, neo-Pythagorean model of cosmic-acoustic accord, such that the meter of his own mid-nineteenth-century alliterations—when united with the harmonic modulations of his music—turn out to merge with the "meter" of the world spirit progressing through time (chapter 4). I then go on to unfold the startling consequences of this Wagnerian harmony theory both for the development of avant-garde poetic principles (particularly in France and Russia), and for the concurrent, late nineteenth-century emergence of a new discipline called "psychophysiology" (a forerunner of modern-day experimental psychology). I argue that the widely received theories of Gustav Fechner, Wilhelm Wundt, and their followers strive to translate the nature-philosophical energy of Wagner's sounding spirit into a natural-scientific dialect of countable psychic vibrations and measurable psychic "tones" (chapter 5).

Part 2 concludes, finally, with an account of the rethinking to which these avant-garde and psychophysiological translations are subjected, in turn, by the phonology (and phonological poetics) of the Russian structuralist linguist Roman Jakobson.[17] The focus rests on a new reading of what Jakobson calls the "linguistic zero," which he defines as the featureless half of a binary opposition between presence and (a kind of) absence.[18] I argue here that this absolutely crucial and much-analyzed aspect of phonological theory can be shown to develop in rigorous, polemical conversation with the systems theories of nineteenth-century language spirit. The zero operates, in essence, as a placeholder for the subtracted soul of structure. It marks the boundary, and thus also the ground, of systemic unity, without ever succumbing to the mysterious, system-transcending dynamic of a telos that gives shape from outside. In the process, it manages to fulfill the seemingly impossible, because seemingly contradictory, condition of infusing by subtracting, grounding by evaporating, unifying by differing, inspiring by deflating. The zero sign can be written down, remainderlessly if somewhat idiosyncratically, and therein lies the source of its system-theoretical priority over the nineteenth-century spirits it reenvisions.

The zero sign cannot, however, be comprehended in isolation from these same predecessor spirits, whose continuous evacuation turns out to constitute the sole "substance" of its all-important systemic function—Jakobson explores this curious dynamic most thoroughly in his work on child language and avant-garde poetry, which is where I end—and therein lies the source, I would argue, of the zero's deep relevance for the contemporary conundrum of historical meaning. Once entelechy disappears from the knowable universe, and the spirit of purpose is divorced from time, the question of what history actually *contributes* to science becomes more pressing than ever before. What epistemic status, if any, should scholars accord to the development of knowledge, and to the various temporal processes by which new theories or concepts emerge? What kind of cognition, if any, is actually generated by studying the relationship of contemporary truths to their predecessors? Once order exists, need we care how it came about, or can we analyze it without understanding its origins? These are not questions, odd as it sounds, to which our age has answers, and *The*

Writing of Spirit is no exception in this regard. I would venture to suggest, however, that the history of modern language science, as reconstructed in the following pages, offers a glimpse of one potentially generalizable instance—the analogies with early twentieth-century mathematics, in particular, speak in favor of a broader paradigm—for which the analysis of a "panchronic" system proves conceptually inextricable from its pasts.[19]

PART I

"The Eternal Etymology":
From Sprachgeist to Ferdinand de Saussure

The whole labor of the linguist who wants to account, methodically, for the object he studies comes down to the extremely difficult and delicate task of defining units.[1]

The search for a linguistic object—the question of what language actually *is*—runs like a red thread through the fabric of Ferdinand de Saussure's otherwise disjointed notes and would surely have dominated the book he so famously never wrote. A science of language like the one he hoped to found requires, after all, an object accessible to scientific investigation. The task of locating such an object, however, turns out to be significantly less tautological than it sounds: to the question "What does the linguist study?" one cannot simply answer "language," as, according to Saussure, the botanist can answer "plants," or the geologist "rocks," for no fixed boundaries separate languages from each other geographically or historically. Transformations occur as a continuous process of minute shifts, making it impossible to say definitively where one language leaves off and the other begins, or to distinguish, except by convention, between dialects and autonomous tongues. Saussure repeatedly declares the unraveling of such language-based confusions the primary duty of every practicing linguist, taking care to point out that it is a duty he believes his (German) predecessors to have honored more in the breach than the observance.

> Our point of view is in effect that the knowledge of a phenomenon
> or of an operation of the mind [*l'esprit*] presupposes the prior defini-
> tion of some kind of term [. . .] which has at some point a founda-
> tion [*une base*] of some kind. This foundation need not necessarily
> be absolute, but it must be expressly chosen as an irreducible foun-
> dation for us, and as central to the entire system. To imagine that it
> is possible in linguistics to manage without this sound mathematical
> logic, on the pretext that language [*langue*] is a concrete thing which
> "becomes," and not an abstract thing which "is," is in my opinion
> a profound error, inspired originally by the innate tendencies of the
> Germanic spirit [*l'esprit germanique*]. (WGL, 17–18/34)

Several crucial elements come to light in this passage—taken from the
collection of manuscripts first discovered in 1996, in the conservatory of the Saussure family's Geneva townhouse[2]—which are absent,
or nearly so, from the better-known and more judicious formulations compiled by Saussure's students in the *Cours de linguistique
générale* (*Course in General Linguistics*, 1916). The link between a
well-defined linguistic unit and the figure of a foundation or ground,
the problem of false concreteness in relation to a theory of historical
unfolding, the rhetoric of depth ("profound"), inspiration ("inspired"),
origin ("originally"), teleology ("tendencies"), interiority ("innate"),
and spirit (*"l'esprit germanique"*) that accompanies Saussure's diag-
nosis of German error—all will turn out to play a substantive role in
a turn-of-the-century linguistic project that demands to be read, so
the following pages will argue, as a battle with the Germans over the
nature of the linguistic endeavor.

The battle begins and ends with the question of the foundational
unit, what Saussure here calls *terme*, and elsewhere, with a dizzying inconstancy that reflects the dilemma he is struggling to name,
seme, *signe*, *unité*, or even, less frequently, *mot*. A precise definition of
the basic building blocks, he contends, would give the linguist somewhere to stand and linguistics somewhere to start. Failure to recognize that such a definition is necessary, on the other hand, undermines
all attempts to discover a common linguistic ground. German linguists of the nineteenth century seek to "explain" the existence of
attested form *a* by linking it etymologically to an older form *b*, such
that both *a* and *b* can be considered manifestations of a single word.
The old Germanic *gagani*, for instance, becomes the modern German
Gegend (region, neighborhood), the medieval Latin *cuppa*, the modern German *Kopf* (head). What remains entirely unclear, however,
in the absence of any explicit definition of the concept "word," is

the status of this transformation. Where, exactly, is the identity that could justify the insistence on the *sameness* of these forms? Each form, after all, is itself a unit, capable of functioning in its own historical moment independently of earlier and later manifestations. The phonetic "traces" that make language historical for the linguist remain largely imperceptible to the average speaker, and play little or no role in the present-tense functioning of language; in a very real sense, they "exist" only for the etymologist, who thus turns out to stand in a dubiously occult relation to his or her chosen object of study.

The implications of the question, as Saussure makes clear, go far beyond a critique of etymological praxis:

> Take for example the series of vocal sounds *alka,* which after a while, passing from mouth to mouth, has become *ōk* [. . .]. Where, at bottom [*au fond*] , is the LINK between *alka* and *ōk?* If we go down this path, *and it is inescapably necessary that we do so,* we will soon find that we must demand an answer to the question where is the LINK between *alka* and *alka* itself, and in this moment we will realize that there nowhere exists as a primordial fact a *thing* that is *alka* (or anything else). (*WGL,* 138–39/200–201)

Etymological relation, no matter how stringently demonstrated, will never *in principle* be able to explain the phenomenon of historical identity because the problem explodes the parameters of existing etymological method. As the condition of possibility not only for language history but also for language *per se,* the continuity of words across time and space poses itself above all as a question for the present tense. To the linguist looking for ground, the dilemma *alka-ōk* is merely a special case of the more general dilemma *alka-alka, ōk-ōk,* for, unlike the flora and fauna of the nineteenth-century organicist analogies, *alka* and *ōk* are not things with bodily contours. Since no two speakers articulate exactly alike, and since even the same speaker articulates differently on different occasions, their respective modes of material existence are necessarily plural and varied. An ostensibly "single" word will always be "represented" by acoustically divergent strings of sounds, pronounced in widely varying moments and places. Meaning, too, varies substantially from one instance to another. And yet, within functional language systems, ordinary speakers will have no trouble recognizing these different utterances as manifestations of a self-identical unit of language; they will align *alka* with *alka, ōk* with *ōk,* and ask no questions about the tie that binds.

This mysterious stability, perceived across a potentially infinite variety of phonetic forms and linguistic functions, confronts the would-be purifier of linguistic terminology with a daunting definitional challenge. While the units must in some sense "exist" in order for communication to occur, articulating their boundaries—the true task of a linguistic notation—proves next to impossible:

> The mechanism of language—always taken AT A GIVEN MOMENT, which is the only way to study the mechanism—will one day, we are convinced, be reduced to relatively simple formulas. For the moment one cannot even dream of establishing these formulas: if we try to fix some ideas by sketching out the main traits of what we imagine under the name of *semiology*, i.e., a system of signs as it exists in the mind of speaking subjects [*l'esprit des sujets parlants*], totally independent of how it came into existence [*de ce qui l'a preparé*], it is certain that we will still, in spite of ourselves, be obliged to unceasingly oppose this semiology to the ever-present etymology [*la sempiternelle étymologie*]; it is certain that this distinction, when inquired into more closely, is so delicate that it draws all attention exclusively to itself, very powerfully, and in countless foreseen and unforeseen cases may well be treated as a subtle distinction; it is certain, consequently, that the moment is not yet near when one could operate with full tranquility outside of all etymology [*hors de toute étymologie*]. (*WGL*, 25/43)

The future tense separating the "one day" of simple formulas from the fruitless fantasies of the present marks the passage above as a manifestation of the very dream it simultaneously pronounces futile. Saussure envisions here, as the precondition of a science still to come, a language susceptible to remainderless formalization. He names the science "semiology," and imagines under this name a sign system so characteristically "Saussurean" that its definition could double as a formulation of the famous arbitrarity thesis. He insists, however, that such a definition must indefinitely remain purely provisional, since the more definitive contours of a truly scientific notation would require the currently unthinkable exclusion of a *sempiternelle étymologie*. Saussure's attempt to accomplish what he himself declares in this passage to be impracticable, by exiling a profoundly German-spirited etymology from the ostensibly neutral territorium of a Swiss course in "general linguistics," transforms the field of language science. The question, however, of how this transformation actually *works*— and of what it *does*, in the process, to the ineradicable dimension of history it acknowledges while refusing to countenance—remains oddly open even today, one hundred years after Saussure's death in 1913. The present study's approach to an answer will require a more detailed investigation into the precise nature of the problem Saussure considers his German predecessors to have posed, which will in turn enable new insight into the solution he dreams about offering.

CHAPTER I

Language Ensouled

GRAMMATICAL LIFE

The nineteenth-century founders of a Germanic "science of language" (*Sprachwissenschaft*) claimed to revolutionize language study by turning their attention from timeless grammatical norms to empirical language change.[1] Pointing to an eighteenth-century tendency to conceptualize language primarily as the instrument of a universal *Logos*—and so as the expression par excellence of a quintessentially human capacity for reason—path forgers such as Rasmus Rask, Franz Bopp, and Jacob Grimm accused their predecessors of neglecting the "real" linguistic object. The rationalist prioritization of *clear and distinct speech*,[2] which required the means of expression to disappear as far as possible into the message, prevented rationalist thinkers, so the claim goes, from treating language in its irreducible, historical particularity. The highest form of expression for philosophers from Descartes to Kant was the one least likely to thrust itself between speaker and listener as an object worthy of independent consideration. Against this classical understanding of language-as-medium, with its countless attempts to exchange the messy contingency of natural languages for the rational purity of an artificial or Adamic one, nineteenth-century linguists insisted instead on the scientific, philosophical, and historical significance of language conceived as *thing*. Language in the de facto multiplicity of its empirical manifestations—which is to say language *in time*—offered for them the only possible path to the timeless essence of a philosophical "language-as-such." The linguistic potential that unites and defines mankind, they argued, can manifest itself only in the form of particular, mutually incomprehensible idioms, since, regardless of the universal grammar that may or may not underlie all language use, the ordinary speaker experiences only confusion when confronted by a foreign tongue.

Historians of linguistics often locate the first indications of this perspectival shift in the sudden Sanskrit fever that swept through educated circles, both linguistic and lay, following the publication of *The Sanscrit Language* (1786) and "The Third Anniversary Discourse" (1788) by the British philologist Sir William Jones.[3] It is in the "Third Discourse" that Jones makes his crucial and much-cited reference to the possibility of a "common source": "The Sanskrit language, whatever be its antiquity, is of a wonderful structure; more perfect than the Greek, more copious than the Latin, and more exquisitely refined than either, yet bearing to both of them a stronger affinity, both in the roots of verbs and the forms of grammar, than could possibly have been produced by accident; so strong indeed, that no philologer could examine them all three, without believing them to have sprung from some common source, which, perhaps, no longer exists."[4] Jones's enthusiastic endorsement of Sanskrit as a bridge to the (specifically European) *Ur*-language made a particularly strong impression on Friedrich Schlegel, who began studying Sanskrit in 1803 and published his *On the Language and Wisdom of the Indians* (*Über die Sprache und Weisheit der Indier*) in 1808.[5] In it he calls for a new form of language study to do justice to Jones's proposal of common ancestry, a "comparative grammar, which will give us quite new information about the genealogy of languages in a similar way as comparative anatomy has illuminated the higher natural history."[6] The first to explicitly heed this call, and take seriously the analogy with comparative anatomy, was his friend and later adversary Franz Bopp, whose *On the Conjugation System of the Sanskrit Language* (1816) won general recognition—with the help of a laudatory introduction written by his better-known teacher, the comparative anatomist and nature philosopher, Karl Joseph Windischmann—as the earliest systematic work of the new comparative, genealogical approach.[7] The idea of a "familial" relationship tying Sanskrit to its European "descendants," and its descendants, in turn, to each other, received in Bopp's thorough analysis of grammatical systems—his "comparative anatomy of language"[8]—its first conclusive demonstration.

Though it is far from clear that the modern discipline of historical linguistics actually originates ex nihilo with Jones and Schlegel,[9] it is certainly true that their respective formulations of the Sanskrit thesis left an indelible mark on the self-understanding of the fledgling field. Due to Schlegel's citation of Jones, and Bopp's citation of Schlegel, both the "common source" image and the comparative anatomy analogy have belonged from the beginning to the internal narrative of the new analytic approach, whose inarguable advances in method and technique, under the guidance of figures such as Franz Bopp, Rasmus Rask, and Jacob Grimm, cannot therefore be disentangled from the particular discursive context of their emergence. The founders of a nineteenth-century language science saw in the notion of shared genetic origins the justification for their revolutionary transformation of the linguistic task, and in the methods of comparative anatomy the model for its execution. Where previously—so the narrative—there had been only Language, a uniquely and universally human *potential* for the articulation of speech and analysis of thought, there was now an *actual* plurality of language families or species, all following their own internal logic of growth and degeneration quite independently of the rational categories and individual whims they had once been thought merely to "represent": "Languages are to be considered organic natural bodies that form themselves [*sich bilden*] according to particular laws, develop bearing an inner life principle within themselves and then gradually die out because they no longer comprehend themselves [*sich selber nicht mehr begreifend*], casting off or mutilating or misusing originally meaningful parts [*Glieder*] or forms, which have gradually become more of an extrinsic mass."[10] The linguist's new mission was thus to investigate these internal logics of development, to uncover the "particular laws" governing the dynamism of language birth and death, by analyzing or "dissecting"—the German *zergliedern* refers to both activities—the bodily structures of particular, developing tongues. Language, in the minds of Bopp and his successors, had come to life.

LIFE SCIENCE

It would be hard to overestimate the significance of this organicist metaphor for the fledgling science of language, and indeed, even the classification of "metaphor" runs the risk of misrepresenting its role. The first historical linguists saw in the notion of language-as-organism neither a rhetorical flourish nor a heuristic device but rather a structural unity that preceded, both logically and chronologically, any conceivable opposition between figurative and literal meanings. And in this they were not alone. By the end of the eighteenth century, a radical transformation had occurred in the notion of organic being, which exerted tremendous influence well beyond its original "comparative anatomical" context—a transformation that entailed a rethinking of organized structure more generally (the modern "organism" comes from the medieval Latin *organizare*, "to organize") and of *temporally* organized structures in particular. Friedrich Schelling's nature-philosophical theory of an unfolding world soul, Friedrich von Hardenberg's Schellingian ruminations regarding the animatory principle of the cosmos, Johann Wolfgang von Goethe's research into the possibility of a morphological world history, Friedrich Schlegel and August Wilhelm Schlegel's commitment to the project of an all-encompassing *Universalpoesie*, Georg Wilhelm Friedrich Hegel's dialectics of a developing world spirit—all were attempts to render productive the new, expanded concept of the organism for a more general account of knowledge and nature. The true significance of the "organicist" approach to language, which endeavored to do the same for the paradigmatic domain of human expression, must therefore be sought against this broader backdrop of a universalized "life science."[11]

The new concept of the organism arose in the course of an eighteenth-century debate about the character of organic causality, and, consequently, about the character of organicist explanation.[12] The two competing perspectives, often referred to as "mechanism" and "teleologism," provided fundamentally different answers to the question of what *kinds* of reasons the life scientist should seek and accept, with the mechanists insisting on the priority of the proximate cause (*Why* does the embryo develop into an animal? *Because* of mechanical interactions set in motion at the moment of conception), while the teleologists maintained the necessity of final ends (*Why* does the embryo develop into an animal? *Because* it contains within itself an inherent tendency to progress toward the "goal" of adulthood). Life grows, from both the mechanist and the teleological perspectives, insofar as it manages to "unfold" a potential that would otherwise have remained "rolled up" in its origins. The question, however, is how the natural scientist should *interpret* this act of unfolding, and with it the meaning of related verbs such as "develop" (from Old

French *des* = off, away + *voleper* = to wrap or fold), "explicate" (from Latin *ex* = out + *plicare* = to fold), "evolve" (from Latin *ex* = out + *volvere* = to roll), "entwickeln" (*ent* = un + *wickeln* = to wrap), and "entfalten" (*ent* = un + *falten* = to fold)?[13] Should growth be understood as a domino effect of mechanical causes and consequences, a chain reaction that merely "activates" what in essence already exists, albeit in a dormant state? Or should it rather be envisioned as a trajectory that ends somewhere entirely *other* than it begins—and in doing so actively *transforms* this beginning into a more encompassing version of itself? Can all change, in other words, be reduced to an analytic reshuffling of matter, and so to a mere permutation of a preexisting totality, or must some change in fact be conceived as a transmutation that *generates* Being?

The traditional, scholastic approach had accepted the priority of mechanical causality in situations of accidental, inorganic flux—like rainstorms or earthquakes—while insisting on the necessity of a teleological purpose to explain the directional arc of growth.[14] With the rise of modern rationalism, however, this bifurcatory account of causation encountered resistance. In the wake of Galileo, Bacon, Descartes, Newton, and Leibniz, with their collective commitment to the dream of a unified theory of nature—and their collective contribution toward its eventual realization in the form of a mechanistic physics of movement[15]—the idea of separate causal principles for the living and the inanimate began to signify to scientists an intolerable deficiency in the foundations of their knowledge. Such a schism in what was permitted to count as "why" entailed for the inheritors of the one-theory mantle an unacceptable schism in what was permitted to count as science, and they therefore demanded a single notion of causation comprehensive enough to explain *all* of natural change. The first widely influential attempts at a purely mechanical account of organic reproduction—the natural phenomenon generally acknowledged to represent the greatest obstacle for a unified mechanistic theory—appeared in this context.[16] Prominent seventeenth-century physiologists such as Marcello Malpighi, and eighteenth-century ones such as Albrecht von Haller, insisted that the transformation from embryo to animal takes place as a deterministic succession of physical effects and is therefore, in actuality, no transformation at all. The structure that the organism appears to spontaneously acquire, in the process of progressing from formless mass to living form, is in fact already present *in toto*, they argued, from the very beginning of its ostensible development. Fertilization occurs when this structural germ (*Keim*)—variously imagined to reside either in the female egg of the mother or in the male seed of the father—collides with the unstructured reproductive material of the opposite sex, which itself does nothing more than *catalyze* the chain of mechanical reactions required for the magnification of the embryo.

Such theories found many supporters among the proponents of an exclusively mechanist science. Other, equally influential voices, however, vigorously disputed the interpretations of this "mechanistic medicine,"[17] and by the end of the eighteenth century, due largely to the work of the physiologist Johann Friedrich Blumenbach, the teleologists had (re)gained the upper hand in the debate. Blumenbach's treatise *On the Formative Drive* (*Über den Bildungstrieb*) marshaled a great deal of persuasive observational evidence for the insufficiency of mechanist accounts (he cites, among other examples, the ability to form hybrid offspring from different species, which would be impossible if reproduction involved nothing more than the mechanical magnification of a preexisting kernel), and a great deal more in favor of his own pronouncedly teleological alternative (here he focuses on the organic capacity to heal, which implies to him the ongoing presence of a purposeful creative power). The resulting thesis is formulated by him as follows: "Everything that I have learned through observation and reflection [. . .] leads me in the end to the conviction: that no preformed germs preexist, but rather, that in the previously raw, unformed procreative juice [*Zeugungssaft*] of organized bodies, after it has come to maturity and reached its proper place, a particular active drive awakens [*rege wird*], which causes the organism to take on a specific shape, then to preserve that shape over the whole course of its life, and, if the shape is mutilated, for instance, to restore it."[18] Blumenbach returns, in this influential passage, to the notion of a specifically organic mode of causality, which would set in motion all the various kinds of changes ordinarily classified as growth. Not only the originary generation of the fetal form but also the continuous development of that form toward a state of maturity, its preservation against the pressure of external influences, and, finally, its *regeneration* in the event of illness or injury all occur, according to him, as the consequence of a single form-generating principle he proposes to call the "formative drive," or *Bildungstrieb*. This drive-to-form accompanies the organism throughout the temporal trajectory of its existence, and can therefore be understood as an energy that animates or in*spires* an otherwise formless matter, along the lines so often figured by a divinity's life-bestowing breath. Blumenbach differs from his vitalist predecessors both ancient and modern, however, in conceiving of the life force as an inherent feature of organic matter, rather than as an immaterial principle infused into living beings from without. New life forms emerge, for him, whenever the formative drive *spontaneously* begins to move (*wird rege*) *within* the previously raw, unformed procreative juices of the parent organism. While the ultimate source of this teleological tendency must therefore remain mysterious, the activity itself can be unproblematically identified with its observable organizational effects, which, much like the observable effects of the equally mysterious force of gravity, can in turn be rigorously, even materialistically, studied.[19] From the perspective of the natural scientist, the principle of life just *is* this principle of *purposeful* organization ("organization," from Greek *organon* = tool, implement), whose objectively accessible, material features distinguish it definitively from nonlife.

Blumenbach's argument for the irreducibility of the teleological model persuaded no less significant a mechanist thinker than Immanuel Kant, whose legendary commitment to the exclusive priority of mechanist explanation—and specifically to the mechanical causality of Newtonian physics—required him to reject the notion of a teleological "force" at work in nature, even as the power of Blumenbach's evidence forced him to concede its necessity for the science of life.[20] Published nine years after Blumenbach's treatise, Kant's *Critique of the Power of Judgment* first extensively unfolded the case for teleology in his "Analytic of the Teleological Power of Judgment" before finally presenting, in the "Dialectic of the Teleological Power of Judgment," his original solution to the conundrum of a specifically organic causality. The "Analytic," which argues along explicitly Blumenbachian lines, provided the era with one of its most influential characterizations of organic uniqueness:

> In a watch one part is the instrument for the motion of another, but one wheel is not the efficient cause for the production of the other: one part is certainly present for the sake of the other but not because of it. Hence the producing cause of the watch and its form is not contained in nature (of this matter), but outside of it, in a being that can act in accordance with an idea of a whole that is possible through its causality. Thus one wheel in the watch does not produce the other, and even less does one watch produce another, using for that purpose other matter (organizing it); hence it also cannot by itself replace

parts that have been taken from it, or make good defects in its original construction by the addition of other parts, or somehow repair
itself when it has fallen into disorder: all of which, by contrast, we
can expect from organized nature.—An organized being is thus not
a mere machine, for that has only a motive power, while the organized being possesses in itself a formative power, and indeed one that
it communicates to the materials, which do not have it (it organizes
the latter): thus it has a self-propagating formative power, which cannot be explained through the capacity for movement alone (that is,
mechanism).[21]

The peculiarity of the organism, for Kant as for Blumenbach, resides
here in its resistance to the *additive* logic of mechanism, which
assumes that a whole will always correspond to the sum of its material parts, and that a change in the arrangement or number of parts
will always imply the influence of an external force. The material of a
clock does not *contain* its own formative principle, or *bildende Kraft*;
it therefore cannot continuously (re)generate the purposive structure with which it was originally, mechanically imbued by its maker.
Clocks do not give birth to other clocks, nor do they have the capacity
to "heal" when broken. They do not bestow on the time they so reli-
ably tick off the shape of a developmental trajectory, because they do
not interact with their environment to protect and fulfill an interior
potential for formal articulation. "Organized natures," on the other
hand, do all of these things, and in the process, they also incessantly
*trans*form their own material, reassigning functions and assimilating
new substance as part of their constant striving after structure. Their
actual physical contours, together with those of their organs or parts,
are determined at each developmental stage by the needs of a purpose
that must therefore be seen to precede, and so to create, the very matter of which the whole is ostensibly composed.

Kant acknowledges, in consequence, that the scientists of living
nature ("the *Zergliederer* of plants and animals"[22]) have no choice but
to presume the presence of such an internal purpose when analyzing,
or dissecting, their objects, since without it they will find themselves
unable to locate the natural joints or organs (Kant's word is *Glieder*,
which translates the Greek *organon* and shares a root with the German *Gelenk*, meaning "joint"[23]) along which to make the proper cuts.
Where the very notion of a body "part" depends for its definition on
the *function* this part is intended to serve (Kant explicitly insists on the
etymological meaning of *organon* as "instrument" or "tool"[24]), the
scientist who ignores the animating spirit of intention automatically

forfeits the ability to cleanly carve up the organic domain. Kant will go on to insist, however—and herein lies his "solution" to the quandary of organic uniqueness—that this presumption of purpose can operate solely at a *heuristic* level: the scientist must assume only so much "spirit," or "formative power," as is absolutely necessary for the determination of material structure, thereafter proceeding as quickly as possible to an investigation of those aspects which admit of a truly mechanistic account. He thereby concedes the inescapability of teleological reasoning in the context of comparative anatomy while carefully denying to such reasoning the status of true, natural scientific explanation. The mechanistically unfathomable "idea" of Blumenbach's drive-to-form provides, according to the famous conclusions of the *Third Critique*, a "regulative" leading-string (*Leitfaden*) to which the scientist may look for guidance regarding the shape of organic life; it does *not* provide an answer to the question of what actually *causes* or "constitutes" that shape.

Blumenbach himself responded with gratitude to this deployment of his treatise by the leading German philosopher of the age, and he was careful, in his later definitions of the *Bildungstrieb*, to incorporate the Kantian caveat regarding its purely heuristic value.[25] Not all of his colleagues, however, were content to consign what they had come to see as the central principle of their domain to the status of a useful fiction. Some, like Blumenbach's student Carl Friedrich Kielmeyer, chose instead to counter the vision of a world divided between phenomenal and noumenal, constitutive and regulative, mechanical and teleological, with the vision of a world united under the banner of exclusively *purposive* development: "I myself would like to derive all variation in the material of inanimate nature from a striving for heterogenesis, analogous to that in the organism, within the *soul* of nature."[26]

The radical impulse to turn the tables on the mechanist champions of a unified theory—by attempting to derive *all* material-mechanical change from the very same structure of soulful striving that governs the development of the individual organism—sprang initially from Kielmeyer's careful, comparative anatomical study of the embryonic trajectory. This study had shown him that embryos pass through distinct developmental stages, or *Stufenfolgen*, and that these stages tend to recall the morphology of less complex species. Thus human embryos, for instance, turn out to share morphological attributes first with plants, then with worms and mollusks, and then with other mammals, before finally culminating in a specifically human-shaped entity.[27] Kielmeyer suggested, on the basis of such parallels, that less complex species should be understood as the developmental precursors of their more complex counterparts, and that the criterion for a true, typological classification of species—historically, since Aristotle, the primary task of all scholars of life—should be sought in the dimension of *time*. If the structure of developmental progression could be definitively established, he theorized, then existing life forms could be defined, and their substantive attributes categorized, by reference to their respective positions within an all-encompassing temporal order. The traditional taxonomy of physical features (hooves, feathers, vertebrae) would turn out to presuppose a deeper taxonomy of developmental change, whose branching, tree-shaped structure could then be seen to articulate, in turn, the form-bestowing flow of a temporal "life juice."[28]

Such a categorization would not require complete empirical knowledge about the actual, genetic chain of bodily descent, or even about the actual, physical qualities of particular organisms or species. It would require, rather, the discovery and demonstration of a general transformational law, in accordance with which natural matter has no choice but to unfold. This emphasis on time as the true ground of body explains why Kielmeyer, in his groundbreaking 1793 lecture "On the Relationships of the Organic Forces among Each Other in the Series of Different Organizations, the Laws and Consequences of These Relationships" ("Über die Verhältnisse der organischen Kräfte untereinander in der Reihe der verschiedenen Organisationen, die Gesetze und Folgen dieser Verhältnisse"),[29] prefers to speak not of creatures but of organizations, conceived dynamically as relationships (*Verhältnisse*) among oppositional forces or powers (in the sense of the original Greek *dynamei*). An organization exists, for Kielmeyer, wherever a multiplicity of antagonistic forces manages to balance one opposition off against another, and in doing so to attain equilibrium: organic relation, in other words, is always *proportional* relation in the traditional philosophical sense of the ancient Greek technical term *analogia*, which refers specifically to relationships of the form "*a* relates to *b* as *c* relates to *d*" or $a : b = c : d$ (in such cases the two relationships $a : b$ and $c : d$ are said to stand *im Verhältnis* or "in proportion"—and thus to be analogous—to one another).

The result of such proportional symmetry is the relative stasis of a developmental level or *Stufe*. This stasis, however, never achieves for Kielmeyer the status of total inertia. In contrast to his mechanist counterparts, who take their point of departure from a passive material substrate, and consequently require for all true change the intervention of an external force, Kielmeyer *starts* from the fundamentally dynamic premise of oppositional relation: "Inequality or relative weakness between two colliding forces is the common condition for all movement in nature."[30] In the beginning, there is a clash of powers that expresses itself organically as a striving to (self-)differentiate (*zu heterogenisieren*), an ongoing primal collision that "punctuates" its diversificatory trajectory with moments of harmonious rest—and in doing so first calls into being the always precarious stability of the various biological phases. Organic development, from this perspective, refers to a particular kind of transformation in the organizational distribution of opposing forces, namely, a transformation that *actualizes* some previously unfulfilled potential for differentiation, some residual dynamic asymmetry, in the original proportional arrangement. The result is always a new, more complex balance of powers, and thus also, but only secondarily, a new, more complex arrangement of bodily materials. The natural joints, or organs, along which the comparative anatomist seeks to cut—the foundational units of the organizational system called life—acquire their definition from the natural joints, or stages, of history.

The observer of natural phenomena, according to Kielmeyer, has more direct access to the shape of such developmental trajectories when they unfold at the level of the individual organization, or embryo, than when they unfold at the level of species (what he calls *die Reihe der verschiedenen Organisationen*). Since the latter unfolding, however, can be seen to follow the same dynamic rules as the former—which just means: to pass through the same proportional configurations or *Stufenfolgen*—he argues that the two kinds of development should be imputed to an identical *cause*: "Since the distribution of forces in the series of organizations [i.e., of organized beings] follows the same order as the distribution of forces in the different developmental states of a given individual, so it can be concluded [. . .] that the force through which the series of the species is produced is in all likelihood identical, in its nature and its laws, with the force through which the different developmental states are brought about."[31] Kielmeyer here phrases his conclusion as though he were merely proposing to widen the applicability of Blumenbach's teleological causality from single organisms to organic species. In fact, however, the implications of the induction he performs extend well beyond the domain of organized structure, since the single force in question causes *emergence* as well as growth. Just as the formative principle succeeds in calling forth the individual organism from a formless matrix of reproductive juices, so also, if the two "series" are to remain truly parallel, must it succeed in calling forth organization per se from the formless flux of anorganic matter. The same laws that govern the development of all life thus turn out, against the claims of partial teleologists such as Blumenbach, to link life to the rest of nature.

The significantly less tentative formulations scattered throughout Kielmeyer's unpublished writings make clear that the ultimate goal is a teleological theory of the universe: "It can therefore be retroactively concluded that the changes our earth has undergone [. . .] were *rule-governed, which is to say developmental changes [regelmäßige, d.i. Entwicklungsveränderungen]*, and it can therefore, finally, be concluded that the developments of our earth and of the series of organic bodies are precisely related to one another, and that for this reason their histories must be connected."[32] By insisting, in this passage from a letter to Bopp's teacher Windischmann, on the possibility of a *developmental* history of the earth—one that would follow the same rules, and thus testify to the same ultimate cause, as the developmental history of species—Kielmeyer makes the moving force of life into the moving force of all terrestrial matter, and the principle of teleological transformation into the origin of all mechanical change. On the basis of this same inductive logic, with its far-reaching equation of growth and emergence, he can speak elsewhere, even more radically, about the possibility of an organicist cosmogony ("I have often had the idea of a cosmogony that would take its point of departure from an *analogy* with the theory of the formation of an organism"[33]), which would theorize the very genesis of the cosmos according to the principles of Blumenbach's *Bildungstrieb*.

Seen from this perspective, the science of comparative anatomy turns out to offer the template for a radically reconceptualized physics of the All. Kielmeyer's "physics of the animal kingdom,"[34] with its return to the ancient Greek concept of a nature defined by development (Kielmeyer gives the original meaning of the Greek *physis*, itself the progenitor of the Latin *natura*, as "emergence, becoming, birth"[35]), paves the way for a unified science of the future that must necessarily remain inaccessible to the mechanist. Knowledge of nature as a single, self-consistent system presupposes, from the perspective here advanced, prior knowledge of a nature systematically unfolding, along the lines of a growth or a birth. The phenomenal world possesses structure and order, of the kind that can be formulated in physical laws, *because* it comes into being as the expression (*Äußerung*) of a singular, purposive tendency. This tendency transcends the laws it grounds as potential transcends actuality; the two are not interchangeable. Neither, however, can they be thought independently of one another, since the originary dynamic principle is for Kielmeyer not a supplemental order of being—not an *An sich* behind and beyond the phenomenal world, nor a regulatory norm imposed from above by the spontaneity of a scientific "I"—but the inner-worldly condition of an essentially *natural* possibility.

KOSMON PSYCHON

Kielmeyer's 1793 lecture, which appeared in print shortly after he delivered it at the Karlsschule in Tübingen, was probably the most radical and certainly the most influential of the texts he allowed to be published during his lifetime.[36] Goethe visited him in Tübingen to discuss it;[37] Alexander von Humboldt declared him to be, on its basis, "the first physiologist of Germany;"[38] Novalis contemplated its consequences for a theory of history;[39] and a generation of comparative anatomists, among them Kielmeyer's most famous student, Georges Cuvier, defined their own central positions with respect to it.[40] From an epochal standpoint, however, the most significant reception may well have occurred in the context of Friedrich Schelling's early, nature-philosophical work *On the World Soul* (*Von der Welt-seele*, 1797), which explores the philosophical potential of an organicist physics very similar to Kielmeyer's, and which concludes with a paean to the lecture itself, "which will mark for future ages, without doubt, the epochal beginning of an entirely new natural history."[41]

Schelling intended his nature philosophy to serve as a corrective to the prevailing Kantian account of knowledge, with its affirmation of a passively knowable, mechanistic nature fundamentally dissimilar to the teleological activity of knowing. The goal was to render philosophically intelligible, as a natural phenomenon rather than a regulative principle, the existence of self-organizing systems like the one Kant had identified solely with the cognizing subject. How, asked Schelling, are we to make sense of empirical realities like life and consciousness from within a conception of the empirical that excludes the possibility of spontaneous self-causation—and so also of goal-oriented behavior—by virtue of its very definition? How can we claim to understand the relationship between our minds and the world they inhabit if the very *dynamic of inhabiting*—and so also of a shared space for necessity and freedom—turns out to pose an insurmountable barrier to comprehension? The elusive common root of subject and object, purpose and cause, cognizer and cognized, philosophy and *physis* must be sought, he suggested, at the point where the two poles actually collide, which is to say, within rather than beneath the natural domain. And this domain must consequently be reconceptualized as *capacious* enough to contain both logics.

Schelling's approach to the problem of the Kantian binaries exerted considerable influence on poet-philosophers such as Goethe, Novalis, and the Schlegel brothers, with whom he became intimately acquainted during his years in Jena.[42] It also resonated among working scientists of the period, many of whom had found their own intuitions about the scientific project difficult to reconcile with Kant's rigorous insistence on a mind-world schism.[43] When, therefore, Schelling declared Kielmeyer's interpretation of the comparative anatomical project—"the idea of a comparative physiology, which seeks the continuity of organic nature not in the transitions of shape and of organic structure, but in the transitions of the functions into one another"[44]—to be the empirical starting point for his own profoundly un-Kantian interpretation of mind-world relation—"for even matter exhibits no other, lesser bond than that of reason: the eternal unity of the infinite [*das Unendliche*] with the finite [*das Endliche*]"[45]—he simultaneously bestowed on Kielmeyer's theory of organic development, in the eyes of many of his contemporaries, the status of paradigmatic natural science. By going on to insist that this paradigmatic science could fulfill its implicit promise of a unified theory of nature *only* on the basis of the proffered nature-philosophical foundation,[46] and by claiming, conversely, that *only* such a truly "scientific physiology" could complete the nature-philosophical system,[47] he also definitively bound the ideal of a new life science to the ambitions of a new, post-Kantian philosophy.

The Schellingian marriage of life science and Idealism occurs by way of a return to Platonic cosmology. Plato's ancient and somewhat cryptic account of cosmic order ("cosmos," from Greek *kosmein* = to order or arrange) emerging under the guiding influence of a divine intention—as articulated in the Pythagorean-tinged dialogues *Philebus* and *Timaeus*, to which Schelling devotes an extended philological study in the years prior to composing *On the World Soul*[48]—already forms the implicit backdrop for Kielmeyer's own tentative reflections on the possibility of a cosmic *telos*, as will become clear in a moment. Schelling, however, takes Kielmeyer's largely unacknowledged Platonism one crucial step further, rendering it explicit and transforming it utterly in the same quintessentially "dialectical" moment: the comparative anatomical notion of a universalized *Bildungstrieb*, inherent to all physical matter, undergoes, at his hands, a generalization in the direction of the Platonic-Pythagorean *Logos*, with the result that the formative principle of nature becomes inseparable from the formative principle of all rational thought.[49]

In Plato's *Philebus*, from which Schelling draws the system-theoretical terminology that anchors *On the World Soul*, the phenomenon of organized structure is investigated and found to arise identically, whether in organisms or in minds.[50] Just as living bodies emerge wherever form enters into a harmonious union with the unformed material of organic existence—Plato's terms in the *Philebus* are *peras*, meaning "limit" or "boundary" in the sense of the Latin *finis*, and *apeiron*, meaning literally the Unlimited or Unbound; Schelling's terms, as cited above, are *das Endliche* and *das Unendliche*—so, too, do concepts come about as harmonious "mixtures" of the finite and the infinite.[51] The divine gift of dialectic method, in which knowledge consists, operates by delimiting, and so de*fin*ing, a previously unbounded domain of investigation, in such a way that an articulated whole can be seen to crystallize out of chaos:

> SOCRATES: It is a gift of the gods to men, or so it seems to me, hurled down from heaven by some Prometheus along with a most dazzling fire. And the people of old, superior to us and living in closer proximity to the gods, have bequeathed us this tale, that whatever is said to be consists of one and many, having in its nature limit [*peras*] and unlimitedness [*apeiron*]. Since this is the structure of things, we have to assume that there is in each case always one idea for every subject of inquiry, and we must search for it, as we will indeed find it there. And once we have grasped it, we must look for two, as the case would have it, or if not, for three or some other number. And we must treat every one of those further unities in the same way, until it is not only established of the original unit that it is one, many and unlimited, but also how many kinds it is.[52]

Socrates goes on to name as exemplary applications of the method the grammarian's analysis of language into letters and the musician's analysis of music into notes, the goal being, in both cases, to carve up an apparently "unlimited" continuity of sound. Such a continuity necessarily admits of an infinite number of possible delineations. Any purely arbitrary partition, however, will yield sound segments incapable of interacting as a harmonious or grammatical whole. True knowledge thus resides in the ability to carve along what Plato elsewhere calls the "natural joints,"[53] and in doing so to uncover the profoundly nonarbitrary "definite number" of purposefully interrelating parts.

In the domain of language, the analytic procedure is as follows:

> SOCRATES: The way some god or god-inspired man discovered that vocal sound is unlimited, as tradition in Egypt claims for a certain deity called Theuth. He was the first to notice that the vowels in that unlimited variety are not one but several, and again that there are others that are not voiced, but make some kind of noise, and that they, too, have a number. As a third kind of letters he established the ones we now call mute. After this he further subdivided the ones without sound or mutes down to every single unit. In the same fashion he also dealt with the vowels and the intermediates, until he had found out the number for each one of them, and then he gave all of them together the name "letters/elements" [stoicheia].[54]

The result is a theory of grammar (from *gramma* = letter, and *grammein* = to write, to draw), conceived as the study of discernible, which is to say *writeable*, letter limits: "And as he realized that none of us could gain any knowledge of a single one of [these letter elements], taken by itself without understanding them all, he considered that this was the common bond that somehow unifies them all and called it the art of literacy [*techne grammatike*]."[55] In the domain of music, a similarly ancient process of distillation yields knowledge of "how many intervals [*diastemata*] there are" together with "the kinds of combinations [*systemata*] they form,"[56] which in turn generates scales of individual notes. The result is a theory of harmony [*harmounia*, from *harmos*, "joint," proto–Indo-European *ar, "to fit together"], conceived as the study of pitch boundaries or joints.

Both examples are necessary because each, taken independently, tells only half the story of what it means, for Plato, to philosophically analyze a domain. On the one hand, the scientist of harmony arrives at the knowledge of what notes there are by studying the intervals (*diastemata*, "what stands between," from *dia*, "between," and *histēmi/histanai*, "to [cause to] stand") and their possibilities of combination (*systemata*, "what stands together," from *syn*, "with, together," and *histēmi/histanai*, "to [cause to] stand"). This procedure makes clear that relationships, in the context of conceptual articulation, take priority over *relata*, such that musical elements turn out to depend for their definitions on an account of the *distances* that separate them, together with an account of the arrangements into which those distances—rather than the entities themselves—can enter. On the other hand, the notion of relations among relations of sound cannot alone account for the kind of privileged structure that allows a definite number of units to emerge out of chaos, since so long as the character of the combinations remains unspecified, the analyst will have no reason to assume that these combinations can in turn relate to *each other*. The result will be a potential for the infinite regress of relations, leading to an uncountable, rather than definite, number of *relata* or elements. To the purely relational understanding of harmonic thinghood must therefore be added the idea of an underlying relational *principle*—a common measure or "bond" (Schelling's words are *Band* and *Einheit*) to which all relations among relations could themselves relate *back*—like the one mentioned in connection with Theuth's discovery "that none of us could learn any one of [these letter elements] alone by itself without learning them all." Only the unifying power of what Plato will go on to call the singular "cause" of combination can turn the *systemata* of acoustic difference relations into the differentiated "systems" of musical and linguistic sound.[57]

According to Schelling's reading, Plato poses and answers the crucial question of what such a relational principle must look like in the *Timaeus*, a late dialogue generally assumed to belong together with the *Philebus*, both chronologically and substantively. The primary speaker of the dialogue, Timaeus, inquires into the possibility of a cosmology, conceived as a unified science of the All, and concludes that the common bond tying together this most capacious of analyzable domains will necessarily take the form of *proportion* or *analogy*. "That is why, as he began to put the body of the universe together, the god came to make it out of fire and earth. But it isn't possible to combine two things well all by themselves, without a third; there has to be some bond between the two that unites them. Now the best bond is one that really and truly makes a unity of itself together with the things bonded by it, and this in the nature of things is best accomplished by proportion [*analogia*]."[58] The example that follows this passage ("whenever of three numbers which are either solids or squares the middle term between any two of them is such that what the first term is to it, it is to the last") makes clear that Timeus here has in mind the technical, mathematical sense of the Greek word *analogia*, which refers to a relationship of equivalence between two ratios. The Greek *logos*, in its originary mathematical meaning of "relation" or "ratio," joins with the prefix *ana*, meaning "up to, toward, back, again" to yield the gesture of "relating back" until an equilibrium among binary relations is reached. (If *a* is to *b* as *c* is to *d*, then the ratio *a* : *b* relates analogously or proportionally to the ratio *c* : *d*.) Such a procedure provides a criterion for defining and distinguishing legitimate relation, since, within the realm of Greek mathematics, two entities can only be said to enter into a ratio at all—and thus to belong among the play of *logoi* that constitute the cosmic Logos—insofar as they are capable of forming an analogical equivalence relation with two other entities. (The numerical entities *a* and *b*, for instance, are considered commensurable if and only if there also exist numerical entities x and y such that $a/x = b/y$, which in turn implies the existence of a common divisor c that can "measure" both *a* and *b* without remainder.)

The sum total of all relations or *logoi* coincides with the domain known in modern mathematical terminology as the system of "rational" numbers. This system represents the largest conceivable collection of entities to which a standard of measurement can be applied. Assuming, therefore, that the purpose of conceptual definition is precisely to formulate such a standard—meaning also a common property or essence—in virtue of which an otherwise heterogeneous multitude of instances becomes thinkable as one *kind* of entity (assuming, for instance, that a proper definition of the concept "white" provides a rule for "measuring" the whiteness of things), then the rational system of numbers must also be assumed to represent the *most general conceptually definable domain*. The "counts" of all particular analyses, performed in accordance with the dictates of Plato's *Philebus*, must be small enough to be contained within this largest of all conceivable conceptual matrices, whose boundaries mark the last stop before the nondenumerable chaos of the *apeiron*.

Platonic diaeresis is possible, from this cosmological perspective, only because the organized structure of the articulated concept *participates* in the organized structure of Being more generally. If "knowing" an entity means discerning the boundaries that define its most basic units, in the sense of rendering them finite or countable, then "being" an entity, it would seem, must entail actually *having* such units in the first place. Not every mixture of the finite and the infinite exists as a countable collection of harmoniously interacting parts; bad weather and bad health, for instance, are mixtures whose components tend to clash with one another, without ever achieving relational equilibrium. Every mixture, however, that can be rationally discerned, classified, named, and so "said" with any kind of temporal stability ("whatever things are said to be"), must necessarily exist in the manner of a Many *cohesively* encompassed by a One, which turns out to mean, also: as a balanced, proportional play of relations or *logoi*, in which all differences get subsumed under the rule of a higher equivalence or *analogia*. To know (a portion of) the All is to reflect (a portion of) the All by uncovering the inherently rational structure of the metaratio that "thinks" it into being. And Plato's Timaeus draws precisely this conclusion when he envisions the universe as a kind of ultimate animal (*zoon*) whose material parts move interdependently under the guidance of an animating mind (*nous*).[59]

It is this ancient presumption of a concept-cosmos correspondence that nature-philosophy must succeed in deductively *demonstrating*, according to Schelling, if Kielmeyer's own inductively established correspondences—between organism and species, between organism and earth—are ever to acquire the status of real scientific knowledge.[60] Only to the extent that the universe itself can be shown to exist as an instantiated concept; only to the extent that the formative principle of *physis* turns out to be identical to the formative principle of thought; only to the extent that the totality of natural phenomena reveals itself to be the All of a unified whole rather than the All of an accidental aggregate: only in this way can the method of comparing and extrapolating find justification, since only in this way can the structural similarities thereby uncovered be assumed to actually *mean* anything about the identity of an underlying cause.

The nature-philosophical anchoring of the comparative anatomical analogies must therefore proceed, as Schelling makes clear in the final lines of *On the World Soul*, by way of a (re)spiritualization of nature: "Now since this principle maintains the continuity of the inorganic and the organic world, and binds the whole of nature into a universal organism, we can recognize in it again that Being which the most ancient philosophy intuitively welcomed as the common soul of nature, and which a few physicists of that time considered to be one with the forming and shaping aether."[61] The passage explicitly identifies Kielmeyer's universalized "formative principle" of systemic organization with the ancient Greek physical concept of a formative aether, and so also with the ancient Greek notion of *psyche* as an animatory breath or wind ("*Psyche*," writes Schelling in his *Timaeus* commentary, paraphrasing Plato, "means nothing other than: *original principle of movement, arche kineseos*"[62]). Schelling thereby underscores the Platonic dynamic of ensoulment he considers the *Bildungsprinzip* to entail. Such a redefinition of the natural scientific project—from physics as the study of a matter accidentally set in motion to physics as the study of a matter inherently ensouled— subordinates the Enlightenment model of truth as adequation, in which a rational subject paints an *accurate* picture of a fundamentally nonrational universe, to a Romantic-idealist model of truth as participatory identity. Scientific knowledge becomes a matter of bringing the individual human subjectivity into line with the generative subjectivity of a world that is more, and yet not other, than the sum of its material parts. The isolated Kantian subject, with its uniquely human faculty of concept formation, reemerges within the context of Schelling's Platonizing nature-philosophy as one modality among others for the Reason that orders existence.

Whereas for Plato, however, these various modalities of mind represent the nodes of an eternally rotating, and thus ultimately ahistorical system, which has been *infused* into chaos fully formed and from above—cosmic time, in the *Timaeus*, is a "moving image of eternity," and the cosmos itself is created in accordance with a time-transcending model by the "intellect of a divine craftsman," or *dêmiourgos*[63]—Schelling follows Kielmeyer in viewing the "joints" of Being as the "stages" in a world-immanent, developmental trajectory.[64] He rejects the "mindless [*geistlose*], yet widespread notion"[65] of a *Geist* that must be retroactively added to matter in order to generate the compound called life, and gives preference, instead, to a comparative anatomical understanding of "the inorganic" as a chaos that essentially *tends* toward cosmos. Every instance of organized structure is for Schelling, as for Plato, a microcosm, and every microcosmic-macrocosmic analogy bears witness to the underlying identity of nature's formative principle. But the sophistication and intensity with which each microcosmic structure manages to symbolize is a function, for Schelling, as for Kielmeyer, of its distance from the generative origin: the further along a given structure turns out to be in the course of the cosmic progression, the more diversity it will succeed in subsuming into its One, and the more of the universe,

conceived here as the endpoint of ultimate Oneness, it will consequently be able to mirror. The essence of nature as cohesive system,
and of the universe as all-encompassing concept, does not therefore
reside in any *existing* state of proportional equilibrium but rather in
the logic of progression governing the passage from one such *anal-
ogon* to another. Nature *is* (a concept) insofar as it *becomes* (a concept), via the dynamic of continuous oscillation—Schelling's word is
Schweben[66]—that binds together the opposing poles of productivity
and product, *apeiron* and *peras*, flux and form.

The following "example" from the *First Outline of a System of the
Philosophy of Nature (Erster Entwurf eines Systems der Naturphilosophie*, 1799) provides this dynamic with its most famous illustration: "An example: a current flows in a straight line forward as long
as it encounters no resistance. Where there is resistance—a whirlpool. Every original product of nature is such a whirlpool, every
organization for example. The whirlpool is not something station-
ary [*feststehend*], but rather something constantly changeable
[*Wandelbares*]—but reproduced anew at each instant. Thus no product in nature is *fixed*, but rather at each instant reproduced through
the force of nature as a whole."[67] Schelling here envisions a directional
flow interrupted, and so also de*limit*ed, by the spiral-shaped redoubling of its own fundamentally forward-tending energy. In doing so,
he effectively reinterprets the traditional etymology of "rhythm,"
which traces the Greek *rhuthmos* back to *rhein* (proto–Indo-European
*sreu), "to flow," across the mediation of repetitively flowing *waves*.[68]
Unlike the moon-controlled meter of the tides, the Schellingian vortices rupture the continuity of their liquid substrate at intervals dictated solely by interior laws of flux. At stake is thus a temporality that
unfolds by imposing rhythmic form on *itself*, a *Schweben* that creates
a cosmos, in the traditional manner of dividing and differentiating—
"and the Spirit of God was hovering [*schwebte*] over the face of the
waters [. . .] and God separated the light from the darkness"[69]—but
one that does so from the inside instead of from above. According to
Schelling, this punctual tendency toward conceptual self-articulation
determines the direction of every natural-historical development. It
is the purpose, and hence also the meaning, of all cosmic happenings, including, paradigmatically, the happening of human cognition,
whose task is to *re*articulate, at the microcosmic level of consciousness, the rhythm of nature's "dynamic series of stages" (*dynamische
Stufenfolge*):[70] "Philosophy," writes Schelling in the introduction to

his *Ideas for a Philosophy of Nature* (*Ideen zu einer Philosophie der Natur*), "is thus nothing other than a natural science of our intellect [*unseres Geistes*]."[71]

HOW INFLECTION UNFOLDS

The list of working scientists who allowed Schelling's reinterpretation of the natural scientific project to influence their approach spanned the gamut of traditional natural scientific disciplines, from biology and chemistry to physics and geology. The theory may well have found its most transformative resonance, however, within a field of study whose object did not traditionally rank among natural entities at all. For in the context of a cosmos progressing toward conceptual form, the emergence of *language*—conceived as an unparalleled instrument of conceptual articulation—became suddenly a critical chapter in the natural history of the universe.[72] Jacob Grimm's account of this emergence, from a midcentury essay titled "On the Origin of Language" ("Über den Ursprung der Sprache") can in this respect be considered paradigmatic: "Every sound," he writes, "is produced by a movement and a tremor of the air, even the elemental rushing of water or the crackling of fire."[73] The world soul thus articulates itself first as aether, and in doing so, begins to make noise. The primal matter of moving air then goes on to articulate itself further in the organized structure of life, and the result is the more rarified wind of breath. "Vocal instruments are proper by nature to both animal and man; by means of these they can produce impressions on the air in a manifold manner."[74] The flow of breath, finally, continues to self-articulate in turn, and the result is a system that *transcends* (mere) biology: "[T]he ordered development [*geordnete Entfaltung*] of sounds means for us organizing into parts [*gliedern*], articulating, and the human language appears as an articulated one, with which the Homeric epithet for humanity coincides (οι μεροϖες, μεροϖες ανθρωϖοι oder βροτοι, from μειρομαι or μεριζω, those who divide or articulate their voices)."[75] Human speech, with its orderly unfolding of the articulatory potential inherent to sound, and so to spirit, marks from this perspective a new phase in the explication of nature. As the form assumed by flux at the highest-known level of biological organization—namely that of human existence, which hereby definitively distinguishes itself from all "lower," less organized life forms—language can justly be assumed to mirror the cosmic dynamic more intensely than any other microcosm, and to lay claim, in consequence, to the status of most natural entity.

To linguists such as Grimm falls the task of making this dynamic *clear*. Such clarity, however, must no longer be a matter of imposing precise definitional boundaries but rather of calling existing boundaries to the fore. Beginning with the very first foreword to his epoch-defining, multivolume study of German grammar, Grimm polemicizes vehemently against rationalist predecessors who seek to "purify" empirical language from a position beyond its purview. The modern grammarian, in his opinion, does not work to bring language back into conformity with some original form-bestowing intention, whether divine or human, by re*form*ing it along more "logical" lines; nor does he seek to *sculpt* young schoolchildren into speakers of an artificially "clarified" language. For the modern grammarian, formation, in all of the many senses of the German *Bildung*, remains rather the exclusive province of language itself, which spontaneously emerges, grows, regenerates, and reproduces according to the laws of an interior *Bildungstrieb*:

> Who could believe that such a deeply embedded mode of growth, occurring in accordance with the natural laws of wise economy, was directed or furthered by the abstractly deduced, feeble, and falsely conceptualized rules of the "Language Masters," and who does not feel aggrieved about the unchildlike children and young people, who speak in a pure and educated [*gebildet*] manner, but feel no homesickness in old age for their youth? One should try asking a true poet, who certainly knows how to command the material [*Stoff*], spirit [*Geist*], and rules of language quite differently from all the grammarians and dictionary-makers together, what he has learned from Adelung, and whether he looks things up in his dictionary.[76]

Grimm's name for this formative principle of teleological development, to which poets and children respond so much more appropriately than pedants, is "language spirit," or *Sprachgeist*: "Language has suffered some injuries, and these it must bear. The only true and beneficial form of reparation stands in the power of the tirelessly creating *Sprachgeist*, which, like a nesting bird, breeds again anew after its eggs have been taken away; poets and writers, however, can perceive this spirit's invisible workings by feel, when they are inspired and emotionally moved [*sein unsichtbares Walten vernehmen aber Dichter und Schriftsteller in der Begeisterung und Bewegung durch ihr Gefühl*]."[77] The consequences for the aspiring student of language are clear. The Logos that drives nature, and with it the All, must somehow be made *explicit* for the speaking but unhearing masses who have been left unmoved by inspiration (from Latin: *in* = in + *spirare* = to breathe), and who consequently perceive in the wind of language spirit nothing other than the product of their own prosaic breath. What the poet discovers by feel (across the *Bewegung* of a divinatory "Be*geiste*rung," in the theophany of "sein unsichtbares Walten vernehmen"), the linguist must therefore rediscover by the more universally accessible means of scientific analysis. The formative principle must be encouraged to "speak" its own interior structure in a language even the layman can comprehend: "If among the people the simple fare of our native language were to find entry, the dictionary could become a household necessity, and could be read with desire, often with reverence."[78]

In the case of the animals of comparative anatomy, "interior structure" means the interplay of various functional organs (*articuli, Glieder*), and the technique of analysis is the dissection (*Zergliederung, Zerlegung*) that lays those organs bare. In the case of a language, "structure" refers to the interplay of various kinds of units—nineteenth-century linguists call this "grammar"—and the technique is an etymological procedure Bopp calls "the anatomical dissection [. . .] of the body of language."[79] In both cases, the real difference between the newer approach and that of its dissecting or etymologizing predecessors emerges most clearly with respect to the question of originating cause, since teleologically inclined comparatists can feel free to *bracket* this issue in a way that their mechanist counterparts cannot. Bopp begins his monumental *A Comparative Grammar of the Sanskrit, Zend, Greek, Latin, Lithuanian, Gothic, German, and Slavonic Languages* (*Vergleichende Grammatik des Sanskrit, Zend, Griechischen, Lateinischen, Litthauischen, Gothischen, und Deutschen*, 1833), which definitively establishes the new field of comparative grammatical analysis, with a pronouncement about the kinds of origins he intends his book (not) to address: "I intend with this work a comparative description of the organism of the languages enumerated on the title page, comprehending everything related thereunto: an inquiry into their physical and mechanical laws and the origin of the forms that distinguish their grammatical relations. One point alone I shall leave untouched, the secret of the roots, or of the reason behind the naming of the *Ur*-concepts [*des Benennungsgrundes der Urbegriffe*]. I shall not investigate, for example, why the root *i* signifies 'go' and not 'stand,' or why the combination of sounds *stha* or *sta* signifies 'stand' and not 'go.'"[80] The linguist who seeks the laws of language life *does* ask, according to Bopp, about the origins of grammatical relation. Such a linguist does *not* ask, however, how and why the original roots of a language come to *refer* to particular extralinguistic activities or objects, whether they do so mimetically or arbitrarily, and as the effect of what external impetus. A particularly precise formulation of the position Bopp hereby intends to reject is provided by the rationalist grammarian Johann Christoph Adelung, whose influence on the behavior of poets Grimm disputes up above:

> Syllables and words did not emerge randomly or by chance from letters, rather the inventors of language worked in accordance with certain basic universal laws [*gewissen allgemeinen Grundgesetzen*], which, however, as with everything in language, they perceived only obscurely. Etymology [. . .], which [. . .] yields to no other science in terms of its importance and fruitfulness, teaches us to discover these basic laws, because it alone is in a position to trace human understanding back to its infancy, to unravel the original reasons behind its concepts [*die Anfangsgründe seiner Begriffe*], and to sketch the whole step-wise progression [*Stufengang*] of its knowledge.[81]

Taken together, the two passages bear witness to a fundamental disagreement about the nature of historical development, which in turn implies a fundamental disagreement about the role of historical *explanation*. Whereas traditional approaches such as Adelung's seek to account for the existence of language by tracing every individual word or name back to its original "reason" for being, self-proclaimed revolutionaries such as Bopp, Grimm, and their Danish contemporary Rasmus Rask insist instead on the explanatory priority of an *interior tendency* toward form, which shapes language "in its Becoming and its developmental trajectory" (*im Werden und ihrem Entwicklungsgange*) irrespective of outside influence.[82] For the new breed of teleologically inclined grammarians, who take their cues from the new breed of teleologically inclined anatomists, the originary "mechanism" of naming no longer represents the goal of etymological analysis, any more than the original "catalyst" of reproduction represents the goal of comparative dissection.

The methodological consequences of this perspectival shift are not immediately apparent at the level of language origins, since Bopp and his colleagues still accept, for the most part, the traditional account of primordial language structure: the oldest roots take shape, for them as for their rationalist precursors, by way of a monosyllabic *Vermählung* between the "feminine" principle of language matter and the "masculine" principle of language form:

> All the sounds of language are divided into vowels and consonants, the former are more fluid, the latter more solid. One may call the consonants the bones and muscles of language; the vowels are that which flows through and animates the solid parts, namely, blood and breath. Consonants seem almost to provide the body, vowels the soul, on the consonants depends the shape, on the vowels the color [. . .]. If, in a certain sense, one can term the vowel the feminine, the consonant the masculine principle of language, then it can also be said that the conception [*Zeugung*] of each root only occurs through the union [*Vereinigung*] of the two.[83]

Despite the pronouncedly nature-philosophical tenor of this passage (from the opening of the third edition of Grimm's *Germanic Grammar*, the polarity it describes would have been equally familiar to seventeenth- and eighteenth-century scholars of language. Just as few anatomists, whether mechanist or teleologist, would argue about the necessity of a gendered coupling at the origin of organic reproduction; and just as few philosophers, whether Platonic, Aristotelian, scholastic, or Cartesian, would argue about the necessity of a matter-form pairing at the origin of organization more generally, so, too, do linguists tend to agree about the necessity of a vowel-consonant cooperation at the level of the originary syllable. What has changed, in the context of Grimm and Rask's Germanic grammars, or of Bopp's more broadly comparative one, is thus less the depiction of the origins themselves—Grimm's emphatic rhetoric of flow and form recalls Schelling, of course, but also Plato, Spinoza, and the Enlightenment language philosopher August Ferdinand Bernhardi[84]—than the role that these origins are being permitted to play. For where the organized structure of the *Ur*-syllable no longer testifies primarily to its "source" in the human capacity for concept formation, an account of language essence can no longer be a matter of turning backward to the "cause" of human reason. Such an account must instead look forward toward the complex shapes that the first seeds, over time, turn out to assume, in an effort to isolate the laws according to which all linguistic development must occur.

The researcher who ignores this dictate risks turning etymology into a pseudoscience of puns, since, in the absence of all further development, the primal "procreative juice" (*Zeugungssaft*) of language remains both acoustically and semantically undifferentiated. Grimm cites as a particularly egregious example the hypothesis of classical philologist Johann Heinrich Voss, who attempts to trace all of Greek, Latin, and German vocabulary back to a series of rhyming roots (*eo heo geo keo cheo neo feo meo beo peo leo reo*) with the single meaning of "to go" or "to move."[85] Such flights of associative fantasy, in which the analogizing mind conjectures away all problematic variations in both sound and meaning to make room for the pleasures of a primordial homogeneity, represent the antithesis of an organicist analysis envisioned as "linguistic comparison that encompasses and animates even the smallest parts of speech."[86] Where comparative grammarians seek the inherent logic of an already articulated language life, striving to "dissect" along existing contours, philologists such as Voss, according to Grimm, return to the chaos of an imagined origin in order to carve up their material at will. They proceed unsystematically—"mechanically," in the polemical rhetoric of the period—because they ignore the specificity of the very system they claim to investigate, substituting consonants and fragmenting words without regard for the unique internal order of their empirical object. Grimm's contemporary, the great Danish linguist Rasmus Rask, expresses the opposition between these two perspectives as follows, in a treatment titled "On Etymology in General" (1818): "[A]lthough language consists entirely of words, separate words do not yet constitute language, unless they are connected to each other in some way. In most languages such connection is established through certain changes in the words, or, as they are called, inflections. Such form changes and more generally the entire structure and system of a language provide a new object of etymology or, as we might perhaps call it in this case, *language explanation*, which may also be used as a general term."[87] The new etymological technique here envisioned, by both Rask and Grimm, as the cornerstone of a teleological linguistics, must explain language as a web of relations rather than a collection of isolated words. It must prioritize grammatical function over lexical meaning. And it must ignore as irrelevant wordplay all sound similarities that do not contribute to this larger context, for its goal is not the original meaning of particular sound combinations (*heo geo keo*) but the structure of the system *within* which those sounds begin to mean.

It is the discovery of a structurally relevant etymology in precisely this sense that leads Franz Bopp to formulate his most significant hypothesis regarding the governing tendency of linguistic development. The hypothesis in question states that the grammatical category of verb conjugation emerges out of the lexical category of personal pronouns: "[I]t is easy to discover that το is the radical form of the Greek article, which is originally nothing more than a pronoun of the third person, and is used as such in Homer. This το, bereft of the final vowel, becomes an essential element of verbs in their third person, singular, dual and plural, as, δίδοτι, δίδοτον, δίδοντι. I have no doubt but it can be proved [. . .] that Sanskrit verbs also form their persons by compounding the root with the pronouns, upon which subject I shall offer a few remarks in its proper place."[88] Originally an independent appendage tacked on to the equally independent verb (the technical term for this combinatory mode is *agglutination*), the personal pronoun eventually fuses with the verb to produce the phenomenon known as *inflection* (Ger. *Flexion* or *Biegung*), in which grammatical functions express themselves through an internal transformation of the verbal stem (*spreche, sprichst, spricht, sprecht, sprechen*). Since such a transformation involves a shift from material contiguity to functional unity—the agglomeration of self-sufficient particles, which retroactively "add up" to an expression of relationship, gives way, historically, to the differentiated oneness of the inflected verb, which subordinates its parts to the larger purpose of a relational whole—Bopp's hypothesis turns out to entail a theory about the emergence of organic form. He can therefore go on to propose, in his *Comparative Grammar*, a new, "naturehistorical classification of languages" to complement the new, nature-historical classifications of comparative anatomists such as Kielmeyer.[89] The history of language life begins, for him, with the inanimate material of languages such as Chinese, which is characterized by utterly unstructured strings of "naked," monosyllabic roots; it then passes through an agglutinative phase, in which the isolated particles begin to cluster loosely together around the "center" of a verbal stem, before culminating in the inflective hierarchies of the Indo-European language family, whose priority resides "in the beautiful connection of these appendages to a harmonious whole, which bears the appearance of an organic body."[90]

Bopp's relatively brief introduction of this natural historical typology in the context of his *Comparative Grammar* sufficed to alienate both Schlegel brothers (for whom the organic unity of inflection represented a linguistic ideal to be rigorously distinguished, in the spirit of Blumenbach, from the inferior "mechanics" of agglutination), and to persuade nearly everyone else.[91] From the 1830s onward, the inflectional grammar so familiar to speakers of modern European languages was presumed by language scientists to emerge from a historical process of condensation (*Verdichtung*), an intensificatory *folding-in* at the level of the verbal root ("inflection," from Latin *inflectere*, "to bend in"), in the course of which the heap of language matter acquired the jointed flexibility of life. Jacob Grimm's midcentury reformulation of this foundational narrative attributes the condensation process to the organizational energy of a form-bestowing, light-and-dark-distributing *Sprachgeist*:

> At the beginning, it seems, the words unfolded, unhindered in idyllic comfort, without any limitation other than the natural succession indicated by feeling. Their impression was pure and uncontrived, yet too full and overloaded, so that light and shadow could not be well distributed [*so dasz licht und schatten sich nicht recht vertheilen konnten*]. Gradually, however, an unconsciously prevailing language spirit [*ein unbewust waltender sprachgeist*] let a smaller weight fall on the subordinate ideas, and they, diluted and abbreviated, were joined to the chief concept as codetermining parts. Inflection emerged from the ingrowth of leading and motivating adjectives [*aus dem einwuchs lenkender und bewegender bestimmwörter*]: these are now dragged along by the main word, which they goad into movement like half or almost fully hidden gears [*triebräder*], and they have passed over from their original, also sensory meaning to an abstract one [*aus ihrer ursprünglich auch sinnlichen bedeutung in eine abgezogne*], through which the former only occasionally shimmers.[92]

The biblical logic of creation via spiritual *in*fusion gives way, here, to the equally mysterious but utterly prefixless dynamic of a *fusion* bestowing form from within. *Coacervatio* becomes *articulatio*,[93] aggregation becomes flexion, word becomes grammar via a spontaneous reallocation of linguistic mass that leaves certain parts "denser" than others, and thus also more capable of operating as relational hubs or nodes. Language unfolds, in the direction of its inflected Indo-European destiny, insofar as it grows inward rather than outward ("from the *ingrowth* of [. . .] adjectives"), coalescing around vortices that implicate rather than explicate (from the Latin *implicare*, "to fold in," and *explicare*, "to fold out"), in order to form meanings that arise internally, as a function of grammatical relation, rather than externally, as a matter of reference.

The progress of such an implicatory unfolding renders the original seeds of language increasingly difficult to discern. Whereas, for example, the "reason" behind the inflectional endings of conjugated

verbs (the *t* of *spricht*, the *st* of *sprichst*) would have been obvious
to an ordinary speaker of Indo-European, for whom they were virtually identical to the corresponding personal pronouns, this identity has long since disappeared from modern European grammar,
and hence from the consciousness of European speakers. Contemporary languages, therefore, require the offices of a historical linguist
to make their latent logics explicit. The linguist responds by reversing the movement of articulatory condensation, tracing known languages back to the moment in which the agglutinative accumulation
of matter first tips over into the purposive intensity of life, and the
originary rationality of the system—as in the case of the comparative anatomist's embryos—still lies open for all to see. Bopp's unprecedented success in this regard makes him the father of the modern,
comparative-anatomical approach to etymology: "Only after the successful analysis of the inflections and derivations, in which Franz
Bopp's perspicacity has performed such a great service, did the roots
come to light."[94]

The resulting reconstruction of the trajectory of language life functions as a record of established etymological conclusions regarding
the development of particular tongues. It takes the notational form of
a branching, tree-shaped typology of nature-historical descent, whose
apex represents the hypothetical "genus" from which all known
inflecting "species" derive. As the universal ground of all flexional
grammar, this apex—called "Indo-German" by Rask and Grimm,
"Indo-European" by Bopp—possesses a theoretical purity that helps
to anchor the empirical investigation of later forms. Bopp alludes to
this notion of an utterly transparent linguistic origin, in which clearly
defined units of sound are presumed to coincide with equally clearly
defined units of meaning, when he speaks of modern languages *no
longer* comprehending *themselves* (*sich selber nicht mehr begreifend*[95]), and assigns to the linguist the task of retroactively *compen-
sating* for this loss. Such primordial purity can be expected to return
toward the end of the Indo-European cultural trajectory, in line with
a teleological interpretation that sees in every origin the anticipatory
figure of its eventual fulfillment. Having passed from the exuberant
fecundity of youth, through a maturity that eliminates the extrane-
ous through a process of controlled decay, inflectional languages will
eventually advance toward a new, more self-conscious level of *scientific* self-understanding, which will allow them to operate once
again as one tongue. Grimm can therefore envision, as a pendant to

his disciplinary creation story, a future day when the scattered Indo-European dialects ("which [. . .] even now may be termed the most powerful tongue on the surface of the earth"[96]) will reunite in reality as in notation, thereby successfully actualizing the anti-Babelian potential of their common genealogical roots: "[W]hereas all human languages are supposed to have departed clouded and shattered from the steps of the Tower of Babel [. . .], so they could someday, in the unforeseeable future, flow purely and clearly back together."[97]

ETYMOLOGY: THE METHOD

The teleological conception of language history, which sees in the Indo-European foundations simultaneously the origin and the end of language study, presupposes a two-tiered notion of its object: modern linguistic phenomena that bear witness to the analytical purity of their beginnings belong for the historical linguist to the living organism of language, while those that resist the etymological "translation," such as words borrowed from foreign tongues or grammatical forms founded on false analogies, belong instead to the bodily decrepitude disparaged by Bopp as an "extrinsic mass." They are the excess that will fall by the wayside when the languages of the world come together to achieve their Indo-European fulfillment. To the schism between those who speak and those who know, in other words, corresponds a schism at the level of the material itself, between the everyday language of the unanalyzed surface and the scientific language of distillable depths. Despite the often polemical assertions of its champions, the true target of historical linguistics is never, strictly speaking, the "naked" empirical idiom, as given to the senses in speech or writing, but a subset distinguished by its link to the (as yet) hypothetical metalanguage.[98] The legitimacy of the new analyses thus depends on the analyst's ability to move from one level to the other *in a consistent and rigorously formalizable fashion*, without falling prey to the arbitrary normativity of more traditional selection processes: "If one plumbs the relationships of individual languages more *frugally* and *solidly* [*sparsamer* und *fester*], and progresses *stepwise* [*stufenweise*] to more general comparisons, it is to be expected that [. . .] eventually discoveries can be made, alongside which, in terms of certainty, novelty, and fascination, perhaps only the discoveries of comparative anatomy and natural history will stand."[99] Though Grimm does not here offer a more elaborate account of the particular practices he has in mind, he clearly recognizes that a rhetoric of "cutting at the natural joints" cannot, on its own, suffice to distinguish comparative grammar from its predecessors. Only the discovery of a language-immanent criterion capable of determining, from the inside out, what it would actually *mean* to plumb linguistic relations more frugally, more solidly, and stepwise can fulfill this demand for a truly rigorous technique, and, in doing so, justify the comparative grammarians' claim to have established a *science* of language.

The turning point in methodological stringency comes with the formulation of the first so-called "sound laws." On the surface, the idea is simple enough. Over a particular, prehistoric period of time, a particular set of speakers began to say *p* where they had formerly said *b*, *t* where they had formerly said *d*, *k* where they had formerly said *g*, and so on for six other consonants. In doing so, they established a new pronunciation that separated them from all other descendants of the Indo-European matrix as speakers of a specifically Germanic language. Some indefinite amount of time later a similar bifurcation occurred within proto-Germanic, generating the phonetic constellation peculiar to Old High German. Proposed by Rask in 1818 and systematized by Grimm (hence the alternative name "Grimm's Law") in the second edition of his *Germanic Grammar* (1822),[100] the dual consonant shift was immediately appropriated by nineteenth-century linguists as a kind of second origin, which enabled them, for the first time, to act meaningfully on the Jones-Schlegel discovery of a common source and on the notion of language "life." The proposed patterns were the product of observations involving innumerable attested forms, and as such they anchored etymological speculation in the ground of the empirical given. But the ability to trace particular consonants back through a stepwise series of phonetic phases, or *Stufen*, also allowed linguists to determine with great precision what portion of a particular language should actually be "counted" as "belonging" to the linguistic organism—and thus to transcend an empirical given that could never completely coincide with its etymological purification. Where consonants conformed to the rules, a sequence of similar-sounding forms could be claimed for science; such words participated in the "life" of language. Where they did not—where, for example, Greek and Gothic forms contained identical consonants—genetic explanations made way for the forces of foreign influence and accident. The Sanskrit *padas* and the Greek *pathos* share no etymological bond, despite the near-identity of their forms, since between the two languages no such shift from *d* to *th* ever took place. Nor, it turns out, is the Gothic *scriban* a direct descendant of the Latin *scribere*, as had often been assumed, since a genetic relation would necessarily have yielded *scripan*; it is, instead, the remnant of a borrowed technical term—adapted from the coloniz-ing Romans and thus every bit as alien to the "originary" Germanic phonetic system, from a nineteenth-century etymological perspective, as the Latinate alphabet it was adopted to describe. "[S]crîban is *scri-bere* itself, fruht is *fructus*, and *consequently un-German*," declares Grimm in his discussion of the sound laws and their consequences.[101] Never before had the linguist been able to draw so definitive a distinc-tion between inside and out, Germanic and non, between the object of investigation and the rest of the world. Etymology, which separates language from the idiosyncrasies of its spoken form, from itself and the encroachments of its others, receives in the sound laws both its empiri-cal justification and its methodological ground.

Grimm's use of a consonant table to represent the two all-important shifts renders readable the true stakes of his codifying exercise:

The whole twofold consonant shift, which has manifold conse-quences for the history of language and the rigor of etymology, is rep-resented by the following table:

Greek	P.	B.	F.	T.	D.	TH.	K.	G.	CH.
Gothic	F.	P.	B.	TH.	T.	D.	..	K.	G.
Old High German	B(V).	F.	P.	D.	Z.	T.	G.	CH.	K.[102]

As the visual synthesis of a demonstrable pattern—Grimm follows up with nine significantly less eye-catching lists of affected words—the table reinforces the overall impression of regularity and reliabil-ity. As the schematic statement of a rule, however, it transcends the subordinate role of mere "illustration" (*Darstellung*) and lays claim to the status of fundament. The lists, readable in one direction only, bear witness to the consonant shift, and the table, read vertically, does the same. Yet the table can also be read horizontally, and doing so yields information—as Grimm's own labeling makes clear—about the *internal structure* of individual languages. The difference between Greek and Gothic, and between both of these and German, boils down, here, to the different roles played in each by certain key consonants (corresponding rules for the vowels are pronounced nec-essary, though "as yet undiscovered").[103] Lined up in their horizontal rows, the consonants thus give graphic form to a wholly new concep-tion of language, defined not as a collection of words, governed from outside by grammatical rules, but as a self-sufficient *system* of *letters* or *Lautsystem*.[104] The fact that these sounds shift together as a group ("in their new positions, the sounds order themselves immediately into the familiar harmony")[105] is proof enough that they relate to one another in some nonarbitrary, deep-structural way, which in turn is proof that a form-bestowing, equilibrium-preserving energy—a linguistic *Bildungstrieb*—governs the phases of language change.

The audacity of Grimm's tabular formulation impressed his contemporaries in a way that Rask's far more modest suggestion of a historical regularity had not, and the contrast between the two representational modes is instructive. For where Grimm pictures the changing structure of consonantal configurations, Rask, from whom Grimm took the rules, still relies on collections of individual words. His "chart" contains an ordered series of examples intended to stand in for other, much longer series, and as such remains firmly within the paradigm of the list. The result has none of the power of Grimm's graphic generalization, which moves beyond the additive model of example gathering to picture the structure implied by the patterns.[106] Grimm himself acknowledges this perspectival shift by drawing attention to the (relative) insignificance of all particular instances, once the general laws that govern them have been laid bare: "Should a few of the given examples still seem questionable and uncertain, the majority may qualify as having been rigorously demonstrated, due primarily to the analogy among the gradation; the correctness of the rule in general is unmistakable."[107] By defending the legitimacy of the table against the age-old claims of the individual word, the rigor of formal notation against the contourless accumulation of the list, Grimm in effect appropriates the apparent chaos of letter change for analysis, and thus for science. His earlier call for a more frugal, more solid, and more stepwise approach to the practice of etymological comparison— one that would subordinate the sameness of sounding substance to the temporal (teleo)logic of a *differentiated* order—acquires in the process a set of rules to correspond to its rhetoric: "[T]he study of words depends less on the sameness or similarity of generally related consonants than on the perception of the historical, stepwise trajectory, which can neither be disrupted nor reversed [*des historischen stufengangs, welcher sich nicht verrücken oder umdrehen läßt*] [. . .] and identity is thus everywhere founded on external difference."[108] Just as, for Kielmeyer, the ability to articulate the various stages of organic development gives access to the boundaries of body parts, and thus to a taxonomy of organic life forms, so also, for Grimm, does the table of linguistic transformation allow for the identification of linguistic units—of *p* as different from *t* but also, etymologically speaking, the *same* as *f*—and thus for the classification of distinct linguistic systems. The nature-philosophical notion of systemic organization as definitionally synonymous with the periodic "standing still" of time (*system*, from Greek *syn*, "with, together" and *histēmi/histanai*, "to [cause to] stand") finds paradigmatic notational expression, here, in the horizontal "standing together" of Grimm's privileged, consonantal *Buchstaben* (*Stab*, "staff" from proto–Indo-European **stā*, "to stand," **stebh*, "to support").

The implications for linguistic praxis were summarized as follows by one of the founders of modern Indo-European etymology, August Friedrich Pott:

> Grimm's historical exposition of the sound changes [*Lautumwandlungen*] in the Germanic languages has alone more value than many a philosophical language doctrine full of one-sided and empty abstractions; for in it is sufficiently demonstrated that the letter—as the tangible [*das handgreifliche*] linguistic element, which, although it is admittedly not stable, nonetheless moves along a comparatively calm track—is on the whole a more certain thread in the dark labyrinth of etymology than the meaning of words, which often jumps boldly around; in it is also shown that linguistic research, especially of a comparative kind, lacks a solid support without precise, historical knowledge of the letter."[109]

Pott, who conducted in his *Etymological Investigations* (*Etymologische Forschungen*, 1833) the first systematic implementation and expansion of the sound laws, here cuts right to the heart of the transformation that revolutionized language science. The shift away from the semantic vagaries of the idiosyncratic word and toward the analytic precision of the law-abiding letter, away from an intuition-based emphasis on similarity—for so many centuries the only available indicator of etymological relation—and toward a reliance on rule-governed difference, combined in Grimm's Law to definitively distinguish etymology, for the first time, from the arbitrariness of mere wordplay. No longer could the study of roots be ridiculed, the way August Wilhelm Schlegel had done in a review of the Grimm brothers' pre-sound shift speculations, as "a science where the vowels count for nothing, and the consonants even less."[110] No longer could the methods of etymology be excluded, as Friedrich Schlegel had advocated in his *On the Language and Wisdom of the Indians*, from the science of a common source.[111] Grimm's 1822 insight into the true "meaning" of the consonants bound the phonic material of lost language forms to the spiritual principle of ongoing language growth, and in doing so, granted dead letters new life: "The individual letters are the organs [*Glieder*] of language; the system of letter-relations [*das System der Buchstabenverbindungen*] forms the body of language, and with this body language spirit is inextricably linked."[112] The resulting revolution in the writing of language laid the foundation for nearly a century of unimpeded etymological triumph.

SPIRIT SUPERFLUOUS?

From the vantage point of Grimm and his contemporaries, the successful formulation of the sound laws bore eloquent witness to the power of a teleological, organicist approach. The hypothesis of the *Bildungsprinzip*, however, ushered in its own set of problems for an ambitious new discipline intent on total analysis, since, while the idea of an interior shaping force did indeed facilitate a dissection of unprecedented precision and depth, it also rendered the essence of language in principle impervious to the scalpel. The underlying flow of a form-bestowing, system-animating *Sprachgeist* was presumed to transcend *per definitionem* the writeable stasis of *Stufen* and *Stäbe*, time periods and consonants, as which it "appeared" for the analyst. It is perhaps not particularly surprising, then, that the most ambitious and thoroughgoing applications of Grimm's frugal, solid, and stepwise techniques eventually came to coincide, in the latter half of the nineteenth century, with a series of markedly un-Grimmian efforts to eliminate this unformalizable spiritual remainder altogether. On the strength of the sound laws, which in theory enabled the linguist to analyze language solely by tracing its empirical forms, the great Indo-Europeanist and Darwin disciple August Schleicher sought to reduce the linguistic organism to the play of its "bodily" letters. Only by subtracting the underlying premise of an unwritable, because indivisible, *telos*, he believed, could linguists begin to approach their ultimate goal of a truly remainderless notation. In the course of his reductionist efforts, Schleicher consciously aligned himself with a midcentury comparative anatomical trend that worked to explain organic development on the basis of purely mechanical and materialist principles.[113] He also, less consciously, radicalized the hegemony of the *Buchstabe* to the point of a discipline-endangering apotheosis. Linguistic writing, at Schleicher's hands, acquired a significance that threatened the very foundations of language science.

The notion of language as an empirically observable (and phonetically transcribable) system of sounds forms the basis for Schleicher's materialist theory of the *Sprachorgan*, which defines the linguistic object as the audible consequence of physical and chemical interactions between a yet-to-be-discovered center of the physical brain and a yet-to-be-physiologically analyzed articulatory apparatus: "Language is the audible symptom of the activity of a complex of material relations in the structure of the brain and of the language organs [*Sprachorgane*], with their nerves, bones, muscles, etc. Of course the material basis of language and of linguistic differences has not yet been anatomically determined. As far as I know, however, a comparative investigation of the speech organs of linguistically diverse peoples has not yet even been begun."[114] The proposed understanding of language effectively reverses the gesture of Friedrich Schlegel's original comparative anatomy analogy—which had expanded the meaning of (bodily) dissection to include that of (linguistic) analysis—by folding the latter back into the former. Language can be analyzed, so the contention goes, precisely because it possesses a bodily correlate that, at least in principle, can also be dissected. Schleicher goes on to admit that such a *literally* comparative anatomical approach to the question of linguistic foundations might well prove unsuccessful, since the relevant material substrate could turn out to involve parts too small to perceive. Such a contingent failure at the level of actual observational capacity would not, however, entail for him a revision of his fundamental thesis, which presumes for all perceptible activity the presence of an underlying "material-corporeal condition."[115]

Where the causal substrate itself eludes perception, the scientist must concentrate attention instead on the chain of its perceptible effects:

> What light is to the sun, so audible sound is to language. Just as in the former case the nature of light bears witness to a material basis, so, here, does the nature of sound. The material conditions underlying language and the audible effect of these underlying relationships relate to each other in the manner of cause and effect, or of essence and appearance more generally. The philosopher would say: they are identical. We therefore consider ourselves justified in viewing languages as something materially existing, even if we cannot grasp them with the hand or see them with the eye, but only perceive them through the ear."[116]

The phenomenon of language as studied by linguists acquires the status of material entity from the material entity that generates it, a relation of causality which Schleicher equates with the relation between appearance and essence, and which he then in turn declares to be, "philosophically speaking," a relation of identity.[117] The pronouncement makes a perhaps unintentional mockery of nature-philosophy's fundamentally antireductionist concern with identity as a oneness, not of matter, but of warring forces, whose conflict powers a dynamic of teleological self-differentiation. Not for Schleicher Kielmeyer's "striving toward heterogenesis" with its implication of real, irreversible change and its rejection of mechanical interchangeability. No difference other than physical distance pertains, here, between the sun, as the object of a purely chemical curiosity,[118] and its sensually perceptible rays. Schleicher's analogy exploits the traditional metaphorical function of sun-as-spirit even in the process of subtracting all spirit from the equation, for his literalization of the sun *figures* his literalization of the language source in the double movement from *Sprachgeist* to *Sprachorgan*, and from *Sprachorgan* to audible speech. Stripped of its ostensibly "figurative" dimension—a dimension that, for linguists like Grimm, had of course never really been figurative in the first place—the animating spirit of language dissolves "back" into the wind of its etymological origins, and this wind, in consequence, renounces its status as a matter *in potentio* ensouled.

Language, for Schleicher, remains a life form—"Languages are natural organisms [. . .] they, too, are characterized by that set of phenomena that we generally associate with the name of 'life'"[119]—but a life form from which the potential for purposeful, teleological activity has disappeared. Life in its new materialist manifestation never progresses in any meaningful way beyond the constraints of its causal origins, and can therefore be remainderlessly equated with the laws that describe its permutations.[120] Such laws, in turn, can be formalized in a variety of notational systems much like the one in Grimm's table of consonants—Schleicher's *Compendium of the Comparative Grammar of the Indogermanic Language (Compendium der vergleichenden Grammatik der indogermanischen Sprache)* offers a tabular overview of sound relationships for every individual language he treats, as well as tables showing the transformations of both vowels and consonants[121]—with the notable difference that Grimm's list of exceptions no longer merits the status of necessary supplement. For both Bopp and Grimm, the existence of the occasional unsubsumable exception had testified to the inexhaustible fecundity of a *Sprachgeist* that must be assumed to transcend, at every turn, the law-governed orders in which it appears.[122] By explicitly consigning all such aberrations to the prenotational purgatory of the as yet unexplained,[123] Schleicher thus arrives at a chain of identifications with the force of a monist-materialist credo: all spirit is identical with the material phenomena in which it "shows itself," and all matter can be described in the form of a notational writing, provided its laws have been uncovered; therefore, all spirit can be written down. *Sprachgeist=Sprachorgan=*audible speech=sound laws=notation. No gap exists here, beyond that of a provisional insufficiency of knowledge, between the essence of language and its etymological representation.

The perfection of linguistic writing for which Schleicher is famous occurs in the context of this commitment to absolute notational adequacy. When, for example, he replaces the traditional practice of citing the oldest attested form of a word with the convention of the asterisk still in use today—writing Indo-European **fathār*, for instance, instead of Sanskrit *pitā*—he does so to make room for hypothetical reconstructions that, according to the sound laws, *must have existed*. These unattested proto-forms have no counterpart in the empirical world of language, either written or spoken, but can nevertheless lay claim, in their notational realization, to a thoroughly positivist reality that Schleicher (following Grimm rather than Bopp) calls "Indo-German." He goes on to demonstrate his commitment to this originary reality by making it the site of a contemporary literary "work": "A Fable in the Indo-Germanic Ur-Language," published in 1868, tells the (very brief) story of two horses and a sheep, in a language derived entirely from etymological reconstructions.[124]

Schleicher's best-known contribution to linguistic notation, the *Sprachbaum* or language tree, stands just as firmly under the sign of a remainderless identity of world and writing (see figure 1). The tree, which represents Indo-European and its "descendant" languages according to their respective genealogical roles, survives in contemporary linguistics as a valuable tool for figuring historical relation. Schleicher, however, for whom differences in idiom necessarily presuppose *physical* differences in the articulatory functioning of the *Sprachorgan*, sees in this diagrammatic mode of writing something rather more concrete.

> [T]he so-called racial differences could only be assimilated into a scientific, natural system with great difficulty. Languages, on the other hand, integrate relatively easily, particularly with respect to their

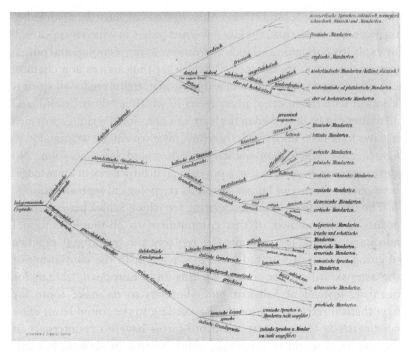

Figure 1. August Schleicher, "The Darwinian Theory and the Science of Language" ("Die Darwinische Theorie und die Sprachwissenschaft"), 71–72 / foldout insert.

morphological side (to their sound-form [*Lautform*]), into a natural system like that of other living beings. For us, then, the externally perceptible form of the cranial, facial, and bodily skeleton is markedly less essential for the human being than that no less material, though infinitely finer, bodily nature, of which the symptom is language. The natural system of languages is in my view simultaneously the natural system of humanity.[125]

The "natural system" of human language, with its ostensible consequences for an equally natural system of race, takes its hierarchical structure from the *temporal* order of historical development ("The comparative anatomy of languages demonstrates that the more highly organized languages evolved very gradually out of simpler language organisms"[126]). The branching course of language life, in its progression from isolated root names ("cells") to inflectional grammar ("higher organisms"), and from Indo-European to the modern European languages, offers the only legitimate standpoint for carving up the linguistic corpus, which in turn provides the only true foundation for an analysis of human kind(s).[127]

True to his mechanist-materialist commitments, Schleicher refuses to interpret this developmental trajectory-cum-typological system teleologically. Both the speech apparatus and the human brain *evolve*, for him, an ardent admirer of Darwin's *On the Origin of Species,* without reference to the mysteries of an interior formative drive: all apparent "striving," on the part of the language organ, toward the incomparable complexity of its (Indo-)German "destiny" can in fact be traced back to countless accidental collisions between individual speakers and their material surroundings.[128] Such collisions add up, over time, to transformations, which are then winnowed, over time, by the "struggle for life" (*Kampf ums Dasein*).[129] The process culminates inexorably, but not purposively, in extinction for the environmentally "unfit" (Schleicher's example is the American Indians), and in expansion for the adaptively endowed (the Europeans).[130] The result is a tree of genealogical relation (*Sippenstammbaum*[131]) that has everything in common with the trees implied by a Darwinian theory of species differentiation *except* for the latter's purely theoretical status. The linguist can actually observe the process that the Darwinian naturalist must hypothesize—owing to a written record more complete than any collection of fossil skulls—and Schleicher can therefore offer the diagram to his comparative anatomist colleagues as a substitute for their famously missing bones: "The linguistic relationships, therefore, as paradigmatic examples for the origin of species out of shared foundational forms, are instructive for those fields of inquiry that, at present, still lack demonstrable cases of this kind."[132] The study of language retains, here, the role of paradigmatic natural science, because it successfully uncovers a particularly useful record of past developments that greatly interest the human scientist. Language itself, however, has renounced the status of microcosm. It develops like a body because it, too, is a (product of) body, and not because it models, according to the logic of a mysterious drive to form, the teleological tendencies of a cosmos-ordering world spirit.

THE DEMISE OF ANALYSIS

By the time of Schleicher's death, in 1868, and due in large part to his notational innovations, the conception of language as a body of writeable sound laws had acquired an utterly noncontroversial place

at the heart of historical linguistics. When, ten years later, Karl Brugman and Hermann Osthoff—both, at the time, colleagues of Ferdinand de Saussure in Leipzig—chose to inaugurate their new journal
by railing against the innumerable arbitrary exceptions with which
their contemporaries regularly diluted these laws, it was therefore
clear to all concerned that the charges were serious. The polemical
first foreword to *Morphological Investigations in the Sphere of the
Indo-European Languages* (*Morphologische Untersuchungen auf
dem Gebiet der indogermanischen Sprachen*) effectively accused
nineteenth-century linguists of defiling their own (and only) disciplinary ground: "Only he who adheres strictly to the sound laws,
these foundational pillars [*grundpfeiler*] of our whole science, has
in his investigations any kind of *stable ground under his feet*. He,
on the other hand, who—unnecessarily and only in order to satisfy
certain desires—allows exceptions to the sound laws governing a
dialect [. . .] necessarily falls victim to subjectivism and arbitrariness."[133] The self-styled "Neogrammarians"[134] of Leipzig were not,
of course, the first to reject what they perceived as an outdated philological residue lingering on at the heart of contemporary linguistic
practice. They were not the first to criticize the widespread prioritization of historical idiosyncrasies at the expense of regularities, or the
unsystematic exploitation of the sound laws as occasionally expedient
tools. Nor were they the first to insist on the exceptionless functioning of these laws as the sole foundation of language science.[135] Indeed,
the actual Neogrammarian approach to concrete problems differed
little, at least at the outset, from the approach of many of their oppo-
nents, for whom the sound laws also formed the backbone of etymological investigation.[136] The tremendous force of the Neogrammarian
polemic, therefore—a polemic that more or less instantaneously
ignited the so-called "Lautgesetz controversy," to which linguists for
decades continued to refer—lay less in the authors' endorsement of
any particular investigative technique than in their rhetorical reframing of an acknowledged linguistic task.

　　This reframing owed its power, in turn, to the Neogrammarian
radicalization of Schleicher's explicitly mechanist perspective ("*First.
Every sound change, inasmuch as it occurs mechanically*, takes place
according to *exceptionless laws*."[137]). The sound laws, by reducing the
chaos of language change to the order of the consonant table, continued to hold out the promise of a perfect correspondence between notation and referent, but the Schleicherian assumption of total adequacy

had run aground on the apparent irreducibility of the empirical surplus. So long as the dreaded exceptions remained, real language would always exceed its linguistic representation in the laws. The first step was simply to make this point, and the first point of attack was therefore a tradition of teleological organicism that, in the eyes of the reformers, concealed the all-important representational dilemma behind a mirage of self-generating structure. Where the question of boundaries is presumed to be synonymous with the question of spiritual extension, as in the notion of a language defined by its *Sprachgeist*, the imperfections of linguistic writing cannot possibly be seen to pose a foundational threat. Empirical excess will simply bear witness to the spontaneity of a spirit *in the process* of bringing ever more matter under its purview; *in time*, the true contours of the system will make themselves known. This practice of ascribing the law-violating remainder to the mysteries of a potential as yet not fully actualized renders inherently imperceptible, and thus also irresolvable, the very dissonance that so plagued the Neogrammarians: the language of a teleological organicism prevented the early historical linguists from tackling the problem of the exception precisely by preventing its emergence *as a problem*. The Neogrammarians therefore condemned, much as Saussure would go on to do a few years later, the "inexact" terminology of a discipline in which "metaphors" of language life could be mistaken for literal reality. They threw the whole weight of their polemical rhetoric against a linguistics "where language and linguistic forms are allowed to lead a life of their own, over and above that of the speaking individuals, and [. . .] where metaphorical expressions are continually taken for reality itself."[138] And they hoped thereby to free their fellow linguists from the fetters of an organicist vocabulary that obscured the "real" linguistic object from view.

Their true contribution to the history of linguistic self-reflection lies in the unparalleled rigor with which they performed this terminological critique. For instead of restricting their polemic to those figures of linguistic body that relied on an animating spirit, as Schleicher had done before them, the Neogrammarians entered into a secondary polemic with the organicist rhetoric of materialists such as Schleicher himself, whom they counted among their forebears. Schleicher's conception of language life, with its empirically unverifiable relationship to a hypothetical *Sprachorgan*, seemed to Brugmann, Osthoff, and their colleagues no less abstract—and thus also no less "metaphorical"— than the pre-Darwinian conception it had claimed to "literalize." The observable phenomenon of speech, which Schleicher had proposed to study *in lieu* of the as yet undissectable substrate, has itself nothing in common with a cadaver, since it offers of its own accord no criterion for determining where (not) to cut. Like the light rays in the analogy of the sun, the sound waves of the acoustic "symptom" flow continuously, without audible delimitation, along a path of gradual, barely discernible change. But where even the fantasy of a future dissection fails to hold out the promise of perceptible boundaries—the bodily substrates, after all, might *also* turn out to differ from one another only in infinitesimal increments—and where no animating spirit "inside" the phenomenal can be called upon to articulate the sound stream, the language tree and letter tables necessarily lose all claim to the status of objective fact. No longer readable as the expression of an empirical *tendency* toward form, the validity of such constructions becomes entirely dependent on the gaplessness of the knowledge they generalize: a single ineradicable exception will suffice to render them illegitimate, or at the very least, premature.[139] From the Neogrammarians' more stringently mechanist, more forthrightly positivist perspective, Schleicher's unattested proto-forms are therefore only ever as real as his theory of sound change is complete, no matter how many fables he cares to compose. Conclusions about empirical languages drawn on the basis of this Indo-German "ground" have for them the character of tautological pronouncements about the current state of linguistic research: "One thus moved, and even today many still move, without knowing it or wanting to admit it, in the most *obvious circle*."[140]

The Neogrammarians do not, themselves, have an alternative theory of linguistic foundations to offer. Indeed, they often appear to be prevented in principle from going in search of such an alternative, by a nearly universal tendency to interpret the vantage point of the critique as itself the site of the solution.

> Only that comparative linguist who once and for all steps out of the hypotheses-beclouded atmosphere [*aus dem hypothesentrüben dun-stkreis*] of the workshop, in which the original Indo-German founda-tional forms [*grundformen*] are forged, into the clear air of tangible reality and of the present, in order to gather knowledge about those things which gray theory can never reveal to him [. . .] only he can arrive at a correct idea of the way in which linguistic forms live and transform, and only he can acquire those methodological principles [. . .] without which any penetration into the time periods that lie behind the historical transmission of language is like a sea voyage without a compass.[141]

The way out of the notational circle, according to Brugman and Osthoff, leads away from the mythical past and into the observable present, where misleading terminology has less chance of concealing empirical realities. Yet the logic of the linguist as natural scientist whose job is simply to observe—here implicitly counterposed to the logic of linguist as philologist, buried among the dusty and disintegrating pages of texts in dead languages, lost in reveries of reconstruction—merely reverses the prioritization of originary language without transcending it. The privileged linguistic object is now no longer the oldest but rather the "youngest" and most alive.

Against the fantasy of a fully reconstructed Indo-European origin ("that nebulous image [*nebelbild*], which simply cannot deny its nebulous home [*nebelheimat*]"[142]), Brugman and Osthoff thus champion the equally fantastic vision of a present-tense language accessible in its originary, "natural" state: "In all living dialects [*volksmundarten*] the sound shapes peculiar to the dialect always appear much more consistently realized throughout the entire linguistic material and maintained by the members of the linguistic community than one would expect from the study of the older languages accessible merely through the medium of writing."[143] The shift in emphasis from past tense to present, which clears the way for an explicit statement of the definitional dilemma and so for a reassessment of linguistic vocabulary, comes to replace, in this passage and elsewhere, the actual quest for a new linguistic ground, as though audible speech given directly to the ear of the linguist could undergird a scientific linguistics merely by virtue of being literally *there*.[144] Etymological transparence, internal consistency, and phonetic purity are all presumed to be available, without the effort of reconstruction, in the linguist's own backyard, if not in his mother tongue then in the tongues of his more fortunate neighbors ("in his native dialect or elsewhere"[145]): the researcher who listens carefully enough to the right idiom should actually be able to *hear* the homogenous region where sound laws function without exception.

It goes almost without saying that this wildly idealized version of an isolated mountain dialect (Brugman and Osthoff cite Johann Winteler's groundbreaking study of a remote Swiss Alps idiom as the paradigmatic example of skillful linguistic listening) does not actually solve the problem of language boundaries. The geographical formations that circumscribe the physical movements of mountain villagers, for instance, cannot be presumed to do the same for their manner of speaking, any more than the bodily limitations that define the single speaker can ensure the self-consistency of his or her individual pronunciation. While the Neogrammarian fascination with local and personal speech (with *Ortsgrammatik* and *Idiolekt*) does indeed prepare the way, then, for increasingly precise general statements about the "mechanism" of language change—as evidenced by the eventual appearance of systematic works such as Eduard Sievers's *Principles of Phonetics* (*Grundzüge der Phonetik*) and Hermann Paul's *Principles of the History of Language* (*Prinzipien der Sprachgeschichte*)—it fulfills this function without ever confronting the conundrum of the changing substrate. To the question "What changes?" and so also to the question "What is language?" the Neogrammarians provide no answer.

Or rather, they provide the answer that there can *be* no answer, and thus also no well-founded analysis. Both Sievers and Paul, whose works, taken together, offer the most comprehensive expression of the mature Neogrammarian perspective, treat definition in general as a necessary but regrettable act of abstraction, and the concept of language in particular as an unavoidable but potentially dangerous fiction. They seek to replace the teleological notion of a formative principle with a mechanistic account of physiological language formation ("under 'phonetics' we understand the doctrine of language formation [*die Lehre von der Sprachbildung*]"[146]) and to build up from this atomic level to an additive account of dialect formation. But they reject the idea that the "dissections" performed by their predecessors have any empirical correlate at the level of individual speech production, and hence also the idea that the results of such dissections—in the form of elementary phonemes, or *Glieder*—could play any constitutive role in the explanation of language phenomena: "A true analysis [*Zerlegung*] of the word into its elements is not merely extremely difficult; it is actually impossible. A word is not a strung-together compound [*Aneinandersetzung*] of a definite number of independent sounds [*einer bestimmten Anzahl selbständiger Laute*], of which each could be expressed by an alphabetical sign, but rather, at bottom, a continuous series of infinitely many sounds [*eine kontinuierliche Reihe von unendlich vielen Lauten*], and letters always only indicate, in an imperfect way, individual characteristic points from this series [*einzelene charakteristische Punkte dieser Reihe*]."[147] Such a perspective, which makes explicit the true consequences of a rigorously mechanistic, materialist approach, sounds the death knell for the nineteenth-century project of writing language. With the principle of linguistic delimitation goes the possibility of a linguistic notation that could ever claim to be more than a distorted approximation of speech. And with the possibility of an undistorted writing goes the promise and purpose of nineteenth-century language science.

CHAPTER 2

Saussure's Dream

IN SEARCH OF THE LITERAL

For Saussure, who received his formal linguistic training in the Neo-grammarian center of Leipzig, all methodological reflection occurs in the context of the definitional dilemma precipitated by the late nineteenth-century shift from a nature-philosophical to a mechanist-materialist perspective.[1] Neogrammarian-style arguments against the organicist hypostatization of language permeate his manuscripts, and his frustration with figures of birth, death, and growth leads him to formulate a general prohibition against metaphor: "There is no object wholly comparable to language, which is a very complex entity, and that is why all comparisons and all images of which we habitually make use regularly end up giving us an idea that is in some way false" (*WGL*, 100–101/152). Radicalized in its turn, the critique of metaphor generates a recognizably Neogrammarian critique of naming badly, of words such as "Latin" and "French" that conjure up the illusion of well-defined givens where no empirical contours exist: "In reality language is not an entity defined and delimited in time; one distinguishes between the French language and the Latin language, between modern German and the German of Arminius [. . .] and then one admits that one begins and the other finishes somewhere, which is arbitrary" (*WGL*, 103/155). Such imprecision obscures the absence of a legitimate, literal object to which linguistic terminology could correspond; a lack of definition on one level hides a lack of definition on another. The book Saussure envisions writing would therefore treat the word as a name that fails to adequately represent the linguistic "reality" of (other) words, a language that gets in the way of investigating language: "There will one day be a very special and interesting book to write about the role of the *word* as the principle perturbance [*perturbateur*] in the science of words" (*WGL*, III/166). The purpose would be preparatory: by exposing terminological flaws, Saussure hopes to clear the way for a renaming that presupposes, as the condition of its possibility, a (re)definition of the actual object. In an environment where the unconscious deployment of metaphor continues to conceal crucial methodological gaps, the disambiguation of linguistic language—the seemingly simple act of defining, and thereby *literalizing*, one's terms—assumes the role of primary task.

The question of Saussure's relationship to the German linguistic tradition will eventually return to the foreground. Here at the outset, it suffices to indicate his teachers' pervasive if subterranean presence in his statement of the problem, for it is only against this background of an absolutely rigorous empiricism, to which Saussure himself subscribes, that the true radicality of the Saussurean "solution" can emerge. Like his Neogrammarian predecessors, Saussure believes (only) in bodies and takes seriously the argument that a lack of perceptible contours deprives linguistics of its ground. Unlike them, he also takes seriously the impossibility of locating these contours. In the definitional aporias brought to light by new subdisciplines such as phonetics and dialect studies—for the Neogrammarians themselves a primarily technological challenge—he sees proof that language exists beyond body not only provisionally but in principle: "This then is our *profession de foi* in the matter of linguistics. In other domains, one can speak of things 'from such and such a viewpoint,' with the certainty of finding solid ground within the object itself. In linguistics, we deny in principle that there are given objects, that there are *things* which continue to exist when we move from one order of ideas to another, and that it is permissible to consider 'things' in several such orders, as if they were given in themselves" (*WGL*, 139/201). While his positivist colleagues, therefore, continue to scour the physical and psychic universe for the measurable given without which they firmly believe no science can function, Saussure, in a breakthrough born of desperation, begins to look elsewhere. The question he inherits— "Where in the world is the real of language?"—gives way to a new query, one that sets him definitively apart from his contemporaries and fundamentally alters the structure of future linguistic reflection: "Where is the real of language," he finds himself forced to ask, "if it is *not* in the world?"[2]

This revolutionary uncoupling of reality and material world, scientific object and sensory experience, paves the way for a confrontation with prevailing methodological wisdom regarding what scientists do and how they relate to the things they study. It thus participates in the same turn-of-the-century reevaluation of scientific principles that marked antiempiricist projects such as Edmund Husserl's phenomenology and Émile Durkheim's sociology.[3] It also, and for precisely this reason, deeply disturbs Saussure, who abandons his empiricist commitments in response to a peculiarly linguistic problem without ever calling into question his allegiance *in general* to the bodies on which those commitments rest.[4] Unlike many of his more philosophically inclined contemporaries, he has no interest in destabilizing the primacy of empirical truth. He sees in the fluidity of linguistic identity not an opportunity to rethink the status of scientific method per se—a project rendered superfluous, he believes, by the enviable *solidity* of the other disciplines ("In other domains, one can speak of things")—but a liability for one particular method in its struggle to become scientific. When he nonetheless takes as his precarious starting point the constructedness of all linguistic phenomena, the task he therefore accepts is not that of transforming science to account for the inevitability of construction but of accounting for construction so as to render it accessible to science. The concept of a transempirical real is for him an unavoidable evil, a necessary response to the idiosyncratic predicament of his own oddly objectless discipline, and it can play no role for him so long as its reality is not somehow attested, indisputably if indirectly, by the self-evidence of unmediated experience. Before Saussure can accept language as a real that transcends the world, in other words, he needs the world to bear witness that such a reality exists.

The world does so, according to his ingenious interpretation, in the *fact* of incomprehension: "First finding among the facts of language [*langage*]: *plurality* of languages [*langues*], geographic *diversity*. [. . .] Even the primitives have this notion; it is this diversity that allows the peoples of the world to become conscious of their language [*langue*]. Perhaps they would otherwise not even notice what they are speaking."[5] Saussure, taking his cue from "the peoples," deduces the reality of language from the universal experience of linguistic

difference.[6] Not language itself but its Babelian opposite, the condition of hearing without understanding, is for him the unmediated
given (*immédiatement donnée*; CLG, 436:6) of linguistic communication. This condition, however, has *la langue*—the ability to speak,
but not to everyone—as the all-important condition of its possibility,
since only where language exists as a public region of mutual comprehensibility can the lack of language appear *as* lack. The notion of
the babbling, "barbar"-spouting foreigner always presupposes its self-
identical (Greek) opposite, a shared mother tongue in which noises
have communally recognized meanings and communally acceptable
articulations.[7] Not sound, then, but *mere* sound is the key to locating
language: streams of speechlike noise that refuse, against all expectations, to dissolve into clearly delineated chunks of meaning, thereby
demonstrating beyond doubt that real language is a matter of more-
than-material analysis for linguist and layman alike. The empirical
event of incomprehension assumes, in this context, the role of originary linguistic fact (*fait primordial*; *Notebooks* III 12a / CLG, 437:5)
because it discloses a realm in which boundaries truly, if transempirically, exist. The result is the reality of the units (and languages)
thereby defined. "In language we have a fact <object> of a concrete
nature" (*Notebooks* III 71a / CLG, 44:5).[8]

Saussure's peculiar choice of the word "concrete," in connection
with the reality of a nonbodily being, announces his conception of
language as a polemical alternative to the Neogrammarian category
of abstraction. In one of his only direct references to what he calls
"the new school," he criticizes his colleagues' positivist critique of
hypostatization for reducing all traditional linguistic categories to the
status of arbitrary mental constructions, devoid of "a real existence"
(WGL, 124–25/183). The practitioners of this useful but ultimately
misguided critique can draw no truly new conclusions about the categories they try and fail to abandon, because their mechanistic paradigm leaves no room for a reality that threatens the duality of thing
and no-thing. If the language of linguists does not refer to a material body, as scholars like Schleicher had assumed, then it must take
its place in the scientific netherworld of disembodied psychological
abstraction. Flesh or (the fantasy of) spirit, nature or (the fiction of)
culture: for the heirs of the materialist mantle, the choices are only
two.

Saussure, armed with his deceptively straightforward analysis of
incomprehension, proposes a third: "*Key principle*: what is *real* in a

given state of language [*langage*] is that of which the speaking subjects are conscious, everything of which they are conscious, and nothing except that of which they can be conscious" (*WGL*, 132/192).
The foundational tenet here so dramatically announced seems at first
glance to offer little more than a new name—and a newly positive
valuation—for the psychological purgatory to which the positivists
had already condemned their object. In actuality, however, Saussure's
"grand principe" adds something crucial to late nineteenth-century
interpretations of mental activity, and this "something" is the plural
of "the speaking subjects," which restores to the concept of consciousness the collectivity implicit in its etymological roots. (*Conscience*,
from the Latin *con* + *scientia*, is a construction parallel to the German *Mitwissen*, meaning "shared knowledge" or the state of "knowing with." *Avoir conscience*: "to have shared knowledge," in this
case, of a language.) "Language [*langage*] is always being considered
within the context of the *human individual*, which is a false perspective. Nature gives us man *organized for articulated language* [*langage
articulé*], but *without articulated language*. Language [*langue*] is a
social fact. The individual, organized for speaking, can only put his
apparatus to use within the context of the community that surrounds
him" (*WGL*, 120/178). What really defines man *as* man, from the
perspective of this passage, is not a natural, individual capacity for
articulated language—Saussure plays here with the traditional Aristotelian definition of the human as *speaking* animal, or *zoon logon
echon*—but a conventional and thus also *communal* knowledge of
one articulated language rather than another: "In this respect, then,
man is only complete by way of what he borrows from his milieu"
(ibid). Since, however, the shared mother tongue is always also con-
strained on all sides by the originary fact of linguistic difference, the
plural of shared knowledge necessarily remains *merely* plural, without aspiring to the quasimystical identity of the All. The euphony of
an originary *Logos* is every bit as foreign to Saussurean science as
the cacophony of uncountable private idioms, and only the no-man's-
land of an awareness *between* one and everyone—an awareness as
arbitrary but unchangeable as the social environment amid which
the speaker dwells—can account for the empirical fact of successfully communicating pluralities, simultaneously multiple and mutually exclusive.

  This double exclusion carves out the peculiarly interstitial space
from which everything else, for Saussure, can be seen to follow. On

the one hand, a knowledge not in principle available to mankind as
a whole can be neither instinctual nor purely empirical. It can have
no objective foundation in nature, human or otherwise, and no privileged relation to any universal truth. It must, in short, be arbitrary,
a question of conventional relations established and acknowledged
only by the group that "knows." The passage above makes such arbitrarity into a fundamental component of the human essence, since it
is only through the *necessary accident* of socialization that the individual can actualize his or her innate potential for a linguistic, and
thus also human, mode of being. On the other hand, and despite its
arbitrariness, the shared knowledge of *conscience* cannot be considered conventional in the conventional sense—as an imposition of (cul-
tured) will onto (natural) matter, or an intentional infusion of physis
with form—because the notion of a nomos-generating naming fails
to account for comprehension as *collective* fact.[9] With no roots in
nature, the sound-sense bond falls prey to individual idiosyncrasies
of pronunciation and meaning, and language therefore changes constantly, unconsciously, and uncontrollably as it passes from mouth
to mouth. Communication, however, continues to occur. In order to
explain this continuity from within the conventional naming paradigm, the linguist would have to presuppose a communal dictionary of sound-sense linkages comprehensive enough to contain every
individual variation. Since the number of such variations is always
in principle infinite—in the absence of naturally existing, empirical
units, the number of possible sound-sense permutations quite literally
knows no bounds—the entries in this hypothetical dictionary could
exist only as so many never-ending lists of admissible alternatives.
Absorbing enough information from such lists to guarantee successful communication with even a few other speakers would be the work
of several lifetimes.[10]

When Saussure refers to the problem of "the role of the word," for
which the German linguistic tradition fails to provide an adequate
solution, he therefore has in mind first and foremost this conundrum
of collective comprehension: the existence of "synchronic" identities such as *alka-alka* and *ōk-ōk* testifies *not* to the past and future-
less simultaneity of an immediate presence but to the bizarre, and
bizarrely bounded, continuity of particular sound-sense relations
across vast swaths of time and space. By locating the literal reality of
a language in the *conscience* of its speakers, he acknowledges such a
restricted continuity as the fundamental fact of linguistic experience,

and dedicates himself to the task of exposing its counterintuitive structure. The mystery of a shared knowledge neither natural nor conventional, neither instinctual nor intentional, becomes the true subject of Saussurean reflection, wreaking havoc in the process on the binary logic of his rather conservative conceptual vocabulary: "By distinguishing thus between language [*la langue*] and the faculty of language [*la faculté du langage*], we see that language [*langue*] is what we may call a 'product' [. . .]. One can conjure up a very precise idea of this product—and thus set language [*langue*] so to speak materially before oneself—by focusing on what is virtually present in the brain of a sum of individuals <belonging to one and the same community> even when they are asleep" (*Notebooks* III 7a / *CLG*, 40–41:5; translation modified). The passage abounds with contradictory rhetoric, employed here as a way of approaching a phenomenon that eludes all readily available categories. One represents to oneself "so to speak materially" what only exists virtually; the site of this material/virtual reality is the singular brain of a plural sum; the singular/plural brain remains conscious of language even while unconscious. The resulting description reproduces quite precisely the definitional recalcitrance of the language space, which unquestionably transcends, despite Saussure's talk of brains, the traditionally inescapable dualities of biology and psychology, nature and culture, flesh and spirit. As the coming together of many (but not all) minds in a single linguistic *(in)conscience*,[11] this meta-physical reality mediates between the universality of instinct and the individuality of consciousness as conventionally understood. The specifically human negotiation of necessity and freedom has its origin in this incomparable collectivity, "without which, in fact, neither the individual nor the species could have in any way aspired to develop its innate faculties" (*WGL*, 95/145).

NEITHER FLESH NOR SPIRIT

Saussure's hypothesis regarding the *structure* of this communal space famously replaces an unwieldy collection of uncountably many imperfectly delimited acoustic impressions with a self-supporting system of differential relations. How is it that native speakers hear phonetically different streams of sound pronounced by different mouths at different times and "know" that they are nonetheless the "same"? The only possible answer, says Saussure, is that the speakers are subordinating an infinitely difficult judgment about sameness to a much simpler, because finite, judgment about significant difference: "*Fundamental principle of semiology* [. . .] Language [*langue*] contains neither *signs* nor *significations*, but DIFFERENCES in signs and DIFFERENCES in signification" (*WGL*, 46/70). Language, as it exists in *conscience*, is for him not a dictionary of unlimited scope but a web of functional oppositions, in which a word simply means everything that the words around it do not: "The *present* state of a form is in the forms which from one moment to the next surround it" (*WGL*, 164/232). The word "cat," for example, will remain "cat" so long as its pronunciation does not veer too far in the direction of "kit," "cut," "cot," "cap," or "cad," and its meaning will fluctuate freely within the limits imposed by words such as "kitten," "feline," "lion," "dog," and "pet." Unlike the set of possible pronunciations or shades of meaning associated with any particular slice of the sound-sense landscape, the number of boundaries needed to keep such a slice distinct from its neighbors is limited by the total number of other slices. Shape rather than substance determines definition for the ordinary speaker of language, since finitude of form alone can explain the empirically verifiable experience of mastery over a system so infinitely complex and varied. The potentially infinite vagaries of positive signification arise only secondarily, in the empty spaces carved out by this finite geometry of negation.

Where units depend solely on their difference from other units for both form and content, the notion of language as representation necessarily comes under attack. The most radical implications of this oft-repeated observation for a Saussurean theory of signification have just as often been overlooked, perhaps because they are largely absent from the published *Cours*. The manuscript pages unearthed in 1996 provide a perfectly concise formulation of the conceptual revolution the theory entails: "*Corollary.*—There is no difference between the literal meaning [*le sens propre*] and the figurative meaning [*le sens figuré*] of words (or: words have no more a figurative meaning than they do a literal meaning), because their sense is eminently negative" (*WGL*, 47/72). Saussure draws here, in the form of a supplementary conclusion, the most significant consequence of a system founded on difference. As a second-order naming that both reflects and, in its emphasis on similarity, corrects the first, the metaphorical relation has offered throughout history a singularly influential model for the conceptualization of language. A literal entity infused with figurative meaning bestows bodily form on the intangible; a thing in the world becomes the vessel for the representation—the *ex*pression—of thought through the marriage of letter and spirit, matter and mind. This marriage, which takes place at the level of the word, presupposes the more profound union it mirrors; before the word can mediate between *sens propre* and *sens figuré*, it must first join sound and sense in the relation of representation that grounds all others, which is to say, in the originary metaphor tying thing to name. The traditional arbitrarity thesis negates this bond of similarity between thing and name but leaves room for a compensatory identity between the *relations* among things and the *relations* among names; it changes nothing fundamental about the structure of representation, which is also, in its most privileged form, the structure of knowledge.

Using the paradigmatic example of the sun to make his point— "(and we expressly choose an example that is relatively [ ])" (ibid.)— Saussure radicalizes the harmless arbitrarity of the unmotivated sign into a principle that explodes the parameters of both reference and metaphor. The fact that in French a person can be "the *sun* of another's existence" has for him nothing whatsoever to do, *even conventionally*, with the "literal" light of the physical sun ("the idea, exterior to language [*langue*], of the SUN"; *WGL*, 48/72), since the various significations attached to the French sun are in fact nothing more than the accidental byproduct of the subtraction governing French signification ("The collection of ideas united under each of these terms will always correspond to the *sum* of ideas excluded by the other terms, and to that alone"; *WGL*, 53/80; emphasis mine). The positive difference in kind between literal and figurative meanings disappears here into a negative identity of number. By thus demoting *le soleil* to the status of one sign among others ("like *star, constellation, light, unity, aim, joy, encouragement*"; *WGL*, 48/72), Saussure rejects both the structure of symbolism in general, which allows language to commune with the other it represents, and the priority of this *particular* symbol, which traditionally represents the heavenly guarantor of all earthly representation. His refusal to privilege the time-honored metaphorical link between the sun and the meaning of existence ("the sun of the existence of another"), between the light of the sun and the light of Truth, goes hand in hand with a more fundamental refusal to privilege meaning and truth *as such*, together with the referential structure they presuppose. Language, for Saussure, does not point outward toward things or inward toward thoughts; it does not try for and fall short of the remainderless identity with reality that is both the origin and the goal of the representational endeavor. It means, but it does not do so in the direction of something one could call truth: "To complain about the inexactitude of language [*langue*] is to misunderstand where its power resides. [. . .] The existence of material facts, like the existence of facts of an other order, is a matter of indifference to language [*langue*]. Language is ever advancing, pressing forward with the help of the formidable machinery of its negative categories, truly free of every concrete fact, and thus immediately prepared to assimilate [*emmagasiner*] any idea that comes to join those that have preceded it" (*WGL*, 51/76). The search for an equation between the sun that is and the sun we see or say—a search that informs parables of provisionally inaccessible suns from Plato to August Schleicher[12]— gives way here to an oddly powerful equation between language and itself, which turns out to *work* precisely insofar as it ignores its own representational "content." Language can say everything, including that which has never been said before, because it expresses nothing but what it says; it mediates perfectly between speakers by mediating not at all between them and their worlds.

As abstract and paradoxical as such claims may sound, the phenomenon they describe remains perfectly accessible to the common sense of *conscience*. In order for the word "tree," for instance, to serve as a sign between two speakers, it must be divorceable from past experiences of concrete trees, which the speakers might not share, as well as from definitions of "tree" (philosophical, botanical) with which one or both may be unfamiliar. At the most basic communicative level, the only condition for comprehending what is meant by "tree" will be the condition both speakers fulfill the moment they start speaking of "trees" rather than *Bäume* or *arbres*: that is, the condition of speaking the *same language*. This language will remain collective precisely to the extent that it continues to signify quantitatively rather than qualitatively, on the basis of internal differentiation rather than representational identity—to the extent, therefore, that it eschews the structure of hierarchical depths inherent in the notion of figurative and literal meaning.

With the structure of hierarchical depths, where levels of signification are determined by their proximity to a meaning-bestowing origin or ground, goes the etymological method, where such distances are calculated in terms of time, and time in terms of phonetic transformations. Language, defined as a public region of mutual comprehensibility, depends on the continuity of its functions across potentially infinite variations in pronunciation and meaning; etymology, on the other hand, operates at the junctures where such continuity can be seen to break down. Dialects that close in on themselves and become incomprehensible for outsiders, languages that evolve until a community can no longer read the testimony of its ancestors: both are cases in which the true structure of language has already dissolved, or rather, divided, rendering once-obvious relationships the exclusive province of the interpretive expert. An etymologist works to reconstruct the bond that once tied these regions together. He or she does so, however, from a perspective alien to that of the speaking communities under investigation. Within these communities, collective *awareness* of the code alone determines its scope, and the opaque connections traced by etymologists—connections either long forgotten or never known by the vast majority of the speaking public—can play no communicative role. For the Saussurean science of language, therefore, they simply do not exist. To do linguistics from a Saussurean perspective is "to systematically ignore all etymological or retrospective factors which are absent from consciousness" (*WGL*, 45/68).[13]

True linguists, in other words, do not concern themselves with levels of meaning whose apparent significance, whether historical or metaphorical, finds no support in a separate sign:

> The synonymist who marvels at all the things that are contained in a word like *esprit* thinks that such treasures could never be therein contained if they were not the fruit of reflection, experience, and profound philosophy accumulated in the fundament [*au fond*] of a language [*langue*] by generations of users. And in some sense such a person might be right, to a certain extent; I shall not examine this point because it is in any event, in reality, a secondary fact. The primary and fundamental fact is that synonymy will establish itself instantaneously in any system of signs that one puts into circulation. To hold the contrary is impossible and would be like refusing to assign opposing values to opposing signs. (*WGL*, 52/78)

Saussure here deploys the word "synonym" in a strictly etymological sense—*syn* "together" + *onoma* "name" = multiple meanings sharing a name—which has more in common with the technical term of its Aristotelian origins than with the conventions of modern use.[14] The neologism "synonymist" designates, in the context of this passage, the scholar who takes seriously a phenomenon more commonly known as polysemy, a position Saussure clearly equates with the teleological understanding of language as a site of evolutionary growth.

The synonymist, like the historical linguist, derives all meaning from a *temporal* relationship between origin and end, in which the latter ("the fruit of reflection, of experience") fulfills the promise inherent to the former by actualizing its potential for becoming ever more explicitly itself. Linguistic explication, understood as a synonymist science, must therefore proceed hand in hand with historical development, understood as the gradual unfolding of truth through time (*explicare*, from Latin *ex* = out + *plicare* = to fold). The task of revealing the spiritual treasures "contained" in a language is simultaneously that of drawing, always proleptically, the teleological arc of its continual, chronological self-transcendence.

While Saussure makes no attempt to argue the synonymist's perspective out of existence entirely, even going so far as to acknowledge the possibility of a vantage point from which such contemplations would not appear absurd, he does insist that this perspective will never be a linguistic one. The admittedly ubiquitous phenomenon of multiple meanings *means*, after all, next to nothing in a system where meaning depends—exclusively—on a perceptible difference of signs, and where sense without sound fails, *collectively speaking*, to register. A comprehensible language always operates with the perfect clarity of a zero-sum game, its tautological, two-dimensional territory unbroken by the crevices of the half-said so often imagined to lurk at its edges.[15] To assume otherwise—to draw linguistic conclusions, for instance, regarding *esprit* as a descendant of *spiritus*, which binds the abstraction of spirit to the concreteness of breath or wind—is to ignore the fundamental arithmetic of one signifier, one signified, without which the negative geometry of difference would collapse. The *synonymiste* who goes digging in the sound-sense plane for the buried treasure of past thought will find little more, in Saussurean terms, than a multitude of empty screens on which to project the play of his (or her) own fantasies, a mirage of positive, private essence in a negative, public space.

BUT RATHER WRITING

Saussure's proposed solution to the Neogrammarian dilemma of an unwritable language does not exhaust itself in his unambiguous rejection of representational paradigms, for not all such paradigms are created equal. While some, including the mechanist theory of conventional naming (together with its late nineteenth-century

Neogrammarian manifestation, the theory of psychological association), deserve to disappear forever from the domain of modern linguistics, others—like the teleological organicist model, with
which Saussure's notion of system in fact has almost everything in
common—merit a more complex form of subversion. This subversion
unfolds, I would argue, according to the principles of *parody*, which
works to subtract the mysteries of a form-bestowing spiritual essence
while leaving the *form* of the parodied original intact. Key figures
and concepts from the nature-philosophical approach to language
undergo at Saussure's hands a transformation that empties them of
their relationship to an animating life force while simultaneously con-
scripting them for use in the structuralist reimagination of system:
sound relinquishes its position as organic body to the psychological
sound-*image*, a differential entity that requires no material manifestation;[16] meaning gives up its role as word-soul to become a function
of quantitative value; *conscience*, as the shared awareness of sound-
images and significatory values, displaces the collective identity of a
historically developing *Sprachgeist*; and the harmonious unity of the
living, growing whole reemerges as "the formidable machinery of its
negative categories" (*WGL*, 51/76).

The result is a negative mirror image of the teleological paradigm:

> The <characteristic> role of language [*langage*] vis-à-vis thought is
> not <to be> a phonic, material medium [*un moyen phonique, maté-
> riel*], but rather to create an intermediary realm [*un milieu intermé-
> diaire*] [. . .]. [W]e must not fall into the banal idea that language is
> a mould [. . .]. <This is not it at all: it is not the materialization of
> these thoughts by a sound which is a useful phenomenon,> it is the
> <in some sense> mysterious fact that the thought-sound implies the
> divisions which are the final units of linguistics. [. . .]
>
> (Comparison of two amorphous masses: the water and the air. If the
> atmospheric pressure changes, the surface of the water decomposes
> into a succession of units: the wave <=intermediary chain which does
> not form a substance!>.) (*Notebooks* II 21a–22a / *CLG*, 253:2)[17]

Explicitly, these lines argue against a mechanistic understanding of
language ("the banal idea"), which sees in sound a passive vehicle
for the causal agency of thought. Implicitly, however, the hyphenated
compound "thought-sound" responds first and foremost to a significantly more sophisticated organicist alternative, for which the coming
together of thought and sound is always also a coming to life, and the
coming to life a temporal process of spontaneous self-differentiation.

The "intermediary realm" carved out by the coupling of thought-sound, where distinctions proliferate according to a "mysterious" principle of definition, could easily be understood as yet another version of the nature-philosophical creation dynamic—where acoustic matter assumes articulated form via the power of a spiritual infusion or *Durchdringung*—were it not for the parenthetical analogy of "the water and the air." Disguised as a meteorological metaphor, this witty and perhaps unconscious retelling of the Genesis story ("The earth was without form and void [. . .] and the Spirit [*ruach*, literally "wind"] of God was hovering over the face of the waters")[18] takes the demystification of spirit to radical, even blasphemous extremes. By recasting the generative spontaneity of the Holy Ghost in the role of low-pressure front, Saussure recalls the literalizations of his materialist predecessor, August Schleicher, who sought to reduce the creative potency of *(Sprach)geist* to the positivist "fact" of articulated air. Saussure, however, has no interest in tracing spirit back to body, since he in fact agrees with the proponents of language spirit that the riddle of system requires both categories. He wants, rather, to envision the origins of structure *as an empty, essence-free version of the organicist form-matter marriage*. What fascinates him is the profoundly *im*material and *un*spiritual mode of being that arises through the confrontation of the voluminous phonic and psychic substances at the dimensionless boundary of their fruit-bearing encounter: "The two states of chaos, in coming together, establish an *order*" (*Les deux chaos, en s'unissant, donnent un ordre*; WGL, 32/51).

Insofar, therefore, as "literalization" remains a legitimate word for the parodic deflation this simile accomplishes, it does so by effecting a transformation in the traditional definition of the "literal," which here turns out to have less to do with material bodies than it does with a lack of substantive depth. Saussure's creation parable in parentheses denies to *both* spirit and wind, both thought and sound, any claim to the status of underlying wellspring or source. It adopts the nature-philosophical insistence on the interiority of the dynamic principle—and thus also on the scientific necessity of a rigorously system-internal account—only to pervert the twin notions of interiority and dynamism by uncoupling them from the profundity of any conceivable animating force. Where Schelling's vortices erupt toward the surface on the strength of a foment that takes place down below, thereby distilling into their self-reflexive maelstrom the whole polarized energy of Being, Saussure's waves come about due to a *drop* in the density

(or intensity) of the atmosphere, and they consequently express precisely nothing about the essential nature of their substrate(s): "the
wave <=intermediary chain which does not form a substance!>").
The waves succeed in giving shapes to the juxtaposition of air and
water, thought and sound, because they *exclude* whatever tendencies, essences, purposes, or forces lie hidden in the depths of *les deux
chaos*. In the process, they establish a rhythm that bears no relation
whatsoever to the direction of any underlying flow.

The schematic illustrations drawn by Saussure's students at the
corresponding point in their lecture notes, which clearly show sound
and sense remaining amorphous except for the border where they
converge without comingling (see figure 2), differ in this respect quite
significantly from the famous and unaccountably misleading *Cours*
diagram (see figure 3), which shows vertical lines descending from a
picture-transcending height—the heavens?—to bestow form on both
air and watery depths.[19] Thought and sound do not interpenetrate, for
Saussure, as in the model of biological reproduction, in order to give
birth to the enspirited body of an organically structured language.
Rather, the empty divisions of a fully articulated *terrain commun*—
the disembodied "organs" of a profoundly life*less* organization—
establish the always-already delimited horizon against which the
"problem" of undelimited thought or sound can first appear as such.
The claim is of course counterintuitive, since Saussure's compound
noun "thought-sound," in the passage above, would seem to presuppose the presence of two more elemental masses called *thought* and
sound. These masses, however, depend on the hyphen, which is to
say, on their own composite product, for the significance attached to
their names. Saussure, who distinguishes sharply between "literal"
understood as the origin and ground of all meaning and "literal" as a
matter of materiality, sees in this hybrid, hyphenated space—though
it rests on no referential link to a world of physical or other facts—the
only legitimately *literal* domain for a theory of linguistic foundations.

A second analogy, following on the heels of the irreverent first
and referring once again to the shape of the thought-sound relation,
brings the discourse of deflation to a head: "<Another correspondence! One cannot cut the back of a sheet [of paper] without also
cutting the front.>" (*CLG*, 253:4). The comparison plays on the shift
from transcendent depths to immanent surface accomplished by a literalization of spirit, envisioning it as an emptying that leaves language *paper*-thin. Deprived of the animatory energy that could open

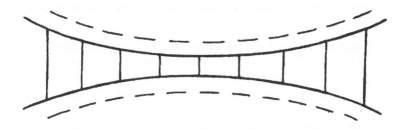

Figure 2. Ferdinand de Saussure, *Cours de linguistique générale*, student lecture notes, *CLG*, 252:2 and 5.

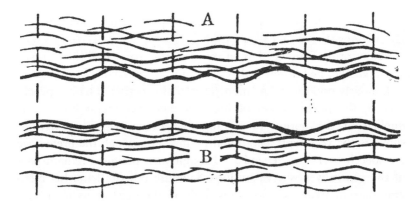

Figure 3. Ferdinand de Saussure, *Cours de linguistique générale,* published version, *Course* 132 / *CLG*, 252:1.

it to the vertical dimension of a temporal unfolding, signification exchanges the surplus value of its bottomless levels for the absolute adequacy of the sound-sense plane, where unmarked meaning exists only as a quite literal contradiction in terms: "*To signify* means [*veut dire*] both to provide a sign with an idea and to provide an idea with a sign" (*WGL*, 74/109). To signify is to be signed, to be signed is to signify, and the result, already suggested by the image of the sound-sense units as so many boundary-drawing cuts (*écrire, scribere,* from proto–Indo-European **sker-,* "to cut"), is an articulation defined by its capacity *to be remainderlessly written*:

> *Articulated* = uttered in a distinct manner. Derived sense. Fundamental sense: a limb or member [*membre*], a subdivision in a series

of things, hence: a subdivision in the syllables, or the division of the meaning chain [*chaîne significative*].

Through Broca's discovery, the faculty of speech has been localized: [this] supports [the idea of] a characteristic founded in nature. But this same [brain] convolution also governs the exercise of writing. (*CLG*, 34–35:3)[20]

The fullness of living language, in which spirit pays homage to its etymological origins by articulating itself as air, gives way here to a language imagined, in the most literal sense of "literally," as the graspable, because *writeable,* space of *litterae*: "Language [*langue*] is tangible, *that is to say*, translatable [*traductible*] into fixed images such as visual images, which would not be possible for acts of speech. (*Notebooks* III 71a / *CLG*, 44:5).[21]

The writing Saussure has in mind in this passage differs drastically, whether he himself realizes it or not, from the one he elsewhere so famously pronounces a "monstrous" deformation of living speech. Indeed, the prioritization of literal letters over literal bodies—as implied in the notion of a tangible-*because*-writeable *langue*— subverts the very representational premises on which the preference for speech over text depends.[22] Though such premises remain part of the generally positivist outlook Saussure inherits from his Neo-grammarian predecessors—whose polemical protestations against an older and insufficiently empirical linguistics he tends, occasionally, to ventriloquize—they can therefore play no role in the fundamental writability of a *langue* which is neither empirical nor mimetic in nature, and which has correspondingly little to do with the faulty orthography Saussure sometimes takes to task.[23] Neither the inaccuracies of conventional phonetic spelling, nor the disintegration of the Neogrammarian phoneme, proffered as phonetic corrective, has any relevance for a writing conceived as the *origin* of the speech acts it definitively does not seek to mirror.

This writing is not serial, like the writing we read, but geometrical, like the web of differences that distinguish a particular language; its space is the two-dimensional plane where sound and sense come together to provide, in the form of marked boundaries, all the information necessary for the construction of linear sentences, without actually pre*scrib*ing the sentences themselves.[24] Two of the student manuscripts contain the following diagram, at the point where Saussure distinguishes the "inner treasury"

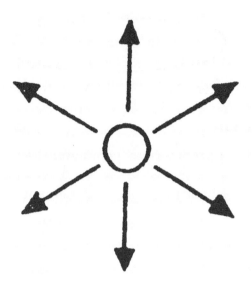

Figure 4. Ferdinand de Saussure, *Cours de linguistique générale*, student lecture notes, *CLG*, 289:3.

(*trésor intérieur*) of associative relation—which is to say, the "reservoir" of potentially usable signs—from the linearity of realized discourse (see figure 4). The schematic representation pictures, as a topographical process of territorial delimitation, the transempirical principle of negation according to which linguistic units emerge and mean: "Every word will fall at the point of intersection of several series <of> analogues: *. This star will vary, but will always be essential for the analysis of the word" (*Notebooks* I 68a / *CLG*, 289:2). The published version of the diagram has tentacles—or branches—rather than rays, and is therefore structurally flawed, but it includes a concrete example of the kind of word analysis Saussure here has in mind (see figure 5).

Taken together, the two diagrams make clear that the units involved in a writing of the "inner treasury" are not identical to the units of conventional grammatical or phonetic analyses, as captured by conventional (grammatical or phonetic) graphic notation: despite Saussure's colloquial use of the term *mot*, the signifying mark of *enseignment* comes into being for a truly linguistic

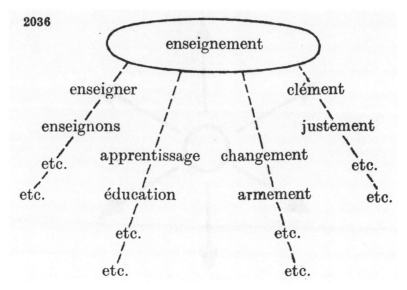

Figure 5. Ferdinand de Saussure, *Cours de linguistique générale,* published version, *Course* 148 / *CLG,* 289:1.

writing, and thus also for language itself, without reference to the traditional category of *word.* Words are composed, according to lexical rules, from parts such as stems, prefixes, and endings; they compose in turn, according to syntactical rules, the linear flow of the thought-conveying sentence. Taken together, such rules bind the raw material of words and the letters that spell them into a hierarchical system of meaningful relation—a grammar—whose branching, treelike structure Saussure here *empties* of all teleological relationship to origin and end. The network he graphically envisions has neither roots nor crown, top nor bottom, no raw materials to be distinguished from the single, negative "rule" of differential relation, and no relations to be distinguished from the spaces they enclose.[25] The actual "content" of these spaces is entirely irrelevant because totally homogenous; "month" is no more *morphologically* linked to "months" (as a *word* to which one might apply the grammatical *rule* of a plural-designating *s*) than it is *lexically* linked to "year." The multiplicity of boundaries dividing the two units from each other, and from every other linguistic slice, render such substantive categories superfluous: "(Sheet of paper [. . .]!) The study of forms and of functions is the same thing. It will therefore not be so easy to establish compartments" (*Notebooks* II 50a / *CLG*, 305:2).[26] Like the astronomical constellations that the diagram, if fully realized, would most closely resemble, the true form of language "writes" itself across an emptiness that would otherwise remain invisible. In doing so, it calls into being the sound-sense shapes that lesser writings can only reflect—imprecisely—as words. Embodied discourse, whether actualized in ink or in air, presupposes as its ultimate ground the stellar writing of a tangible but transempirical *langue*, where only position counts, and "word" is just another, less accurate, word for the empty space in the center of the *enseignement* star.[27]

The Saussurean linguist who goes looking for the signs beneath the words, and thus for the writeable ground of language, winds up, like the Neogrammarian one, playing catch-up with the speaker, whose participation in the shared knowledge of *conscience* assures an awareness of "real" units more subtle than any linguistic notation. Since the speakers themselves, however, cannot actually *say* what they know— the fictitious unity of the word, conceived as a positive "thing" to be spoken, will always conceal, at the level of consciousness, the emptiness of the sound-sense bond—a Neogrammarian-style transcription will not suffice. Only by folding the unreflected *analyses* of (falsely) conscious speech *back* into the collective geometry of an unsayable *conscience*, where the writing of the "inner treasury" resides, can the linguist generate the unambiguous notation required to ground a language science. This notation will then prove true "to the extent that it accords with the analysis of language [*langue*] as attested by neologisms or analogical formations" (*WGL*, 125/184), that is ("of language" being understood here in the strong sense of the *genetivus subiectivus*),[28] insofar as the boundaries of its units correspond to the transempirical analysis that language has always already performed on *itself.*

The project relates to the explicatory efforts of the first historical linguists, like the barometric pressure parable to the story of divine (self-)creation, through an evacuation of spirit that is also, simultaneously, an evacuation of time. Saussure's critique of the traditional etymologist's *conscience*-transcending methods, and of the synonomist's deep-structure "discoveries," appears in this context as part of a broader critique aimed at the paradigm of projection per se: "The ANACHRONIC, artificial, volitional and purely didactic point of view that PROJECTS from one morphology (or one 'past

language state' [*état de langue ancien*]) onto another morphology (or
onto another, later state of language [*un autre état de langue posté-
rieur*]) [. . .] this point of view is the ETYMOLOGICAL point of
view" (*WGL*, 7/21–22). The etymological point of view illegitimately
imposes an ideal continuity of meaning where in reality there is none;
the "explanation" it cobbles together by mixing language states
exceeds not only the consciousness of the average speaker but also
the realm of science.[29] The result is for Saussure, however, an illusion
that can be *emptied*, like the organicist prioritization of genesis it represents, into a formal structure commensurate with legitimately linguistic categories. He accomplishes this transformation with a gesture
that makes perfectly explicit, for perhaps the only time in his frag-
mented corpus,[30] exactly what he means by linguistic "explication":
"These observations allow us to deepen [*approfondir*] the sense contained in the word *etymology*. Instead of saying that it is 'the study of
the origin of a word,' we could say with more truth that it is the study
of the relationships of the word to others, which amounts to exactly
the same thing as saying that it is 'the explication of a word.'" (*CLG*,
433:2). As though infiltrated by the associations of profundity that
cling to this particular subject matter, the passage turns Saussure's
redefinition of explanatory unfolding into a "deepening" of the sense
the word *etymology* "contains." This deepening, however, carves a
direct route to the two-dimensional surface of language, and a direct
route away from the seminal Raskian rethinking it otherwise recalls.
Against an analytic *act* that dissects the word-body in the quest for a
single, organic "connection," Saussure presupposes here the analytic
fact of multiple inorganic relations, without which no words would
even exist. Rask's notion of a historical grammar to be developed
by way of the history of (parts of) words, and of an etymology that
would thus be equivalent to an explanation of language—"etymology
is like the spirit of this dead mass [of words], the bond which holds
the individual parts together"[31]—finds its negative mirror image in a
Saussurean "explication of a word," which subtracts the animatory
energy of grammar from the definition of definition. Etymology, etymologically speaking the science of true sense, reemerges here in a
sense Saussure calls "more true" (thereby simultaneously performing
the very practice he describes) as an explicatory mode that seeks true
sense in the writeable constellations of *la langue*. Saussurean explication is thus every bit as much a matter of "etymology" as its historicist counterpart. The etymology in question, however, is no longer the

study of *historical* origins, and its insights no longer distinguish lay-men from linguists. Where nineteenth-century etymologists aspired to increase knowledge by drawing historical conclusions unavailable to the average speaker, the Saussurean linguist strives only to make clear on one level what is already obvious on another—an obstinately nontranscendent mode of explication that generates, when successful, a nonnarrative technique of articulatory repetition.

When Saussure subjects the confused terminology of his profession to the explicatory disambiguation he considers his primary task, the issue of chronological precedence thus rarely arises. Key terms such as *articulation, signifier, conscience,* and *étymologie* are all reduced to their "literal" foundations—foundations that turn out to *literally* des-ignate the literal, and hence writeable, essence of language—without reference to the history of their "actual" usage or the historical mean-ing of their component parts. Instead, *conscience* becomes the shared knowledge of *con + science,* and *signification* the state of being *signed,* through an analytic procedure sanctioned by the negative geometry of the system in which they both—originally—participate. Only the presence of *other* forms containing "con" or "sign" allows speaker and linguist to recognize these sound slices as meaningful subunits, and hence to deconstruct words in a way that simultaneously clarifies and literalizes their sense: "It is to the extent that these other forms hover in the vicinity of *défaire* that we can analyze, break down *défaire* into units" (*Notebooks* II 56a / *CLG,* 293:2). One could, of course, perform the same operation on *enseignement* (education) where surrounding forms such as *enseigner* (to educate) or *changement* (change) make possible an analysis into *enseign* and *ment.* This example from the published *Cours,* however, lacks the dimension of self-reflexivity that characterizes Saussure's actual choice of *défaire,* since the latter, with the literal meaning of "to dismantle, to undo," turns out also to *name* the very process of decomposition it here undergoes. Such deconstruc-tions in no way contravene the arbitrarity principle, despite their ten-dency to seek meaning in form, since they assume no relationship between sound and sense per se, but only between a unit and its fel-lows. They allow for an explication that goes beyond the word as it exists in the (false) consciousness of everyday speech, without impos-ing an alternative (positive) meaning on the emptiness this explication thereby opens up. In the notion of a word that only ever wants to say itself by saying its neighbors—"to signify *wants to say* [*veut dire*] a sign," "articulations, *that is to say* [*c'est-à-dire*] articuli"—Saussure thus discovers a legitimate path linking *le mot* to its own *pre*historical preconditions. His redefinitions take seriously the ancient punning technique of the *figura etymologica*—in which two words with the same etymological derivation come together to double, and potentially transform, the interpretation of the "root"[32]—as the only potentially permissible "etymological" mode.

The most significant exception, in this case, serves primarily to confirm the rule. Saussure contends, in one of the most expansive terminological reflections from his notebooks, that the word *seme*, a transliteration of the Greek σῆμα (sign), better designates the basic sound-sense units of language than the amorphous French *mot*. He defends his preference by referring to the history of the Greek word's usage, which includes the Greek tradition's earliest known reference to written characters, namely, the famous *semata lugra* (dire signs) of Homer's *Iliad*: "In this difficult question of adopting a word that does more or less violence to [ ], we must at least not forget, among those things that confirm the word *seme*, the σήματα λυγρά, graphical *semes,* but we know of their origins [*parenté*]. (We will at least accord with the oldest word used by the poet for [ ])" (*WGL*, 72/108).[33] The *semata lugra,* appearing as they do in the work of an oral poet, where they are said to be *inscribed* on a tablet, pose one of most famously intractable problems of classical philology, with a commentary count to match.[34] Saussure, however, treats the phrase less as a lexical idiosyncrasy to be explained philologically than as the logical consequence of a transcendental apriority, an apriority here *figured* historically as a genealogical kinship relation (*parenté*). For him, the "oral" reference to a writing technique not yet empirically available bears witness first and foremost to an atemporal truth about the structure of the sign in general, and thus also, by extension, about the properly differential meaning of the particular sign *seme*. The argument is in line with his more general insistence that a careful investigation of chronological trajectories can bring language to light as the differential condition of possibility "beneath" and "before" historical existence: "In practice, and, secondarily, [. . .] we recognize that the labor of the historian may incidentally throw vivid light on the conditions that govern the expression of thought, mainly by affording proof that it is not thought that creates the sign, but the sign that primordially guides thought (and thus in fact creates it in reality)" (*WGL*, 28/46). Historical narrative, like the historicist meaning it ostensibly reveals, plays the role of metaphorical afterthought within the Saussurean system, but this afterthought carries within it the structure of a "before" beyond all time, and can therefore be interrogated for information about the emptiness from which it emerged. The historical "fact" embodied in the *Iliad* verse thus acquires linguistic significance insofar as it manages to testify to its own belatedness, and to the belatedness of all positive content.

POSTMÉDITATION

The model of a tautological illumination has stylistic consequences that Saussure considers awkward yet unavoidable: "The division of this book into miniscule paragraphs has something a bit ridiculous about it, which I would have liked to avoid" (*WGL*, 65/95). The promised book about the word, still present only as an idea ("there will one day be a book") in the notes about what it should contain, here leaves the hypothetical far enough behind to become the object of methodological reflection. The result is a defense of definition, conceived as the only conceivable rhetorical form, but it is also—and herein lies the source of Saussure's instinctive distaste—a submerged reflection on the *effort* of articulatory repetition. On the one hand, to write otherwise than tautologically would be to deny the nonnarrative nature of language science, which explicates by articulating and articulates by repeating, at the level of consciousness, a preexisting *analyse de la langue*. Such an explication cannot be expected to *unfold*: "For if this book is true, it shows above all that it is deeply false to imagine that one could produce a radiant synthesis of language [*la langue*], by starting from a determinate principle that develops and incorporates itself [*qui se développe et s'incorpore*] [ ]" (*WGL*, 65/95). On the other hand, this mode of writing, with its relentless focus on terminological clarification, appears to Saussure to entail a disturbing degree of analytic violence, for the stream of spoken language does not, of its own accord, make reference to the star-shaped network on which it depends: "It is therefore a terrain in which each paragraph must be set like a solid object sunk into the swamp [*le marécage*], with the capacity to indicate [*de retrouver*] the route both backward and forward" (ibid.). The image recalls the story of Ariadne and the string, and the trope of a textual labyrinth to be negotiated with the help of (land)marks. In the Saussurean version, however, it is the text itself that functions as marker, and the markers that *generate* the path. Saussure's pillar-like paragraphs must actively introduce navigable structure into the ever-shifting topography of the discourse swamp, and in order to do so they must perform the etymology of definition (*de* "off, from" + *finere* "to limit") in all its way-clearing, chaos-differentiating force: "They are simply, and in the pure etymological sense, aphorisms, *delimitations*" (*apo* "off, from" + *horizein* "to limit or bound," *de* "off, from" + *limitare* "to limit or bound"), "but limits between which the truth constantly appears (*se retrouve*)" (*WGL*, 82/121). The price, of course, is the very explicatory transparency that Saussure intends these aphorisms, *as* definitions, to provide. His "miniscule paragraphs," with their rejection of synthesis and first principles, do not reflect their subject matter so much as they transform it—through a process simultaneously etymological and ahistorical—into the scientifically accessible space it (ostensibly) already transempirically is. Theirs is the task of making clear by *making* clarity, in a domain where repetition occurs together with the all-too-proactive production of truth.

This distinction between the *fact* of *langue*-as-writing and the *project* of linguistic notation—a gap that for Saussure can be measured in units of effort, frustration, and nausea—complicates the antiessentialist explanatory paradigm to the point of near-total subversion. The language scientist who wants to decompose *défaire*, for instance, must first gather together the "neighbors"—*décoller, déplacer, découdre, déplacer; refaire, contrefaire*; and so on—with which the target word can be seen to have something *in common*. ("Every word will fall at the point of intersection of several series <of> analogues"; *Notebooks* I 68a / *CLG*, 289:2.) The result is a tautological explication that depends, despite its purely differential ambitions, on a nebulous, criterionless intuition about what it would mean for certain words to be considered in some respect "the same." Linguistic analysis entails an associative synthesis, the *figura etymologica* forges new identities, definition distinguishes but also creates, and the real question, of course, is why any of this should be the case. What is it about the attempt to *re*write the structure of language that leads inevitably away from the oppositional emptiness of *conscience*?

On occasion, Saussure seems to suggest that the culprit is a kind of inherited false consciousness, a historical accident of poorly chosen terminology that obstructs the language scientist's progress toward what would otherwise be obvious truths. There are other moments, however, in which he assigns to the dilemma a status far more fundamental, and thus also more commensurable with the depths of his despair ("But I am completely disgusted with it all"[35]). The predicament of the linguist mirrors quite precisely, according to this second interpretation, the structure according to which humans, from the beginning of time and the origins of the sign, have interacted with the world: "*Misc.* While out walking, I make <without saying anything> an incision [*encoche*] in a tree, as if for pleasure. The person who is with me retains the idea of this incision, and undeniably associates two or three ideas with this incision from that moment onward, when in fact I myself had no other idea than to mystify them or to amuse myself.—Every material thing is for us already a sign: that is to say an impression we associate with others, <but the material thing seems indispensable>" (*WGL*, 79/115). In this tongue-in-cheek story about the birth of language-as-writing, which appears midway through a series of aphoristic reflections devoted to the terminological conundrum of the sign, Saussure tackles the deadly serious problem of projection as an unavoidable component of every linguistic scenario. No matter how hollow the initial incision, the intending mind of the observer immediately "fills in" the emptiness with the positive substance of signification ("two or three ideas"), thereby transforming the originarily contentless mark into a meaning-bearing material object ("every material thing is for us already a sign").[36] The real obstacle to a rewriting of language, the thought experiment suggests, must therefore be located in the ahistorical and profoundly noncontingent tendency of the *seme* to tip over into its opposite, and near-homophone, the *sôme* (from the Greek σωμα, meaning "body"): "*Misc.* [. . .] Even a term like *sôme* [σωμα] would become, if it had a chance of being adopted, in very short time synonymous with *seme,* which is meant to be its opposite. Linguistic terminology here pays tribute to the very truth we are establishing as a fact of observation" (*WGL*, 77/113).[37] The hybrid, hyphenated structure of the thought-sound relation, to which the word *seme* bears such emphatic, Homeric witness, has precisely nothing of substance in common with the structure of an enspirited body. No amount of terminological rigor, however, can prevent its various empirical manifestations from being continually mistaken for a cadaver.[38]

On one of the loose sheets of paper discovered among the conservatory notebooks, Saussure goes on to radicalize his interpretation of this most elemental of impulses under the title "Postméditation-reflexion":

Thus in a language composing a total of two signs, *ba* and *la,* the totality of the mind's confused perceptions [*des perceptions confuses*

de l'esprit] will NECESSARILY be pulled either to *ba* or to *la*. The
mind [*l'esprit*] will find, due to the simple fact that there exists a
difference *ba/la* and that there exists no other such difference, a
distinctive feature [*caractère distinctif*] that allows for regular categorization under the first or the second of these headings (for example
the distinction between *solid* and *non-solid*). In this moment the sum
of positive knowledge will be represented by the common feature
which happens to have been attributed to the *ba* things, and the common feature which happens to have been attributed to the *la* things.
This feature is positive, but in reality it has only ever sought the negative feature which makes it possible to distinguish between *ba* and
la; it has never tried to unify and coordinate, it has only sought to
differentiate. And, finally, it has sought to differentiate only because
the material fact of the presence of the different sign invited it, indeed
forced it to do so. (*WGL*, 60/88)

The passage operates once again with the notion of projection, which
Saussure here holds responsible for the inauguration of all positive
knowledge. A primordial "esprit" imposes its own confused perceptions onto the orderly emptiness of linguistic delimitation. In the
process, it "discovers" the concept of the concept, here figured etymologically in the example of "solid and nonsolid" as an entity with graspable contours (*concept*, from Lat. *con+capere*, "to grasp together").
Such an intermingling of mind and language matter, which proceeds
according to a logic of investment and incorporation, must necessarily remain a secondary affair within the realm of language origins, and
the titulary prefixes of "post-" and "re-" ("*post*méditation-*re*flexion")
would seem to uphold this expectation. The development of the passage, however, shifts the terms of the debate in such a way as to render these prefixes almost meaningless, for what shows up elsewhere
in Saussure's manuscripts as the symptom of a strictly human spontaneity reappears here as the *inevitable consequence* of the differentiating cut. The "esprit" that fills the emptiness of a *ba/la* opposition
with the substance of its own experience does so only at the unignor-
able behest ("invited it, indeed forced it"; ibid.) of material signs whose
very negativity thus appears to *call forth* ("MECHANICALLY" and
"NECESSARILY"; *WGL*, 60/87–88) the positive postmeditation that
blocks access to their writeable purity. And Saussure's bizarrely liberal
use of reflexive verbs—"every sign in existence thus INTEGRATES
ITSELF [*s'intégrer*], post-elaborates itself [*se postélaborer*], a determinate value" (*WGL*, 60/88)—suggests that the retroactive supplement of
reflexion takes place also, and perhaps even primarily, at the level of a
language that *does* positivity to *itself.*

The tension surrounding the question of priority and postmediation comes to a head in the following two-part definition:

> There can be no analogy between language [*langue*] and any other human thing for two reasons: 1) the internal nullity of the signs; 2) the ability of our mind [*la faculté de notre esprit*] to attach itself to a term that is in itself nothing.
>
> (But that was not what I initially wanted to say. I have digressed [*J'ai dévié*, literally "I have deviated"].) (*WGL*, 74/109)

The characterization of linguistic uniqueness occurs according to the traditional, Aristotelian formula of category ("human thing") plus specific difference (the "two reasons"), whereby the second of these two reasons contains an apparent contradiction: an attribute of non-linguistic being ("the ability of our mind") is here pronounced to be one of the defining characteristics of a linguistic structure ("the internal nullity of the signs") to which it is simultaneously clearly opposed. The implications are explosive enough for the parenthetical disclaimer to carry the weight of real regret. For what Saussure explicitly does not *want* to say, but what he clearly cannot quite bring himself to disavow, is that the structure of projection belongs, with its supplement of spirit, to the differential essence of language—and thus to the essence of an explication whose task is to hold the two apart. To the extent that this "deviant" definition can be taken seriously, the attempt to write *langue* can therefore be expected to proceed in constant conflict not only with its own superfluity (the writing of *langue* is always already written) but also and more crucially with its own impossibility (the writing of *langue* does not exist apart from the projections of spirit that render it unwritable).

As the simple consequence of an even simpler observation ("no one disputes the principle of the arbitrariness of the sign"; *Course* 78 / *CLG*, 153:1; translation modified), the emptiness of the linguistic origin should almost go without saying. The real problem, however, is that it also almost *must*: "In all the domains of linguistics, it is remarkable that whenever a proposition acquires a general character, it expresses either, as one prefers, the most banal thing, such that one feels somewhat embarrassed to enunciate it—or the most paradoxical thing, which will then be blindly opposed by the very people who had previously laughed to see the same truth spoken in a simpler form" (*WGL*, 39/61). The difference between banality and paradox corresponds here to the difference between definition conceived

as the repetition of an original writing, and definition conceived as
the repetition of an original writing performed *in the face of constitutive, language-internal resistance.* The latter is the source of the
famous Saussurean frustration. It is also the reason that the *figurae
etymologicae,* for all their synchronic emptiness, cannot help but participate in an essentially diachronic tradition of positive rather than
differential meaning: why, in other words, a word like "articulation,"
which has the negative structure of a sign determined by its fellow
signs, must also quite literally *mean* its own nullity, according to a
logic that subordinates body (the articulating mouth) to writing (the
articulatory boundaries of *langue*) across the "metaphor" of an etymological literalization. "[N]o generalization is possible if one continues to consider each product in terms of its genesis and its essence
simultaneously [. . .]. Any type of expression which seems at times
to establish a connection between [vertical] facts and horizontal facts
is without exception an image; the other source of disgust [*l'autre
cause de dégoût*] is that one can neither do without these images nor
resolve to accept them" (*WGL,* 152–53/218). The first part of the passage essentially repeats the insight of a specifically Saussurean etymology, which makes the explication of the word dependent on the
explication of the purely differential system, and the understanding of
historical development (here the "vertical facts") a mere supplement
to the understanding of the atemporally articulated surface. If this
distinction between essence and temporal beginnings is not strictly
maintained, Saussure claims, terminological chaos threatens to ensue.
("It would be impossible as a result to discuss a single one of the terms
of linguistics currently in practical use without taking up again, from
the very beginning, the entire question of language"; ibid.)

The problem is—first of all—that the ahistorical essence of language can be approached only backward, through the language of
history and touchable "things"; one must begin with existing nota-
tions and the surface phenomenon of "words." And the problem is—
secondly—that one can move backward only with the help of images,
which reside illegitimately at the intersection between history and
system. In Saussure's case, these images have been transformed into
silhouettes of the sought-after imageless signs, but his "etymological figures" remain, for all their empty purity, mere *representations*
of emptiness. The Saussurean literalizations project the substance of
a spirit-emptying, nothingness-affirming *profession de foi* onto the
incisions of the originary *litterae,* in a gesture of protective redoubling

intended to shield them from the encroachment of other, even less legitimate material. The effect is repugnant—a "source of disgust"— but inescapable. In the face of a perpetually affirmative threat to the negativity of the linguistic "essence," Saussure takes a card from the enemy's deck and makes language repeat its own writable purity at the level of its original, "etymological" content. He thereby defines a sign*like* alternative to the inescapable hegemony of the word, proleptically perturbing the "principal perturbateur" in preparation for the *truly* negative linguistic units about which he dare not yet even dream.

CHAPTER 3

Verse Origins

THROUGH THE LETTERS WAFTS THE SPIRIT

The transformation of literary studies for which early nineteenth-century German scholars are famous, a transformation often cited as the genesis of an academic *Germanistik*, in particular, and of nationally based philological disciplines, in general, occurs as a direct result of the early nineteenth-century reconceptualization of language. A new interpretation of the linguistic object prepares the way for a new approach to linguistic artifacts, and thus also for a new understanding of what it means *to read*. This new notion of reading, in turn, according to a dynamic less widely recognized, works to extend the very theory of systemic spirit that conditions its emergence, such that the philologically reconstructed domain of an originary Indo-European poetry becomes—for the linguists of *Sprachgeist* as well as for Saussure—the hypothetical site of a fully realized linguistic writing. In the (dis)enspirited letters of a rediscovered poetic *arche*, so the always only implicit claim, language scientists can hope to encounter the purest conceivable expression of the primal language essence, which is to say also of the formative principle or *telos* that drives the development of grammar.

One of the first theoreticians of the nature-philosophical approach to reading, the Romantic poet Friedrich von Hardenberg (Novalis), includes among his *Teplitz Fragments* (1798) the following reflection on the concept of philology:

> Concept of Philology—Sense for the life and the individuality of a letter-mass. Soothsayer from ciphers—*Letter-augur*. A supplementor

[*Ergänzer*]. His science borrows much from *material tropes*. The physicist, the historian, the artist, the critic, etc. all belong to the same class. / Away from the singular [*vom Einzelnen*] to the whole—from semblance [*Schein*] to the truth and *sic porro*. All is encompassed by the art and science of progressing from one thing to another—and so from one thing to the All—rhapsodically or systematically—the art of spiritual travel [*die geistige Reisekunst*]— the art of divination.[1]

Composed during a period of intense preoccupation with Schelling's *On the World Soul,* this fragment sounds virtually all the chords of the ancient, Pythagorean-Platonic model—cosmic system as language, cosmic elements as letters, cosmic scientist as grammarian— into which nineteenth-century nature-philosophy breathes its new temporal, teleological energy. Novalis focuses on philology as a mode of *progressing* from multiplicity to unity, from part to whole, from ones to One, and thus as a quest for the formative principle that turns heaps of letters into text ("sense for the life and the individuality of a letter-mass"). In doing so, he takes seriously the etymological root of the German *lesen*, "to read," which derives from the Old High German *lesan*, "to gather, to collect." But he also makes the gathering of reading into the origin of organized structure per se, on the strength of the fact that Logos itself has its roots in the activity of collecting (*logos*, from *legein*, "to say, to relate," from proto–Indo-European **leg*, "to collect, to gather"). The force of attraction that causes certain letters to *stand together* in the shape of a literary system is for him the same one that binds the disparate bits of existence into the harmonious accord of the cosmos ("*All* is encompassed by the art and science of progressing from one thing to another—and so from one thing to the All"). The reading of texts mirrors, in its attempt to relate singular marks to each other, the primal "reading" that establishes the common ground or measure of the all, and the philologist's love affair with language, when properly interpreted, turns out to imply a divinatory science of world-spiritual, analogical affinity.[2]

This notion of reading as an etymological return to the cosmic principle of collection, so ambitiously if vaguely prescribed in Novalis's fragment, acquires for linguists committed to Franz Bopp's theory of flexion—with its emphasis on language development as a process of grammatical *gathering*—a concrete significance to be bolstered by a reconceptualized etymological technique. The new language science seeks to unfold the generative infolding at the root of all

Logos, and so also to systematically explicate the dynamic of an originary, language-establishing *poesis*. Jacob Grimm and his colleagues
thus reject the "either-or" dynamic that still separates, for Novalis,
the systematic from the rhapsodic modes of relating things to things
("rhapsodically *or* systematically"), and insist instead on the intuitive syntheses of a prehistoric poetry ("rhapsodic," from Greek *rhap-
tein*, "to stitch together," but also from *rhapsodius*, "reciter of oral
poems") as the wellspring of a modern linguistic analysis.

The particular disciplinary techniques of an etymologically founded
philology first emerge, then, at precisely the same moment when etymology itself first acquires its language-scientific underpinnings and
the German language its linguistically formalizable boundaries: in the
second edition of Jacob Grimm's *Germanic Grammar* (1822).[3] Grimm
readers such as Karl Lachmann quickly discovered that by applying the
tools of a newly law-abiding linguistics to the study of literary texts,
they could transcend classical text-based scholarship in the direction
of pretextual poetic origins. Layer after paleographic layer could be
stripped away from existing documents to reveal ever more originary
versions "hidden" beneath, with the aim of arriving, eventually, at the
oral foundations of a truly national literature. These rediscovered foundations were of course no more empirically verifiable than the starred
forms of August Schleicher's Indo-European *Ur*-language; as "records"
of an oral age untouched by the technologies of text production, they
acquired a written dimension only at the hands of the philologist himself, who made them available to readers in the shape of a critical edition. And yet, (re)constructed on the basis of rules that had been, in
their turn, induced from the study of existing and thus verifiable texts,
such philological representations of an otherwise inaccessible origin
could lay claim to a positive truth. Insofar as poetic creativity expressed
itself in language, so the logic went, it participated in the collectivity
of *Sprachgeist* and obeyed the laws governing the "growth," or historical development, of natural (language) life. The original utterance
behind the transmitted word could therefore be divined with an effectively mathematical certainty so long as the diviner—"a transcendental
researcher who knows how to establish certain facts above and beyond
the transmitted tradition [*Überlieferung*]," in the words of Wilhelm
Dilthey's philologist brother-in-law, Hermann Usener—possessed a
sufficiently sophisticated "language-consciousness."[4]

Though the celebration of a new, linguistically grounded approach
to texts spanned the length of the philological century, as the

examples of Lachmann and Usener suggest, it reached a self-reflexive high point in Jacob Grimm's own *Germanic Mythology* (*Deutsche Mythologie*, 1835–44), which brought the science of German letters into contact with the history of German gods. Attempting to extrapolate from the remnants of an oral, "hero-legend" (*Heldensage*) tradition the outlines of an indigenous Germanic theology, Grimm offered up a reconstructive tour de force that made the true stakes of the linguistics-literature relation explicit. Etymological trajectories conceived as phonetic evidence (the name Siegfried, for instance, is said to derive from the Old Norse Sigfödr or *Siegvater* ["father of victory" or "victorious father"], one of the titles of the Norse god Odinn/Wuoton)[5] alternate here with etymological trajectories conceived as metaphors for the temporal development of German spirit(s):

> Just as all the sounds of language go back to a small number, from whose simplicity all the rest come about, the vowels by means of gradation [*ablaut*], a-mutation [*brechung*], and dipthongization, the mute consonants through the division [*zerlegen*] of their three rows into three levels [*stufen*] [. . .]; so also in mythology do I trace the many different divine phenomena [*die vielfachen göttlichen erscheinungen*] back to their unity, allowing variety to emerge from this unity, and it can hardly be wrong to assume also for the gods and heroes such unification, mixing, and shifting [*einigung, mischung und verschiebung*], in accordance with their characters and particular qualities. [. . .] If this transformation took place haphazardly, there would be nothing to it; but it seems to proceed according to definite stages [*nach bestimmter stufe*], and without leaps.[6]

This condensed story of alphabet origins, cited by Grimm as an analogy for the emergence of the gods, rehearses a process described at great length in his work on the structure of the German language. The section titled "Von den Buchstaben" (On the letters), with which he opens the seminal second edition of the *Germanic Grammar*, maintains that not all currently available letters play a role in the beginning of speech. Some, like the "pure" vowels *a*, *i*, and *u*, contain within themselves the seeds of others (*e*, for instance, is in fact a mixture of *a* and *i*), and these others therefore appear only later, as language gradually evolves toward ever more explicit actualizations of its latent articulatory potential. The nonarbitrary nature of this unfolding is the true subject and discovery of Grimm's entire grammar, and reason enough for the treatment of letter origins to take pride of place. The laws according to which the sounding substance of language "matures"—it is in this section that Grimm introduces his all-important consonant shift—render readable an otherwise inaccessible reality beneath the empirical surface of transmitted texts. In doing so, they tie the "otherwise wildly wandering etymology" to a practice of systematic analysis that definitively distinguishes the new organicist science of language from its speculative forebears.[7] They also, however—as Grimm's daring simile in this passage suggests—lay bare the logic governing the progress of *Logos* in general: the letters of the divine names can be presumed to change *in tandem* with the holy spirits they designate ("but Siegfried *is* also Wotan"[8]), because the transformations of the mind can be presumed to run parallel to those of language.

The "just as" with which the above passage begins gestures, then, toward a fundamental if largely implicit presupposition of nineteenth-century philological scholarship. No amount of regularity at the level of phonetic change can help the scientist of literature pass from writing to writer unless the poets themselves—or rather, the life-giving energies of their intending *minds*—turn out to lie dormant inside the transmissible body of their works. The hypothetical *Ur*-text must therefore be presumed to obey the natural laws of language life *while* adequately expressing the spontaneity of an authorial will, in an originary synthesis of spirits and letters. This requirement of a reanimatory reading explains the traditional philological prioritization of archaic over modern modes, since the primitive forms subsumed by the Grimms under the names of "natural poetry" (*Naturpoesie*), "epic" (*Epos*), and "hero-legend" (*Heldensage*) are assumed to *bypass*, in their orality and collective naiveté, the dilemma of individual artistic freedom. In an age where the poem merely rearticulates, according to a principle of emphatic condensation, the glories of the national *(Sprach)Geist,* there can be no question of an unbridgeable gap between the poets and their linguistically trained interpreters, between meaning and derivable forms. A rigorous textual critique will render the formative principle of a primordial *poesis* accessible to scientific systematization, and this systematization will lay the groundwork for a philological communion with the spirits of epochs past: "Through the letters wafts the spirit that wrote them down, and among the leaves laugh the golden fruits of the true, of the spiritual life."[9] He who picks these fruits harvests from trees of historical knowledge.

2 L, 2 P, 4 R (= 2+2)

This nineteenth-century German understanding of the etymologizing philologist as "letter-augur" or "soothsayer from ciphers," capable of reading his way backward from the structure of language to the structure of the world, will be transformed to the point of paradigm-shifting parody, around the turn of the century, by a much-discussed but little-understood Saussurean poetics. Saussure's first attempt to affect such a transformation, which involves a polemical reinterpretation of the *Ur*-Germanic *Nibelungen* material, fails because it seeks simply to eradicate the presupposition of language spirit from the foundations of textual analysis. The collapse of the *Nibelungen* project, however, prepares the way for a second, more successful attempt, which involves a radical rethinking of the origins of poetic meter. This latter project has traditionally been fundamentally misconstrued, I will argue, precisely because it has never before been seen in the context of Saussure's anti-*Sprachgeist* crusade. Reinterpreting Saussure's theory of ancient verse against the backdrop of the Germanic tradition it simultaneously rejects and retains will thus turn out to entail a new understanding of his structuralist model of reading, and of the role played by this model in his reconceptualization of language science.

At some point shortly after 1900, Saussure began work on a book he planned to call *Histoire et Légende: Etude sur l'origine des traditions germaniques connues sous le nom de Heldensage* (History and legend: Study of the origin of the Germanic traditions known under the name of Heldensage).[10] It was to focus, as the title conveys, on the geographical and historical origins of Germany's national epic, the *Nibelungenlied*, a topic that had dominated the discipline of *Germanistik* from its early nineteenth-century inception onward, and which continued to cause controversy among Saussure's German contemporaries.[11] The goal, for German scientists of language from Jacob Grimm to Eduard Sievers, was a reconstruction of the ancient oral tradition—the *Heldensage*—which was presumed to have served as source material for the author of the medieval poem. Saussure's proposed contribution grew out of his conviction that German philology, like its linguistics, remained in dire need of a new, nonmystical foundation. Without some objective, external criterion for separating essence from accident and origin from scribal error, textual reconstruction threatened to degenerate into textual denigration: the nineteenth-century notion of a "critical edition" (*kritische Ausgabe*), first exemplified by Karl Lachmann's seminal *Nibelungenlied* editions, remained for Saussure both a fiction and a farce in the absence of any empirical touchstone of truth.[12]

When, then, he claims in his notes to have discovered just such a touchstone, or rather its remnants—in the historical chronicles of Gregory of Tours!—he clearly means to challenge, first and foremost, the methods according to which prior conclusions had been reached. The results are typically provocative: by locating the foundations of the *Heldensage* in the sixth-century Burgundian kingdom of Lyons, as described by Gregory, rather than the fifth-century kingdom of Worms, as represented by the medieval poem itself, Saussure effectively renders a quintessentially Germanic material geographically *French*.[13] Nor does he stop with geography, for his notes also posit significant Latinate influence at the level of epic content: according to his reading, the Huns of the *Nibelungenlied* stand in for the historically more proximate Franks, while the Siegfried cycle, as recorded in the Old Norse *Edda,* tells a version of the classical Roman Theseus myth. The Romanization of the *Nibelungen* legend serves, however, no purely nationalist end. Rather, it provides a platform from which Saussure can launch, once again and with slightly different emphasis, his more general polemic against the teleological model of language history. While his nearly contemporaneous manuscripts on general linguistics seek to suck the spiritual substance from a nineteenth-century German tradition of etymological *Sprachwissenschaft,* or language science, the even more copious *Nibelungen* notes work to undermine an intimately related, and equally German, tradition of etymological philology.

A series of strikingly Grimmian analogies between letters and mythological entities informs the methodological reflections scattered sporadically throughout Saussure's polemical *Nibelungen* notes. The Saussurean versions, however, carefully deconstruct precisely those categories—lawful transformation over time, organic unity of form and meaning—that constitute the nonnegotiable foundation of an etymologically informed philology:

> It is true that in going more deeply into things, one notices in this
> domain, as in the related domain of linguistics, that all incongruities
> of thought spring from inadequate reflection on the nature of *identity*
> <or the characteristics of identity>, when one is dealing with a non-
> existent entity like a *word*, or a *mythic character*, <or a *letter of the*

alphabet>, which are merely <different forms of the SIGN>, in the
philosophical sense. (1)
(1) Poorly discerned, it is true, by philosophy itself.
<Perhaps to put into the Preface>
A letter of the alphabet, for example a letter of the Germanic runic
alphabet, manifestly possesses no other *identity*, from the very beginning, <than that which results from the association>
<of a certain> phonetic value
<of a certain> graphic form
through the name <or of the nicknames> that can be given to it
through its place <(its number)> in the alphabet
If two or three of these elements change, as occurs in every
moment <and all the more rapidly as one change often entails
another>, one no longer knows, after very little time, *literally and
materially* what is understood, or rather [ ]
<The graphic individual and, also in general the semiological
individual will not, unlike the organic individual, have a means by
which to prove that it has remained the same because it rests from the
ground up [*depuis la base*] upon a free association.>
[. . .]
*Not <a creation> which is more or less fragile: but <a creation>
which is radically devoid [*dénuée*] of any unity principle; it is only
the relative duration of certain traits that gives [the former] illusion,
and it is a daily lesson for the one who studies [these phenomena] to
see that the association—which we <some>times cherish—is nothing
but a soap bubble, is not even a soap bubble, since the latter at least
possesses a physical and mathematical rather than an accidental and
ignoble (*indigne*) unity. (*LEG*, 387–88)

In place of the gods, with their traceable names—for Grimm the "pillars" (*Pfeiler*) upon which he can construct the "scaffolding" of his
mythology[14]—and in place, also, of the letters, with their knowable
laws—for Saussure's Neogrammarian colleagues the "foundational
pillars [*Grundpfeiler*] of our science"[15]—Saussure offers only the traditional iconographic vocabulary of *vanitas,* and then retracts his
offer. *Not even a soap bubble.* The articulatory units that structure
and support the German reconstructive project (in accordance with
the etymology of (Buch)*Stab,* from proto–Indo-European **stebh,* "to
support"), in its quest for the articulating breath of a long-past life, are
in his view nothing, and less than nothing, for the simple reason that
they do not, coherently and continuously, *exist.* Even the famously
ephemeral soap bubble possesses, despite its utterly unpillar-like fra-
gility, a rule-governed integrity to which the accidental agglomera-
tions of epic and alphabet, divine names and runes, cannot hope to
aspire.[16] Mere disembodied qualities, recombining at random into

forms with flexible contours, can never lay claim to the *literal* and *material* reality that Saussure, as empirical positivist, feels he needs to demand from his linguistic and philological "elements."

The critique is a Neogrammarian one, with the exception of its object, for Saussure here brings his Leipzig-trained, fin de siècle-flavored skepticism to bear on a domain that the Neogrammarians themselves had traditionally spared: the German epic as sacred reservoir of national and linguistic spirit receives, at Saussure's foreign hands, the *de*animation his colleagues had already performed on their language. The process has disastrous disciplinary consequences that Saussure, like his predecessors, refuses to draw, preferring to remain instead within the bounds of the paradigm he nonetheless defiantly destabilizes. Faced with a discipline rendered undisciplinary by the ubiquity of lawless change, he proposes not a *new* understanding of history but an ostensibly purificatory intensification of the old. A more rigorous implementation of accepted methods—a pedantic exactitude at the level of empirical praxis—will compensate, he hopes, for the absence of a methodological ground, and allow him to accomplish *in fact* what his theory deems *in principle* impossible: "Therefore, in principle, one should simply abandon the attempt to track [these phenomena], seeing that the sum of their modifications is incalculable. In fact, [however], we see that one can have some relative hope of tracking them, even over large intervals of time and <distance>" (*LEG*, 368). The resulting technique takes the notion of historical reconstruction to fantastic and unprecedented extremes. The epic is assumed to originate as a "chronical in rhymes," and thus to reproduce, with the highest possible degree of precision, a society's collective memory of historical facts and events; only later, once crucial elements have been forgotten and imagination called in to fill the gaps, does Saussure allow this serial recitation of actual events to rearrange itself ("of its own accord") into a "literary work" (*LEG*, 375). Philological reconstruction thus becomes a matter of peeling off the poetic projections of later epochs in an effort to reach not the literary *Ur*-text but the historical *Ur*-event—a kernel of extrapoetic, bodily truth that has less to do with the spirit-infused letters sought by an organicist philology than with the despiritualized *literalness* of Saussure's teachers' positivist facts.

Given the ultimate fate of the letter in the Neogrammarian discourse of the literal ("a true analysis [*Zerlegung*] of the word into its elements is not merely extremely difficult; it is actually impossible"[17]), it is perhaps not surprising that Saussure's Neogrammarian-inspired attempt to anchor the epic meets with less than unqualified success. In the *Nibelungen* notes, the outside world still plays the role that the notes on general linguistics so ingeniously assign to a transempirical *conscience*—and it does so poorly.[18] The textual "evidence" necessarily distorts the historical "real" in ways that can never be established with any certainty, while the relationship between this evidence (here, the Gregorian chronicles) and the epic it ostensibly "explains" amounts to little more than an article of philological faith ("And why not? Is it not the simplest supposition?"; *LEG*, 374). The philologist who professes such faith can defend it only by painstakingly comparing the two texts, gathering coincidences in the hope that the burden of proof will eventually shift to the doubters' camp.[19] Saussure believes he has accomplished this much, though he also freely admits that he will never actually know. His tone fluctuates in accordance with the precariousness of his position. One moment he is launching an aggressive polemic against proponents of the "mythological thesis," whose readings of the *Nibelungen* texts he deems insufficiently historical,[20] and the next he is acknowledging the inevitability of his own ignorance with a retreat to the subjunctive tense: the link between epic and history *would* solve all philological problems, so the alternative and immeasurably weaker claim, *if* it existed: "In this case—which we are in no way positing as expected, but rather as unexpected—there would in effect be an *unforeseen* possibility of tracking the legend <There would be an exterior hook upon which the legend would hang>" (*LEG*, 427–28). We are not far, here, from the not-so-hidden subtext of the soap bubble simile, with its insinuation of a reconstructive futility to which Saussure never gives more explicit voice. And indeed, by the time of the first course on general linguistics, in the winter semester of 1906–7, the problem of history in the *Nibelungenlied* has already been definitively displaced by other concerns. Despite his obvious affinity for the kind of positivist, particularist approach at work in the *Nibelungen* notes,[21] Saussure never returns to the notion of an external anchor.

Around the same time that he abandons the *Nibelungen* project, however—and concurrently with his first lectures on general linguistics—he begins to take an almost obsessive interest in another, largely German, largely nineteenth-century philological debate.[22] This time the question is the metric structure of the most originary Indo-European verse. The German starting point was the role of

alliteration in ancient Latin poetry, and its relationship to the alliterative practices of the Old Norse, Old English, and Old High German
poets. Midcentury German philologists had noted the fidelity of the
so-called Saturnian verse form—the form of the oldest known Latin
lines, preserved only in monument inscriptions and quotations from
more recent works—to the rhythmic structure of the ancient Latin language. They then contrasted this fidelity, presumed to spring organically from an oral folk tradition, with the Greek-influenced artifice
of later Latin poetry, which subjugates the natural rhythms of living
speech to the "tyranny" of borrowed, conventional rules.[23] They went
on, finally, to draw parallels between Saturnian verse structure and
the oldest *Germanic* rhythms. Where the ancient Greeks, together
with their educated Roman imitators, counted syllables and measured vowel lengths, both the Germans and the most original Latins
were thought to have *alliterated* key words, according to a principle of
composition that left the timing of ordinary language free to unfold
without constraint.

Such a parallel could then be exploited, in combination with the
evidence of Iranian and Sanskrit sources, as a springboard for sup-
positions about the rhythm of the Indo-European *Ur*-verse. Following the example of August Schleicher, who relied on the hypothetical
forms of a reconstructed *Ur*-language to generate the content of his
Indo-European fable, philologists such as Rudolf Westphal and Hermann Usener traced the various (Indo-)European poetic trajectories
back to their origins in a common source, in an attempt to determine
the metrical structure that such a fable might have assumed.[24] They
argued for a "comparative metrics of the Indo-German peoples" that
would stand alongside the more established "comparative grammar"
at the foundation of a truly rigorous philology.[25] And they proposed a
hypothetical *Ur*-line of sixteen syllables and eight accents, separated
by a caesura, to be represented graphically as follows:

$$\acute{-}\,\acute{-}\,\acute{-}\,\acute{-}\ |\ \acute{-}\,\acute{-}\,\acute{-}\,\acute{-}\ {}^{26}$$

Saussure clearly has such ambitious reconstructions in mind when
he, too, moves from theories about ancient Latin meter to a theory
of originary Indo-European poetry—by way of comparisons with
Greek, Sanskrit, Germanic, French, and Italian material, and in a
proliferation of manuscripts now known (somewhat misleadingly) as
"the anagram studies." His hypothetical origin, however, replaces a
discussion about the type and location of accentual emphasis, and

thus about the relative primordiality of various accentual modes, with an entirely unorthodox understanding of verse structure. Poetry, Saussure will come to insist on the basis of his Saturnian "discoveries," is originally and essentially a matter of combinatory arithmetic rather than accentual exhalation. It requires neither a spontaneous rhythmic "feel" for the proper "movement" of a natural phrase, nor a rule-governed proficiency in the arrangement of measured syllables but rather a calculating analysis of lines into even *pairs* of letters: "2L, 2P, 4R (=2+2), 6A, 2O, 4U, and so forth, without there being any particular importance whatsoever to the initial position."[27] In this model, every letter in every poetic line, irrespective of its position within a given word, effectively calls forth its own reflection, according to a law of literal symmetry that radicalizes the principle of alliterative repetition—with its exclusive focus on *first* letters of *key* words—to the point of absolute erasure:

> The whole phenomenon of alliteration (and also of rhyme) which one has observed in the Saturnian verse is only an insignificant part of a more general phenomenon or, rather, of an *absolutely total* phenomenon. The totality of syllables in each Saturnian line obeys a law of alliteration, from the first syllable to the last; and in which every single consonant, as well as every single vowel, as well as every single *vowel quantity*, is scrupulously counted. The result is so startling that one wonders how the authors of these lines [. . .] would have found the time to indulge in this kind of brainteaser: for the Saturnian verse is a veritable Chinese game, *quite apart from anything pertaining to metrics.*[28]

Where every syllable "weighs" the same and every letter "counts," no combination of heavy and light, stressed and unstressed, can play a role in the constitution of verse. Having raised this arrhythmic and antimetrical game to the status of candidate for general Indo-European principle by "discovering" it in other ancient sources—including, most significantly, the Vedic hymns of the Aryan priests, or *Kavis*—Saussure then goes on to draw the conclusion that German-style alliteration is derivative. In the primitive simplicity of alliterative verse, so prized by German philologists as an indigenous link to the Indo-European lineage, he sees rather the barbaric perversion of a prior intricacy, a degenerate misunderstanding characterized primarily by the absence of a more original, more universal symmetry.

Saussure considers this "consequence" of his Saturnian studies significant enough to bear repeating, because he believes it to entail a revolutionary new theory about the origins and essence of phonetic letters. The following passage, probably composed in 1907, is one of at least four nearly identical versions of the theory in question.[29] I cite it here at length, since only in its totality can it be seen to introduce— via a particularly pregnant etymology of the German word *(Buch) Stab* (letter)—the thoroughly spiritless philological ground that Saussure had tried and failed to envision in his work on the origins of the *Nibelungenlied.*

Germanic Alliterative Verse

Although nothing links the facts of Saturnian Latin alliteration to the rhythm of a line—even supposing a state of the Latin language in which the initial syllable is stressed—it is certain, on the other hand, that the alliterative initials of the Germanic languages (Old Norse, Old Saxon, Anglo-Saxon, and one or two High German texts) cannot be separated from the rhythm of the line because (a) the line is rhythmic and is based on the accentuation of its words; (b) the words are accented on the initial syllable; and (c) consequently, if, by an equality of consonants one emphasizes the initial, one is simultaneously emphasizing the rhythm.

But historically one can wonder if, instead of taking German alliteration as an original model—which set the standard by which one could judge Latin *alliteration, rhythm, and accentuation*—one could construct an entirely inverse argument by which, to the contrary, it is the Germanic tradition, through changes which are, moreover, understood, that arrived at the form [which has become] celebrated in its own tradition as a general model for versification [. . .].

It is also in setting out from this given of an Indo-European poetry, which analyzes the phonic substance of words (to make of it either an acoustic series or a series which acquires significance through allusions to a particular name) that I thought I could understand for the first time the famous German *stab,* in its triple sense of (a) rod; (b) an alliterative phoneme of a poetry; (c) letter.

As soon as one so much as suspects that the elements of a line had to be *counted,* one is faced with the objection that such counting is extremely difficult, given that we, who have writing at our disposal, are obliged to take great care to be sure of counting properly. Also, one can see from the outset or, rather, foresee, that if the craft of the *vātēs* was to assemble a predetermined number of sounds, this would only be possible, so to speak, by means of some external sign like pebbles of different colors, or *rods* of different shapes, which, representing the sum of *d*'s or *k*'s etc., which could be used in the *carmen,* passed successively from right to left as the composition progressed and rendered a certain number of *d*'s or of *k*'s unavailable for the subsequent lines. (One must begin with short poems of 6 to 8 lines, of which the *Elogia,* or certain Vedic hymns, or Germanic magic formulas, give us an idea.) Thus it happens that even *a priori* the link between a rod (*stab*, or *stabo*) and the PHONEME appears absolutely <natural and> clear if the poem *counted* the phonemes, whereas I have never been able to make any sense of *stab*, *stabo*, the alliterative letter, or the letter in the usual conception of alliterative poetry. Why, then, would a letter have been designated by a rod?? A mystery.

<The entire question of *stab* would be clearer if one could avoid the unfortunate intermingling with *buoch* (the bark of the beech tree, on which one could trace *characters*). These two objects of the vegetable kingdom are, in the matter of Germanic writing, entirely separate; as demonstrated in my earlier analysis, I consider *stab* = *phoneme* to be anterior to all writing and absolutely independent of *buoch*, which precedes it in the current German compound, *Buchstabe* (apparently "beech-rod").

The whole to be considered for the interpretation of the passage of Tacitus which I am here leaving to one side.> (*WW*, 24–26/38–40)[30]

The notion of an alphabetic analysis anterior to all writing, from which all Indo-European poetry can ostensibly be derived, presupposes the inherent divisibility of the speech stream at a level more fundamental than words, on the basis of units too elemental to carry an independent semantic meaning. At stake is thus (also) the originary significance of both the Greek *gramma* and the Latin *littera*. The fact, however, that Saussure's hypothesis finds its most elaborate expression in an argument about the etymological origins of the German *Stab*, suggests that he is here once again polemicizing against the German predecessors and contemporaries he otherwise nowhere mentions. And the fact that his appended note situates his reflections on the German letter in the context of an interpretation of Tacitus's *Germania* suggests, in addition, that he has once again framed his polemic as a battle of Teutons and Romans, Germans and French. This polemic plays almost no role in the reception history of his anagram studies.[31] The polemic, however, is the point, for only a reading that pays close attention to what Saussure does *not* say about the German *Stab*—to what he, in fact, quietly *refuses* to say—will be able to make sense of his seemingly incidental obsession with the prehistory of poetic meter.

LITTLE STICKS, LETTER RHYMES

Although Saussure mobilizes a rhetoric of discovery in order to report on his *Stab*-related findings, he has no real illusions about the

belatedness of his position. By the time he experiences his eureka
moment regarding the common ground of *Stab* as stick, as letter, and
as poetic device, a generally accepted solution to the conundrum of
the three meanings had been circulating through German philological circles for close to half a century, with the rather predictable result
that the discussion itself had all but ceased to exist. It is this generally
accepted Germanic account of *Stab* origins, therefore, that Saussure
implicitly negates on his way to an assiduously un-German theory
of originary poetic essence. The etymological "breakthrough" of the
Stab conceived as calculating tool presupposes a different and more
spirit-filled *Stab* story as the foundational error from which it departs.

The German story, like its Saussurean counterpart, grows out
of an encounter with Tacitus's *Germania*. In a passage devoted to
German practices of divination, the Roman observer speaks of little sticks, or *Stäbe* (*surculi*), on which the Germans inscribe small
marks (*notae*). These signs are then "read" by a priest or patriarch:
"[The Germans] attend to auspices and lots like no one else. Their
practice with lots is straightforward. Cutting a branch from a fruit
tree, they chop it into twigs and, after inscribing them with certain
marks, cast them completely at random over a white cloth. Then a
civic priest, if the consultation is official, or the head of the family,
if private, prays to the gods and, gazing up at the heavens, draws
three separate twigs: these he interprets by the previously inscribed
mark."[32] The passage is cited early on by the Grimm brothers, as
part of a pre-sound law attempt to establish the etymological identity of language and plant life. Extravagant enough in its associative
reach to have aroused the ire of August Wilhelm Schlegel,[33] the brief
"Significance of Flowers and Leaves" ("Bedeutung der Blumen und
Blätter") plays the role of programmatic metatext for the collection
of etymological reflections—the appropriately titled *Old German
Forests* (*Altdeutsche Wälder*, 1813–16)—in which it appears. It also
contains the first of several nineteenth-century attempts to tackle
the question of the *Stab*:

> Writing was, like speaking, according to the meaning and the activ-
> ity involved: the cutting and integrating [*Einfügen*] of branches.
> *Feadha* in Irish and *gwydd* in Welsh mean both tree and letter
> simultaneously, the Welsh *coelbreni* (from sing. *coelbren*, stick of
> omen, *Reiß der Weissagung*) served in prophecies and the casting of
> lots, exactly as, according to Tacitus, *surculi* did for the Germans.
> Whence, still, our letter games and our practice of telling the future
> by letting a book fall open to an arbitrary page [*Wahrsagung aus*

Bücheraufschlagen]. Because, however, in ancient times only a few needed to comprehend the art [of writing] and a rare abstraction went along with it, there developed also in this respect a necessary, unavoidable connection between writing, secrecy, song, and magic.[34]

The Grimms here pronounce writing, like speaking, to be a matter of sticks, inscribed and thrown. Having already bolstered their position with the sequence of dubious etymological "examples" that so horrified Schlegel, they clearly feel no need to elaborate further. *Surculi* imply *notae*, and *notae*, *surculi*; letters and sticks-of-omen share a name but also, on some level, a being. How exactly they do so is less important than the undeniable fact of their sameness. "The following hypothesis [. . .] belongs to the realm of general etymology, and can therefore, as is there typical, be rejected as a whole but not entirely disproven in its particulars."[35] The identity of *Stab* (letter) and *Stab* (stick) is for the brothers an article of organicist faith, a codetermination that transcends in its deep essentiality ("a necessary, unavoidable connection") any distinctions of kind, since metaphor, metonymy, genealogy, and causality all apply in equal measure at the matrix of the common origin.[36]

Despite their evident reluctance to waste time disentangling the complex of writing and magic, secrecy and song, the Grimms do manage to suggest a direction for future research in their brief treatment of the German rune:

"Rune" thus means both the word and the sign with which the word is written; "to write" [*schreiben*] means, in turn, the same as *rita, rista, ritzen,* to inscribe [*einschneiden*], i.e., to cut rune-sticks [*d.i. Runstäbe schneiden*]. The Nordic runes, which are certainly related to other modes of writing, for example the Roman and the Greek, from which, however, they just as certainly do *not* descend, have consequently received their shapes (and even in some cases their names) from sticks and sprigs [*Stäben und Reisern*]; the a. from the ash, the b. from the birch, the th. from the thorn. [. . .][37]

The exemplary imbrication of this *indigenous* letter—a theory attributing the invention of the runes to the influence of the Greek or Roman alphabets is here, in passing, rejected—mediates not only between letter and stick but also between letter and language, across the connecting link of the runic names and shapes. Runes come from trees, etymologically and actually, but they also *mean* trees at the level of linguistic content (their tree-derived names), and *represent* them at the level of linguistic form (their tree- and stick-shaped signs)—all of

ᚠᚢᚦᚨᚱᚲᚷᚹ : ᚺᚾᛁᛜᛃᛈᛉᛋ : ᛏᛒᛗᛗᛚᛟᛞᛉ

fuþarkgw : hnijëpʀs : tbemlŋdo

Figure 6. The oldest-known runic alphabet with its transliteration. From Karin Fjellhammer Seim, "Runologie," in *Altnordische Philologie: Norwegen und Island*, ed. Odd Einar Haugen, trans. (from Norwegian) Astrid van Nahl (Berlin: De Gruyter, 2007), 147–222, here 158.

which equips them, in the eyes of the Grimms, with a truly unique capacity to *figure* the language-nature bond (see figure 6). At stake is the notion of a natural writing—of a graphic convention that accords with natural laws but also, in a sense still to be clarified, of a nature that *writes*—and the possibility of its particularly elegant actualization among the Germanic peoples. Though this passage leaves the implications of such an actualization largely undeveloped, the question of runic writing will return to the forefront a few years later, together with Tacitus's *surculi*, in an appendix to Wilhelm Grimm's *On German Runes* (*Über deutsche Runen*, 1821).

Grimm accepts as the ideological burden of his treatise the responsibility of arguing for an originary Germanic alphabet. For him as for his fellow historical linguists, the alphabet is a privileged mode of writing, since it assigns graphic marks to only the most elemental *articuli*; it has the structure of an organic system rather than an ideographic inventory of names, and seems therefore to penetrate more deeply into the nonarbitrary nature of a language that is itself more life than list.

> This [inner, fundamental view of language] feels the soul of language [*die Seele der Sprache*] and sees in it something intrinsically grounded and living [*begründetes und lebendes*], something independent of human arbitrariness; in the single word it recognizes the organ [*Glied*] that belongs to a whole, not merely an inherently lifeless form that has been furnished with meaning. From this perspective there emerges the need for true letters [*das Bedürfnis der eigentlichen Buchstaben*], which indicate the living sounds, and in turn allow the sounding word [*das tönende Wort*] to sprout [*hervorwachsen*] and acquire form [*sich bilden*].[38]

Phonetic writing of this kind is for Grimm the common inheritance of peoples with *(Sprach)Geist*, and reason therefore dictates that the Germans, as the people of *Geist* par excellence, must have written

phonetically. Having thus established the probability, if not the logical necessity, of an indigenous German alphabet before the advent
of Roman influence, Grimm goes on to dismiss the absence of pre-
Christian evidence with an argument about the impermanence of the
runic *material*: "[T]he wooden letter tablets [*Brief-Tafeln*] on which
one cut the runes [. . .] were easily destroyed, and could not be preserved beyond a certain amount of time."³⁹ Only the wood, in other
words, did not survive. The runes themselves go all the way back to
the time of the migration period, when the Germanic tribes split off
from their Indo-European brethren to descend on northern Europe;
they may even have been invented by the hero-god Odinn (German
Wuotan), as a famous poem from the medieval Norse text known as
the *Edda* contends.⁴⁰

Given Grimm's commitment to the notion of a pre-Roman writing, it is hardly surprising that he ends an important appendix by
harnessing the Tacitus passage to the runic cause. On the basis of
etymological evidence for the Germanic equation of letters and lots—
some of it familiar from the *Old German Forests*, some of it new—
the section "Divination from Tree Branches" ("Weissagungen aus
Baumzweigen") comes to the conclusion that Tacitus's *notae* most
likely refer to runic letters, and that the passage therefore testifies to
the early existence of German writing.⁴¹ The implications of such an
interpretation, however, transcend the question of an indigenous Germanic alphabet. Grimm's rather dubious logic, which suggests that
the *notae* are letters because they are runes (thereby conspicuously
ignoring the possibility of runes that are not also letters), turns the
Tacitus passage into an Indo-European origin scene of the first order.
Here again, and far more explicitly than in *Old German Forests*, the
rune plays the role of missing link, binding *Stab* to *Stab* across the
mediation of a hybrid nature that expresses itself, for Grimm, in a
double etymology: *ritzen* (to carve, ancestor of the English "write")
and *raunen* (to whisper mysteries or magical incantations) come
together in *runa* to yield a writing that captures both phonemes and
future events.⁴² Simultaneously letter and mystical symbol, Grimm's
secret-filled inscription can be said to write nature insofar as its free
fall escapes the interference of human intention. Like the free-flowing
blood of an animal sacrifice, which was also used for divination, the
dark sticks scattered randomly across a white cloth form a graphic
pattern that belongs to the inhuman vocabulary of fate (from *fatum*,
past participle of *fari*, "to speak, to utter"): "The underlying idea is

that in the living, trembling movement [*lebendigen und zitternden Bewegung*] of the falling twig, or of the downward pouring blood— since both are free of all human influence—the divine will is active and must reveal itself."[43] God speaks, in other words, the language of spontaneous movement, which is also, not coincidentally, the language of nature and life. To the extent, however, that this "vital, trembling movement" can afterward be interpreted as a particular constellation of *letters*, the runes also render the divine will accessible to the language of human convention. The significance of the Tacitus passage resides in its portrayal of this all-important transitional moment during which the two languages—historically—intersect.

Grimm himself does not explore the problem of what such a moment would mean for a theory of writing or a treatise on runes, with the result that *On German Runes*, like *Old German Forests*, comes to no definitive conclusions on the subject of the *Stab-Stab* relation. In place, however, of an answer to the question he nowhere articulates, stands once again the runic name, to which his treatise tentatively ascribes the mysterious capacity to mediate: "If one takes the *notae* on the twigs to be actual letters, one is led to the further presumption that the name of the rune located on the selected twig provided the desired answer or decision. I cannot believe that these names were attributed to the runes through some kind of accident [. . .] for all of them designate the most immediate environment of that time, the desirable or the ominous."[44] The language of the gods and the language of men come together—somehow—at the level of letters-turned-names, which are also, according to Grimm, the most originary kind of words. Taken as a group, such names constitute the foundational vocabulary of a pivotal, primal epoch for which writing is always also a divinatory practice, and divination a mode of natural writing.

A corrected and completed version of this same hypothesis— fruitful, clearly, in spite of its vagary—plays a crucial role in the study that, by virtue of its incomparable cogency, eventually puts an end to the nineteenth-century discussion of the *Stab*. The philologist Rochus von Liliencron, in his brief article "On the Doctrine of the Runes" ("Zur Runenlehre," 1852), assigns to the runic names the status of oracular content because he believes, with Grimm, that they encompass "the entire conceptual domain of the oldest times."[45] This small but comprehensive "alphabet" of *Ur*-notions is for him the elemental matter out of which the divinatory interpretation must emerge. Matter alone, however, cannot generate its own interpretation, and Liliencron, therefore, does not content himself with observations about the runic substance. Instead, and unlike his more illustrious predecessors, he focuses on the neglected question of runic *relation*, seeking the rules of transformation that bind *surculi* to (the writing of) fate. He begins by disentangling the one from the other—the metamorphosis he wants to investigate cannot even appear as such, much less submit to analysis, under the conditions of identity assumed by the Grimms—and both from the modern *Buchstabe*. The German runes, he claims, were originally and essentially "mystical signs," *not* letters, and as such they required *actualization* in order to function; it was not movement alone, as Grimm had thought, that turned natural materials such as blood and wood into the natural writing of the gods, but movement in conjunction with the pronunciation of interpretive *spells*. Liliencron defends his thesis by marshaling the many descriptions of Germanic lot-casting practices that make mention of either *incantationes* or *carmina*. He concludes from this list that divination for the Germans is originally a matter of poetry, though not, of course, in any modern or specifically literary sense.

> No expert [*Kundige*] could today still believe that the term *carmina* means actual lyric or epic songs, which would have been written during the time in question with runes, or even, as Saxo describes in his preface, hammered into rock; there must, therefore, be something else behind this phenomenon, and what this something else is cannot be doubted if one takes the former passages together with the latter. The expressions *carmina* and *incantationes* must be grasped, every time they appear, as belonging together to one concept, and where only one stands, the other must be thought as well. The referent of this double expression is: poetic (i.e., alliterative, letter-rhyming [*stab-reimende*]) formulae of incantation or better, religious formulae and such. [. . .] In short: one inscribed [*schnitt ein*] the runes [*notae*] as mystical signs, from which the expert [*Kundige*] could form and combine religious formulae [*carmina*].[46]

The practice of lots casting, or *loosen*, Liliencron here describes is less a kind of writing than of reading—less *schreiben* than *lesen*—but in a sense that is also less conventional than etymological, and which in its emphasis on gathering and sorting can hardly be distinguished from the combinatory poetics of a *dichten* (to poeticize) conceived "literally" as *Verdichtung* (from *sich verdichten*, to concentrate or intensify).[47] "The expert," whose duty is to "read" all manner of mystical signs, interprets the divinely ordered inscriptions by weaving them into the incantatory formulae he then recites rather than writes.

These formulae open up a new space in the German discourse on the *Stab*, which corresponds to the peculiar transitional status of the Tacitus passage. For despite clear structural similarities, the form of inscription practiced among the oracle-loving Germans is of a different, and, for Liliencron, more originary order than the alphabetic mode employed by the description itself. What Tacitus writes for his fellow Romans to read is the record of a moment that precedes, developmentally if not chronologically, both his writing and their reading—a moment governed neither by beech sticks nor *Buchstaben*, but by a third and hitherto unutilized meaning of *Stab* as a pillar of poetic "(d.h. alliterierende, *stab*reimende)" technique.

> If, then, it was possible to discover and recite "carmina et incantationes" by following the lead of the mystical signs, then the signs must necessarily have expressed something that constituted [*bildete*] an essential part of the "carmina." If we now ask what constituted, formally speaking [*formell*], the foundation of the ur-germanic verse, we see that this is the *Stabreim* [alliteration], i.e., the same initial sound within two or three words of a verse that consists of two half-lines. This same initial sound, however, shares its name in ancient poetics with the runes; both are called, with one and the same word, *Stafr* [*Stab*]. What we must explain is the fact that rune and verse hang so tightly together, such that the expert [*der Kundige*] could form [*bilden*] the latter from the former; and here we now find a formal [*formellen*], essential component of the verse, which quite simply *is* a rune, and which, therefore, from ancient times carries its name. The rune *Stab* [rune-stick or staff] was spoken or sung into a verse *Stab* [the poetically alliterating sound].[48]

The interlacing of the *Stäbe* according to a rigorous, predetermined pattern of alliteration turns a jumble of randomly thrown sticks into the skeletal structure of an oracular verse. Like the aphoristic sign-posts Saussure hopes will give discernible shape to the empirical language swamp—like, too, those other law-abiding *Stäbe*, which Pott called the etymologist's "solid support" and Brugman/Osthoff "the foundational pillars of our science"—the runic vertebrae here perform an articulatory function more fundamental than that of the empirical writing to which, among other things, they give rise, lending the stability of nonarbitrary form to the amorphous magic of meaning and the legitimacy of system to the explicatory project of prophecy.[49]

They do this, however, at a level so primal as to transcend the very dualism they thereby appear to negotiate: "Thus it is proven by plenty of passages that treat the carving [*das Ritzen*] of magic runes from later times, that likewise in these later times the dead sign in itself counted for nothing; it became vital [*lebendig*] and effective [*wirksam*] solely through the singing and speaking of the verse, whose *Stab* it was."[50] Form and meaning, matter and spirit, language and life unite in this version of the pagan rune as though to proleptically counter the famous Pauline dictum ("For the letter kills, but the spirit gives life"; 2 Cor. 3:6), which a Christianized Germany will later take for gospel truth. Liliencron supports his claim about a poetry of animated signs with a reference to the sign-animating god Odinn (German Wotan), who is portrayed in the medieval Norse epic, the *Edda*, as the primordial inventor of runes—"I then bent downwards / pondering the runes, I learned them, sighing [. . .] / and drank a drink of the precious mead"[51]—and to the magic liquid the god consumes, "according to a quite standard poetic periphrasis,"[52] as a metaphor for the power of verse. "I peered downwards. / I took up the runes, screaming I took them [. . .] and I got a drink of the precious mead." The result is a vision of runic origin, which is simultaneously a parable of poetic inspiration, conceived, according to the Romantic model of Be*geiste*rung, as an infusion of (linguistic) spirit. The protoletter turns out to be inseparable from the protopoetry of a pagan godhead, which in turn prepares the way, with its alliterative protophonetics ("the recognition and isolation of the initial sounds of words was brought about, in a practical manner, through a fundamental requirement of poetry"[53]) for the protofulfillment of a nineteenth-century linguistic fantasy regarding the foundations of letter science.

By anchoring the nature-philosophical notion of divinatory *lesen* in an actual, datable practice of alliterative *loosen*, Liliencron thus retroactively accomplishes the "historical" realization of a quintessentially German explicatory ideal (Novalis's "concept of philology [. . .] soothsayer from ciphers—letter-augur"). The theory of a poetry-infused *Stab* at the origin of German letters posits the originary reality of a common ground presupposed, explicitly or not, by every significant linguistic work of the period. It hauls into the realm of literal fact that crucial corner of the language science imaginary where stick intersects with symbol, nature with sign, and where, consequently, the organism of language gets *(re)articulated* as a writing that moves the world: "Petersen in his 'Nordisk Mythologi' [. . .]

has very nicely demonstrated how the runes that adhere to all things came to be understood as their essence [*die Wesenheit der Dinge*]; insofar, then, as one breathes life into the rune—which has been as it were 'scraped off' the things—through a magic spell, one thereby also sets the essence of things in magically effective motion [*zauberkräftig wirkende Bewegung*]."[54] The theory also, however, traces this vision of world-spiritual movement back to the pan-Germanic particularity of its literal, runic roots, for the *Stabreim* is an Old Norse, Anglo-Saxon, and Old High German form, with no equivalent among the Romance languages. The dream of a linguistics that would be always also a poetics (and of a poetics that would be always also a physics of *Geist*) becomes, therefore, in Liliencron's midcentury actualization, a reality of strictly Teutonic provenance, in spite of its originarily *Roman*tic pedigree.

THE RHYTHM OF *GEIST*

Liliencron's gesture of nationalist particularization anticipates, and perhaps also helps motivate, a midcentury paradigm shift in the realm of German philology, which "discovers" the verse-structuring technique of alliteration—called *Stabreim*, or "letter rhyme"—as an indigenous alternative to the classical tradition of Greek metrics. Treated only occasionally and in passing by the Grimm brothers,[55] and with industrious perplexity by Lachmann (whose careful but inconclusive treatise on Old German prosody remained unpublished until 1990[56]), alliterative verse did not acquire central disciplinary status until the revelation of its metric uniqueness raised the all-important question of a uniquely German poetic mode.[57] This relatively late revaluation, with its implications for a uniquely German understanding of language "life," proved a welcome springboard for philologists of Germanic literature, who saw in it an opportunity to distinguish their object from the dead, dusty letters of the classicists, on the one hand, and the superficial elegance of the Roman(ce) poets on the other. For such scholars, the presence of a verse style so different from the classical, and hence foreign, conventions of meter and end rhyme testified to the necessity of a national poetics. The project of articulating such a poetics assumed, in the years following Liliencron's seminal contribution, and in the absence of *political* nationhood, a programmatic significance rivaled only by the question of epic origins. Like this latter domain—with which the question of an indigenous poetic mode in any case intersects, since the earliest *Nibelungenlieder* were assumed to have alliterated—the *Stabreim* discourse proved controversial enough to elicit contributions from countless prominent philologists and linguists. More important, however, than the details of the controversy itself, with its focus on the still-unsettled issue of proper scansion, were the tacitly accepted premises that structured the debate for all sides: premises that turned Liliencron's "historicization" of the living letter into a full-fledged theory of spirit-based versification, and against which Saussure found it necessary to polemicize by repeatedly *re*-solving the three-*Stab* riddle.

The premises in question concern the fundamental shape of the originary German verse in its relation to other, more familiar models of poetry, and they constitute the (implicit) German argument for the uniqueness of alliterative verse. The basic rules of line construction, which remained essentially uncontested throughout the scansion debates, are as follows: two half-lines, each with two accented syllables and an unstipulated number of unaccented ones, are separated by a caesura; they are bound together by the alliteration of the third accent with one, or both, of the first two:

> *bā́t* under *béorȝe.*　　*Béornas ȝéarwe*
> on *stéfn stíȝon:*　　*strḗamas wúndon*[58]

Each consonant alliterates only with itself; a few select consonant groups (*sk*, *sp*, *st*) operate as single letters and alliterate only in combination. All vowels alliterate indiscriminately: what actually alliterates, in this case, is not the vowels themselves but—according to one widely accepted theory—the *consonantal* glottal stop that precedes them, unwritten:

> *í* siȝ ond *ū́tfūs*　　*ǽ ðelìnȝes fœr*[59]

In accordance with the terminology established by the thirteenth-century Icelandic poet Snorri Sturluson—author of the text known as the *Prose Edda* and its concluding treatise on verse construction[60]—the first letter of the third accented syllable is called the *Hauptstab*, from the old Icelandic *hofuðstafr*, meaning "main staff," while the alliterating syllables from the first half of the line, which are governed by their relationship to the *Hauptstab*, are known as *Stollen*, from the old Icelandic *stuðill*, meaning "staff" or "support." The verb *staben* refers to the actual act of alliterating and, by extension, to the process of verse construction per se: *staben* is the function the poet performs, and as such is equivalent in resonance and scope to both the modern German *dichten* and its classical Greek counterpart. The technique it describes, however, could hardly be more different from the technique prescribed by a classical poetics, since the "poem" it produces has no fixed meter in the conventional, syllable-counting sense. Unlike the Greeks and Romans, who built verses by combining long and short syllables into preestablished patterns such as iambs and trochees, the *Stabreim* effectively ignores its syllables in favor of alliterating accents. By refusing to regulate the total number of syllables in a line, it makes possible a "freedom" in the expression of poetic "feeling" that compares favorably—in nineteenth-century German eyes—to the ultrarational *measurements* of the ancients.[61]

This freedom has nothing to do with the arbitrary chaos of *ungebundene Rede* (literally "unbound speech") or poetic prose. To the Germanists, lawlessness in poetic practice remains the special province of the French, who inherit, via the Romans, the Greek poets' fetishization of number without their respect for rules, resulting in "verse" forms defined solely by a set quantity of syllables per line. The absence of articulatory units, or feet, in the French (and late Latin) conceptions of poetry entails an absence of structure that can be mitigated only by recourse to the twin medieval innovations of assonance and end rhyme: "[T]he assonance of the Romance languages, and the rhyme that emerged from it [. . .] functioned essentially to make palpable for the ear the otherwise only weakly suggested verse rhythm."[62] The notion of a historical genesis motivated by metric insufficiency strips sound parallelism of its relationship to poetic essence. End rhyme, in which the early Romantics had still seen a legitimate technique for transcending classical, syllable-measuring pedantry,[63] appears here, together with its weaker, vocalic counterpart, as an unsatisfactory substitute for lost metrical rigor. The makeshift unity it confers on the otherwise nonexistent form of French verse is but a pale echo of the deeper, more systematic articulations of the ancients.[64]

Not, then, the unbounded freedom of an essentially arbitrary French poetry but the seemingly paradoxical freedom of an organic necessity: such is the alternative the Germanic *Stabreim* presents to the law-abiding elegance of the classical metric form. The Greek articulation, like the German one, takes place as a function of stress, or ictus, which breaks up the continuity of the poetic line in conformity with certain incontrovertible rules. The difference between the two traditions, from the perspective of the *Stabreim* scholars, lies in the relationship of these rules to the question of poetic meaning—or rather, in the fact that, for classical poetry, such a relationship does not and cannot exist. The "accents" that structure ancient Greek and Roman verse mean nothing, not because they are insignificant, but because the laws that determine their placement have nothing to do with the accents of ordinary speech, or the lexical meaning of the accented syllables. They have to do instead with the purely quantitative length of the syllabic vowel—long ones take accents, short ones do not—which in turn explains why the classical "foot" never contains more than one long-voweled syllable, and why classical meter is so often referred to as a poetry of "quantity." No matter how elaborate or aesthetically pleasing these patterns of long and short might become in the hands of a poet with perfect technique, they will never bear any *meaningful* relation to the poetic content, which belongs by definition to the other side of the terminological equation: "quality," like *Geist*, is a German specialty, and the true contribution of the *Stabreim* to the history of poetic form.

The conditions of possibility for this qualitative innovation are to be found, according to the philologists, in the character of the (proto-)German language itself. The story of German uniqueness in this regard begins with a reconstructed linguistic event known as the accent shift, which is thought to have occurred some time after the first sound shift, when proto-Germanic diverged phonetically from the Indo-European matrix, and before the second, when it splintered into mutually incomprehensible dialects. The accent shift fixed the originarily flexible Indo-European accent to the word stem, with the result that the Germans, whose Indo-European forefathers had accented "meaningless" endings as often as "meaningful" stems, now stressed only the bearer of the etymological root. The result was the subordination of their language to what the metric scholar Rudolf Westphal terms *das etymologische Prinzip*: "In the Germanic languages [*im Germanischen*] it is always the foundational element of the word in its actual materiality [*das eigentlich materielle Grundelement des Wortes*], i.e., the root syllable, which draws the accent to itself, and only then releases it when it unites with a second root word or with a meaningful, separable preposition to form a composite."[65] When, then, the proto-German poet refuses, in line with the peculiarly intimate bond he shares with his language, to ignore the accents of ordinary speech, he also ties the basic articulatory unit of his form

to the (etymological) meaning of his words.[66] The most lexically significant syllables unfailingly receive the poetic stress, which manifests
itself not as a quantitative increase in vocalic length but as a qualitative intensification of expressive sound. What is thereby accented or
emphasized is thus not an arbitrary aesthetic form imposed on the
matter of language but the spirit-infused matter of language itself:
where the Greek poets distort the "natural" accent patterns of spoken Greek in the service of a sophisticated but ultimately alinguistic
aesthetic ideal—namely the rhythmic and tonal symmetries of wordless *music*—the Germans draw their inspiration from the etymological, and uniquely German, correlation of structure and meaning
that governs their every linguistic activity. The alliterating consonant
marks the ictus with a "rhyming" letter and, in doing so, gives material shape to the dynamic intensity of the accent, which in turn articulates the hierarchical play of poetic meaning and, *simultaneously*,
the etymological architectonics of language. The poet, in this case, is
quite literally enspirited by language—a *Sprachbegeisterter*, in Jacob
Grimm's sense—infused with the animatory power of a medium he
never conceives as mere tool, and capable in his inspired state of distilling the energy of everyday speech down to its explicated, poetico-
etymological essence.

Beneath a terminology that distinguishes between the quantitative
aesthetics of the Greek "song verse" (*Gesangvers*) and the qualitative
aesthetics of the German "speech verse" (*Sprechvers*) pulses, then,
the central premise of nineteenth-century *Stabreim* scholarship. An
ostensibly classical model of poetic inspiration, according to which
the poetic mind infuses its linguistic matter, Pygmalion-like, with a
foreign structuring principle of metrical but meaningless sound, goes
head to head here with a model of meaning-based verse infused by
a particularly pure distillation of the collective linguistic spirit. The
purity stems from the refractory rather than disruptive influence of a
poet who functions as prism rather than source, and who, in doing so,
allows language to articulate itself anew on the higher than ordinary
plane of poetry. The implicit hierarchy governing the two inspiratory
models recalls the Grimm brothers' prioritization of "nature poetry"
(*Naturpoesie*), which opposes to the individual articulations of a more
sophisticated, more *will*ful "art poetry" (*Kunstpoesie*) the spontaneous self-expression of an entire language-sharing people, or *Volk*. The
seemingly oxymoronic notion of a wholly natural literary mode, conceived as a compositional art that eschews all compositional artifice,

performs in its Grimmian context the familiar nature-philosophical gesture of transcending the traditional matter-spirit divide. In the space opened up by the midcentury *Stabreim* scholars—a space that the Grimms themselves never thought to search seriously—it performs this same gesture at an entirely new level of formal, and historically "verifiable," specificity. The poetic interchange between an inherently meaningful nature and an inherently natural language, as imagined but not studied by both Grimms, acquires in alliterative verse a concrete metric shape that bears retroactive witness, in the minds of its proponents, to the validity of the original hypothesis.

That *Naturpoesie* should turn out to be, historically speaking, a German category, realizable only in accordance with Germanic alliterative rules is, on the one hand, for the language scientists in question, an unavoidable consequence of the all-important accent shift that linked German pronunciation to the source of the sound-sense bond. Unlike other peoples, who were free to obscure the energy of the etymological origin by emphasizing subsidiary syllables, the Germans had no choice but to turn, every time, to the life-giving, meaning-bestowing root. This compulsory reverence for the animating principle, however, is also no mere historical accident, since, according to Westphal, the punctual event of the shift simply gives outward expression to a preexisting and specifically German intensity of spirit: "It seems almost as if the ancient German, during the period when this persistent accentuation of the root syllable became a fixture within his language, also fixed the type of his own being therein—his thirst for movement and deeds [*seine Bewegungs- und Thatenlust*]—by elevating this syllable, which denotes activity and movement, to the center of the whole word and sentence. For what he accentuates in his language is quite simply the moment of action [*das Moment des Handelns*], of movement; the whole energy of speaking is directed toward this end."[67] Westphal makes oblique reference, here, to the theory, prevalent among historical linguists from Bopp and Grimm onward, that the grammatical stems of all speech parts derive exclusively from the roots of *verbs*, since verbs provide the privileged medium for the formative principle of inflection.[68] To accent such stems is thus to emphasize the dynamic activity of infolding—or gathering—that gives rise to the organism of language. The Germans alone have access, so the argument goes, to a truly living language in which to compose their truly natural poetry, because they alone express themselves precisely by expressing the energetic origins of all

truly articulated speech. In other words: their very *essence* is etymological, in the new sense of the term defined by the founders of a
nineteenth-century language science.

The *Stäbe* of the *Stabreim* just *are*, then, the living letters about
which these founders could otherwise only fantasize. August Pott had
claimed for the "tangible" but meaningless consonants of Grimm's
sound shift the ability, within the realm of language science, to
replace the vagaries of the meaningful word.[69] He had also, however, called for a science that would end by bringing letter and word
back together, a science that would eventually find itself in a position
to reinvest the sounding stuff of language with the significance of
its etymological origins: "[T]he researcher has the responsibility of
unleashing the life that is bound up within the letters."[70] The rhyming runes, with their primal unity of matter and meaning, relate to
their law-abiding, sound-shifting cousins as the figure of the latter's
eventual fulfillment. They embody, in a preliminary and anticipatory
form, the principle of a natural, oracular writing capable of revealing, through the patterns of poetico-linguistic movement, the *fatum*
of a self-articulating, world-spiritual will—a *fatum* that can then be
"read," in turn, by poet-priests trained in the letter-sorting techniques
of a divinatory philology. Liliencron's little sticks, which are not (yet)
Roman letters, articulate the intricate sound figures, or *Klangfiguren*,
of a harmoniously vibrating world, in perfect attunement with the
natural rhythms of gods and (German) men.

THE CULT OF CANCELLATION

When Saussure repeatedly contrasts his notion of a total Saturnian
symmetry with the Germanic ideal of alliteration, he is therefore
implicitly but forcefully declaring his hostility, both as French speaker
and as linguist, to a nineteenth-century ideology of living letters that
prioritizes all things German. The fact that this Germanocentric ideology makes room from the very first for the possibility of a Roman
exception does little to bridge the national divide, since, according
to the most prominent students of the Saturnian, the ancient Italians
were to be commended for their commitment to a language-faithful
folk tradition of "naturally" accented poetry, but *not* for the discovery
of the one truly natural, alliterative form, which they are presumed
to have inherited from the etymologically accenting Germans.[71] (The
same scholars who commend the Saturnian thus go on to condemn

a more modern Roman(ce) poetry for its fall from the grace of [German] *Geist*.) Saussure resists this logic with a hypothesis that turns the ethnic tables: his theory of verse origins arrives, "in starting from the Latin facts,"[72] at a *Stabreim* deprived of all priority and a Saturnian poetics purged of all dependence on the crutch of the spirit-filled *Stab*. At stake in this reversal, however, is something more than mere parochial pride, for the structure of a self-articulating flow of animating energy, as implied by the German organicist paradigm, threatens the very foundations of Saussure's transnational linguistic vision: "The simple expression will be algebraic or will not be at all" (*WGL*, 165/236). So long as letters live, or at the very least maintain a connection to their original, poetic ability to do so, they will never allow themselves to be satisfactorily—which is to say remainderlessly— *written down.*

In place of the living letter, Saussure thus proposes a theory of absolute articulation and writability "without remainder."[73] "The ideal verse," composed in accordance with the "first principle of Indo-European poetry," is defined not by its alliteration of accented initials but by its even numbers of each appearing "phonic element," such that every line can be analyzed into a series of self-cancelling pairs. A second rule, introduced to compensate for the enormous difficulty of a strict, line by line cancellation, dictates that "extra" phonemes may be carried over to the following line, where they can then be paired with an identical remainder. A third rule allows for the possibility of an intentional "miscount," which would generate enough "remainder" phonemes to spell out the names of gods or heroes.

Saussure will eventually uncouple this last rule from the rigor of phonic symmetry to produce an independent theory of poetic anagrams. Together with a relatively conventional understanding of the relation between divinities and their names—according to which the anagrams serve to "rivet" the god in question to the prayer of a superstitious poet-priest[74]—this latter theory will then come to dominate his analyses to the exclusion of all other considerations. By that time, he will also have given up on the fantasy of a perfectly remainderless articulation, turned away from the question of Indo-European origins, and lost faith in the redemptive potential of the poetics he describes. He will go so far as to pronounce the anagrammatic method "by nature deplorable, but inescapable" (*WW*, 102/134), and to declare himself dissatisfied with his own conclusions—"I would be delighted if someone could demonstrate to me, for instance, that there are no anagrams but simply a repetition of the same syllables, or elements, in accordance with rules of versification which have no connection with proper names or with a specific word" (*WW*, 93/124)—before finally abandoning the project altogether in frustration over lack of proof.

Here at the outset of his investigations, however, and still under the influence of the elation called forth by the suddenness of his Saturnian insight, Saussure dares to hope that he has located, at the origin of Indo-European time, a truly *differential* poetics. For, undertaken in consonance with the higher law of compensatory cancellation, the technical flourish of the anagram rivets not god to word but word to letter, turning names, and with them the divinities on whom those names call, into a subsidiary function of rule-governed phoneme play. The total number of phonic elements minus the total number of those properly paired yields a residue which is simultaneously generated and excluded by the empty purity of a "closed arithmetic,"[75] and which, in its status as "THEME or TITLE" (*WW*, 13/25), can then be considered *quantitatively equal* to the meaning of the poem.

In the hypothesis of a poetry that subordinates expressive spirit— "whether the critic on the one hand, and the versifier on the other, desire it or not" (*WW*, 17/30)—to the negative precision of an articulatory analysis, Saussure clearly imagines himself to have identified a mode of expression capable of doing justice to the Janus-faced structure of language as such. Here, too, the divisions of a differential writing *call forth*, according to a logic both inevitable and essentially linguistic, the substance of a positive meaning that is in this case the content of the anagrammatic names. Here, too, the substance eventually vies for priority with the nothingness from which it emerges, for the anagrams that start out as ornament end up as a mandatory rule, and thus as an integral component of the definition they originally seemed only to supplement. The difference—and the reason, perhaps, for at least some of Saussure's Saturnian excitement—is that in the case of a phoneme-counting poetry, the challenge posed by positivity has no real chance of success. So long as the rules of phonic symmetry remain firmly in place, the anagrams can never infiltrate the emptiness they help to define for the simple reason that this emptiness exists on an entirely different plane, not only from the words for gods and heroes but from the problem of words in general. Whereas the syllable *ba* can mean, simultaneously if not originally, both "everything that is *not* la," and "things that *are* solid"—like a coin with both negative and positive sides—the phoneme *b,* as counted by the Indo-European poet, has only a single, differential definition: "Phonemes are above all oppositional, relative, and negative entities" (*Course* 139 / *CLG*, 268:1; translation modified). It is identical to the set of acoustic characteristics that distinguish it from all other sounding segments of language, across infinitely many variations in pronunciation, and *it is nothing beyond these distinctions*. Attempts to reify it into something more break down at the most fundamental level of articulation, since *b*, empirically speaking, does not actually, independently exist. In order to be heard, it must be preceded or followed by the vowel that gives it voice, and only in this syllabic combination—a combination that remains poetically impossible so long as the Indo-European poet remembers to double his *b*'s—can it open itself to an influx of human-centered sense.

The doubling operation that protects Saussure's *Stäbe* from the encroachments of reference thus bears a strong resemblance to the repetitions of his *figurae etymologicae*, which identify key linguistic concepts with *themselves* in order to hold all (other) meaning at bay. Here, however, the strategy of self-reflection does not falter at the level of the "self," for the units being repeated, and in repetition being affirmed, do not thereby acquire any illegitimate kernels. Repetition, within the realm of phonic symmetry, does not fulfill or emphasize or intensify on the way to its profoundly non-Hegelian cancellation of language sound; rather, the folding over of the phonemes reproduces in (rounded) shape but also in (absence of) essence the paradoxical power of a language-generating *zero*. The result is a philological fantasy of the ideal linguistic sign, from which everything emerges and to which nothing, whatsoever, can return—the fantasy of a sign that can be articulated, written, and defined without acquiring positive meaning, and that therefore runs no risk of confusion with respect to the words it grounds. A sign that can affirm nothing without affirming anything and whose empty enclosure deflects all penetration attempts to which it simultaneously gives rise. A sign, therefore, that, technically speaking, is no sign at all, since Saussure's official definition of the sign still requires that a sense coincide with a sound: the notes on general linguistics do not seek to articulate a concept of sense-less elements because they consider sense-less sound to be inarticulable by its very nature. The divisions that determine the basic units of a language ("articulations, that is to say *articuli*, small limbs"; *Notebooks* II 22a / *CLG*, 253:2) can only ever occur, according to a Saussurean model of signification, at the level of the sound-sense plane ("the thought-sound implies the divisions which are the final units of linguistics"; *Notebooks* II 21a–22a / *CLG*, 253:2).

By making the imaginary letter-sign into the foundation of his Indo-European philology, Saussure effectively presupposes, in other words, a notion of the unit that would have been impossible to formulate within the realm of his linguistic writings, and he does so in full consciousness of the outlandish conclusions that such a notion entails. His rejection of the beech tree half of *Buchstabe*—"I consider *stab* = *phoneme* to be anterior to all writing and absolutely independent of *buoch*" (*WW*, 25–26/40)—and with it all reference to the raw material of an empirical inscription, commits him quite explicitly to the transempirical reality of an alphabet beneath and before all (other) writing. And a related terminological clarification, which claims for this alphabet a (chrono)logical priority over the bodily wind of speech itself, goes on to give voice to what the *Stab* story itself leaves unsaid: "Neither anagram nor paragram means to say that poetry is arranged [*se dirige*] for these figures in accordance with written signs, but to replace *gram* by *phone* in either of these words would result precisely in the suggestion that we are dealing here with an unheard of species of thing" (*WW*, 18/31). Saussure here pronounces the terminological shift from "anagram" to "anaphone" inadvisable, despite the fact that the poetry he interprets is an aural one, because such a shift would suggest—or rather admit—that the "things" in question are in fact *inaudible sounds*. The Indo-European poet plays with acoustic units whose boundaries make possible the articulation of ordinary speech without themselves making noise ("the acoustic image is not the material sound"; *CLG*, 149:2), and the "phonemes" Saussure equates with the originary meaning of the German *Stäbe* ("I consider *stab* = *phoneme*"; *WW*, 25/40) are thus never, strictly speaking, *phonetic* in kind.[76]

Only in this entirely unprecedented sense of the term "phoneme," which Saussure nowhere overtly theorizes but which clearly anticipates the transempirical object of a differential phonology—as first elaborated in the 1920s by Nikolai Trubetzkoy and Roman Jakobson—does anagrammatic poetry turn out to coincide with a "phonic analysis of words": "I affirm in effect (as my thesis from this point forward) that the poet devoted himself to the phonic analysis of words and considered this his basic trade [*ordinaire métier*]: that it is this science of the vocal form of words which is probably responsible, from the most ancient Indo-European times, for the superiority, the distinctive quality, of the Hindu *Kavis*, the Latin *Vātēs*, etc." (*WW*, 22/36). The Indo-European poet, together with his Indian and Latin descendants, raises himself above his fellow mortals by practicing a mode of sound writing that figures here as the origin of verse but also of language science. (Saussure will go on to posit a causal connection between Indo-European phonic play and the rise of the Sanskrit grammatical tradition that had so fascinated Schlegel and his compatriots.)[77] The superiority of these rule-governed and compulsively counted lines—over the incomplete analyses of a meaning-based *Stabreim* but also, one assumes, over the phonic laxity of an "expressive" modernity—rests on a technique that makes conscious the transempirical divisions of *conscience* at the level of a wholly empirical craft. Where the alliterating Germans content themselves with emphasizing the rhythms of ordinary speech, "according to a principle that in no way implies, on the part of the poet, the analysis of the word" (*WW*, 22/36), their more sophisticated Indo-European predecessors, according to Saussure, poeticize in precisely the opposite direction, mysteriously managing to transcribe the conditions of linguistic possibility into the lines of a poetic *parole*.

It is therefore as "specialist of phonemes,"[78] rather than as worldspiritual divining rod, that the poet becomes a seer for Saussure (the Sanskrit *kavi* and Latin *vātēs* both refer to a caste of soothsaying priests), and as a story of linguistic, rather than cosmic, revelation that the question of *Stab*-origins gets resolved. Instead of the little Tacitan sticks, falling freely in consonance with the divine writing of fate—a writing that must therefore be imagined to infuse and inspire the alliterative unfolding of its priestly-poetic interpretation— Saussure insists on the empty precision of his counting twigs ("passing successively from right to left as the composition progressed"; *WW*, 25/40), which have only as much to do with the German runes as they do with all the other alphabets they prefigure. Poetry and philology come together for him in an originary ritual of explication that subordinates everything, and especially holy spirit, to the meaningless letters of an empty *langue*:

> With respect to that which specifically concerns the Vedic text itself, and the spirit in which it has been transmitted from inaccessibly ancient times, this spirit would find itself eminently in conformity, *via its attachment to the letter*, to the first principle of Indo-European poetry, <as I now conceive it>, apart from all specifically Hindu, or specifically hieratic factors, invoked in relation to this superstition of the letter.

I will even reserve my opinion as to whether the *Pada-pâtha* text of
the hymns is designed to safeguard phonic correspondences whose
value is traditionally understood and is, <in consequence>, relative
to the *line,* although the text seems to want to establish the form of
words, apart from the line. There is still a need for a study I haven't
made, and which is, by all evidence, immense. (*WW,* 24/38)

The commentator-priests charged with correctly interpreting, and
thus with protecting, a protolinguistic poetry can stay true to the
Indo-European "spirit" of their sacred object only by reading it without reference to its meaning-filled words, in which all German-style
spirit resides. Rather than reanimating the written letters by linking
them to the "living movement" of their (runic) names, they must seek
to acknowledge, through the "simple" act of repetition, the poetic
articulation of an essentially wordless ground.

Saussure's own procedure in the Saturnian notebooks provides a
perfect example of this priestly *attachement à la lettre*:

For example, in:

Subigit omne Loucanam opsidesqve abdoucit

one sees	2 times *ouc*	(L*ouc*anam, abd*ouc*it)
	2 times *d*	(opsi*d*esqve, ab*d*oucit)
	2 times *b*	(su*b*igit, a*b*doucit)
	2 times *–it*	(subig*it*, abdouc*it*)
	2 times *ĭ*	(sub*i*git, ops*i*des-)
	2 times *ă*	(Louc*a*nam, *a*bdoucit)
	2 times *o*	(*o*mne, *o*psides-)
	2 times *n*	(om*n*e, Louca*n*am)
	2 times *m*	(o*m*ne, Loucana*m*)*

The principle remainders are found to correspond precisely to those
the preceding verse left in abeyance:

In effect the *p* of *opsides* (last line)
= the *p* of *cēpit* (penultimate line)

each remains without a correspondence in its own line: but together
they are compensated, from one line to the other.

*Why not *omneM* Loucanam? This is precisely what I think I can
prove by a long series of examples: that the imprecisions of form
which have sometimes been taken as archaisms in epigraphic Saturnian poetry are *deliberate,* and in accord with the phonic laws of this
poetry. *Omnem* would have made the number of *m*'s uneven! (*WW,*
20–21/33–34)[79]

If, within the realm of linguistics, Saussure's dramatic rethinking of language science forces him to leave the security of Grimm's consonant table definitively behind, then the breakthrough of his Saturnian philology must be sought in the return it permits him to perform—at the higher level of a more originary *Stab*—to the security of a tabular ground. The fascination with phonic structure, and with the phoneme in the unsurpassable purity of its alliterative annulment, generates here a mode of reading that does away with the philological problem of poetic language, conceived as a vehicle for spiritual meaning, but also with the problem of language as such, conceived as a collection of words. The Saturnian line *expresses*, in accordance with "phonic laws" more fundamental than Grimm's, nothing other than the insubstantial nothingness at the origin of all expression. The philologist, in turn, makes this nothingness explicit by rewriting it as a table of cancelled pairs. Critique becomes a matter of precise arithmetic notation, and the clean count a methodological ideal to which Saussure can devote himself with unabashedly fanatical vigor. The fact that he sees in this profoundly un*inspired* mode of reading ("apart from all specifically hieratic factors") the literal foundations of a numinous Logos, and in his own role as letter-counting commentator the only originary cultic mode, is the true implication of the peculiar motto he inscribes across one of the anagram notebooks: "NUMERO DEUS PARI GAUDET" (God loves even numbers; *WW*, 11/23).[80]

The fact that this literalization of spirit occurs by way of an eminently Saussurean literalization of the literal itself—and of a century's worth of privileged German letters—is the true implication of his idiosyncratic etymology of *Buchstabe*.

PART II

Tending toward Zero: From Runes to Phonemes

The great Russian linguist Roman Jakobson—one of the inventors of contemporary phonology[1] and the ingenious fulfiller of Saussure's wildest differential dreams—pays frequent homage to the published work of his most famous predecessor, whose reflections on language structure so decisively inform his own. The early nineteenth-century theorists of *Sprachgeist*, on the other hand, play almost no explicit role in his oeuvre, for the deceptively simple reason that they are at once too close (in substance) and too far away (in time). Like Saussure, Jakobson polemicizes vigorously against the more recent mechanist-materialist approaches he considers fatal to any real understanding of linguistic structure ("For the naturalistically inclined researcher, the sound inventory of language disintegrated into an uncountable number [*eine Unmenge*] of unstable motoric or acoustic atoms, which he painstakingly measured, but whose purpose and meaning he consciously renounced"[2]). Like Saussure, Jakobson also remains tellingly silent on the subject of a spirit-drenched system theory more intimately related, and hence ultimately more threatening, to his own.

What the phonological battle with *Sprachgeist* lacks in rhetorical prominence, however, it recuperates in radicality and depth, for while the mechanistic perspective of Saussure's Neogrammarian teachers can simply be summarily rejected on the way to a demonstrably more effective method of analysis, the teleological tradition of the *Bildungsprinzip* must instead be painstakingly, if quietly, transfigured. The following pages will argue that this transfiguration—which turns out to chart a surprising course from the harmony theories of Richard Wagner to the experimental psychophysics of Wilhelm Wundt to the poetics of the European avant-garde before coming to language-scientific fruition in Jakobson's disenspirited phonemes—just *is* the true content of the phonological revolution. "Definition by difference," according to the logic here unfolded, must necessarily remain an essentially meaningless phrase so long as its agon with a historically prior and systemically still powerful "differentiation by spirit" continues to be ignored, because the agon, not the slogan, makes the point: only a structural emptiness reconceived as a structural emptying, and thus also as a *process* of eliminating past selves *from within*, can do justice to the real conceptual power of the structuralist linguistic paradigm.

Jakobson accomplishes the task of a system-internal elimination by applying the principles of differential definition to a level beneath and before the conventional signifier—as Saussure had tried and failed to do—according to a new logic of analysis made possible by a breakthrough he nowhere thematizes (and which consequently remains unthematized throughout the scholarly literature on Jakobsonian phonology).[3] At stake is still, for him, the quintessentially Saussurean conundrum of what it could mean for two or more distinct, empirical phenomena, separated by changes in time, space, and/or material appearance, to be considered "the same," or in this case, to be considered a linguistic unit. And the breakthrough in question involves a corresponding revolution in the system-theoretical concept of analogy (Greek *analogia*) under which the most problematic instances of sameness have so often historically been thought.

As a technique for establishing relations among relations in the form of a ratio between ratios—if *a* is to *b* as *c* is *d*, then the relation *a* : *b* stands in a one-to-one relation with the relation *c* : *d*—analogy appears to hold out the promise of a similarity founded on form rather than substance: sameness of material substrate can but need not be present in order to account for an analogical equivalence of structure. (To cite some famous Aristotelian examples: the analogy of "fish spine is to fish as bone is to animal" rests on the identity of an underlying osseous matter, while analogies such as "hand is to human as claw is to bird" and "feather is to bird as scale is to fish" do not.[4] In the latter two cases, the similarity refers only to the structure of the relationship that obtains between the hands, claws, feathers, scales, and their respective organisms, which is to say, to the *function* of these organs within the context of a living body.) Analogy thus allows for the idea of a sameness that need not be sensorially accessible to be known, a mode of belonging together that goes beyond what can be seen, felt, or heard "as one." In doing so, it paves the way for a broader understanding of conceptual substrate or essence without ever appearing to challenge the idea that a substrate *of some kind* must exist. To the extent that a proportional correspondence can be assumed to count as knowledge, it *must* turn out to rest on a real foundation of substantive Oneness just as, arithmetically speaking, every equation between numerical ratios entails the existence of a common unit.[5]

It is this more capacious yet still fundamentally "rational" notion of conceptual and systemic unity that so strongly appeals, as previous chapters have demonstrated, to early nineteenth-century scientists concerned with domains such as life and language, where identities of function seldom coincide completely with identities in the perceptible material. And it is this nineteenth-century reinterpretation of analogy, in turn, that Jakobson so powerfully subverts when he subtracts, via a procedure that remains to be analyzed, the spirit of unity from the One. The project sounds paradoxical, since to empty the ratio among ratios of the common ground that constitutes the analogy would seem to imply the end of all cohesive collection. It does not. Nor is Jakobson alone in his attempt to reconceive the rationality of system, "meta-analogically," as a relation among *ir*rational relations: A comprehensive treatment of the origins and consequences of the new, antiessentialist system theory in question would require a much broader discussion, one capable of assimilating nineteenth- and twentieth-century developments in domains such as set theory, mathematical logic, group theory, information theory, and cybernetics.[6] Amid this proliferation of new system-theoretical disciplines, however, the status of linguistics remains unique, since only the linguist can (and must) insist that such fundamentally transrational systems are actually *intuitively* accessible, which is to say also relevant for all human(ist)s. To know a language is to understand it effortlessly, almost without thinking, as a series of articulated units rather than an undifferentiated stream of sound. To know a language is thus always already to engage, in a way that phonological theory sets out to explicate, with the peculiarly beautiful logic of an irrational, because incommensurable, unity.

CHAPTER 4

Wagner's Poetry of the Spheres

PHILOLOGY + HARMONY

Ferdinand de Saussure may never have known that the spirit-drenched poetics he hoped to counter with his combinatorial *Stab* story had escaped early on from its philological confinement, becoming toward the end of the nineteenth century an independent poetic phenomenon of peculiarly international influence. For many of his contemporaries, however, the *Ur*-Germanic technique of alliteration was, in fact, *primarily* the province not of Norse bards but of the paradigmatically "modern" composer Richard Wagner, whose rehabilitation of the *Stabreim* for the purpose of a future-oriented, operatic art led him to a dramatic reinterpretation of language spirit.

Wagner was not the first nineteenth-century poet to attempt a revival of alliterative verse. Friedrich de la Motte Fouqué's early dramatic trilogy *The Hero of the North* (*Der Held des Nordens*, 1810), for instance, which took as its foundation the Norse version of the *Nibelungen* myth and sought to preserve the alliterative structure of its source text, the *Völsunga Saga*,[1] may well have influenced Wagner's decision to prioritize the Norse *Nibelungen* strain over the German.[2] The alliterative *Edda* translations of Ludwig Ettmüller (1837) and Karl Simrock (1851), which Wagner certainly knew and used, combined the creativity of a literary "adaptation" intended for a lay audience with concern for textual authenticity and philological expertise. (The former contains a preface explaining the fundamentals of the *Stabreim* technique, the latter a knowledgeable scholarly apparatus, including commentary on the Liliencron theory of the rune as original *Stab*).[3] A Wagner contemporary named Wilhelm Jordan, traveling around throughout the 1860s and 1870s in epic-poet garb, recited the alliterative lines of his own two-volume *Nibelungen* epic to large and occasionally even international audiences—a practice Wagner claimed to have learned about from the newspaper reports, and which he clearly considered a distressing parody of his own attempts to bring the *Nibelungen* material to the stage.[4]

All of these otherwise quite disparate efforts at modern-day alliteration had in common the desire to "translate" the language of the Germanic origins into the dialect of the present tense. Ancient, indigenous conventions of poetry formation were to be updated, but otherwise obeyed, and philological study, which had uncovered those conventions, deserved on this account the final word. Only Wagner, whose respect for authenticity and origin clearly rivaled that of his alliterating contemporaries, had the artistic courage—and intellectual arrogance—to reverse this order of precedence by transforming, rather than merely reflecting, the terms of the philological debate. At his hands the *Stabreim* became, once again, a viable because poetically independent compositional technique, and its newly regained independence was confirmed, not compromised, by its relationship to a second but nonsupplementary compositional mode: *Music*, which lay beyond the reach of all traditionally text-based philology, was Wagner's utterly unique contribution to the definition of German verse. It was his answer to the scholarly controversy surrounding the meter of the *Stabreim* line, to the problem of differentiating poetry from prose, and to the age-old question of language origins. More importantly, it was also the reason that his alliterative experiments acquired, across the medium of the operatic stage, a significance for subsequent theories about language that no (other) philological model could hope to rival.

The philological controversy over the measure, or meter, of alliterative verse grew out of the difficulty of distinguishing the *Stabreim* rhythm from the rhythms of ordinary speech, which it was presumed, in accordance with nature-philosophical principles, to distill but not distort. So long as the accentual patterns of prose could not give way, as in Greek poetry, to an external order of regular "beats" or "tacts"; and so long, also, as German versification had to remain distinct from the lawlessness of French free verse, which tacked onto its "lines" of everyday language the occasional ornamentation of end rhyme or the

superficial "structure" of the stanza, students of the *Stabreim* needed
a new way of understanding—and articulating—the formal structure of poetic time. Uncountable theories came and went throughout the second half of the century, but none managed to strike the
peculiar balance, required by the ideology of alliteration, between a
meter that poetically transcended the ordinary and one that remained
faithful to everyday rhythms. It is no coincidence, therefore, that the
only proposal to gain widespread acceptance during this period made
no claims at all upon the title of general "theory": Eduard Sievers's
"five types" paradigm, which distributed all known examples of alliterative verse across five different categories of nonbinding accentual
patterns—without making any attempt to reduce these categories to
a single, rhythmic principle—offered, by his own account, no more
than a provisional, "statistical" solution.[5]

Beginning already in the 1850s, Wagner proposed a different
approach. Like his philologist contemporaries, he considered the discovery of a unifying principle behind the *Stabreim* to be the highest
priority of a serious poetics. Unlike them, however, and in defiance
of all scholarly discipline, he chose *not* to search for this principle in
existing examples of alliterative verse. Text alone, he reasoned, can
never give reliable information about what is after all finally and essentially a question of *sound* and, more specifically, of the rule-governed
unfolding of sound over *time*. To argue that such a question has nothing to do with music is to play a pedantic game of redefinition at the
expense of true understanding, particularly since true understanding
must always involve, for Wagner, the *sounding* essence of meaning.
Nevertheless—and herein lies the crucial, problem-solving gesture of
his proposal—the musical answer need in no way conflict with the
fundamentally linguistic understanding of the *Stabreim*, since the
concept of music has *also* been misunderstood, according to Wagner,
so long as it does not encompass—and experience its fulfillment in—
the etymological language of alliteration.

Wagner first developed his theory of the intertwining of language and music in the lengthy treatise *Opera and Drama* (*Oper
und Drama*, 1852), three years after completing his first alliterative
composition (he would eventually rework this composition, the opera
libretto *Siegfried's Death* [*Siegfrieds Tod*], for the final installment of
the alliterative *Ring* cycle). The structure of the theory is determined
by the nature-philosophical ideal of organic emergence, which dictates that the text-music interaction take place in such a way as to

avoid the additive logic of a text *plus* music scenario; each mode must achieve its highest fulfillment in accordance with its own laws, at the precise moment of its greatest openness to the other. Opera, as the only significant artistic genre that brings music and language together in a single form, is uniquely situated to explore this potential for total interpenetration. Its traditional failure to do so testifies, for Wagner, to the deficiencies of its historical manifestations, and these deficiencies, importantly, permeate both sides of the music-language divide. Where modern poetry has forsaken the strictures of rhythmic structure in a misguided attempt to counteract the artificial impositions of Greek dance meters, modern music, it turns out, has erred in the opposite direction by fetishizing contentless form. And where the loss of rhythm thus results in a poetry that cannot be distinguished from prose (Wagner cites an anecdote about actresses who have their iambic Schiller lines recopied in the form of ordinary speech),[6] modern compositional technique culminates in the rigorously rule-governed but ultimately empty architectonics of "absolute music." Such music subordinates the expressive potential of melody to a harmonic development devoid of determinate *sense*, and in doing so, renounces the originary relationship between development and explication, time and teleology.

This relationship, whose absence from modern art Wagner bemoans, is in fact the central topic of his treatise: his idiosyncratic interpretation of the compositional process ties the unfolding (*Entfaltung*) of language to the unfolding of music across a series of intensificatory identifications that culminate in the explication *of* language *through* music—and vice versa.[7] The starting point is a position that takes seriously the aesthetic medium of linguistic sound, both as a thing to be perceived with the senses (*aisthesis*) and as a matter with which to make art. Sound is, for Wagner, as his metaphors make clear, a substance in the nature-philosophical sense of the *apeiron*: an undifferentiated *Ur*-swamp of potential being that simultaneously lacks and implies its own actualization as form.[8] Music, *per definitionem* the category of structured sound, becomes, once distilled from the acoustic chaos, a new substance in its turn, with the potential to assume the even more particular, *signifying* structure of language. There can be no question, here, of a supplementary meaning added on retroactively to a previously purely musical sound, since the nature-philosophical understanding of substance/structure relation necessarily excludes the combinatorial mode of mere aggregation. That which spontaneously emerges, like signification, as a distinguishable component of actualized form, must also already potentially exist at the level of its less differentiated preconditions ("For even matter exhibits no other, lesser bond than that of reason: the eternal unity of the infinite with the finite"[9]).

Wagner presents his argument to this effect—an argument that is no less an argument for its lack of a discursive frame—as a story about the origin of language in music, envisioning the self-transcendence of matter into form as an inexorable, and thus logically necessary, evolution from the sound of unarticulated feeling to the sounds of complexly structured thought. The musical medium ("that maternal *Ur*-melody") out of which language grows, or rather is born, is itself the organic outgrowth of an undifferentiated emotion that for Wagner *cannot be distinguished* from its immediate and unmediated expression in sound ("Melody [. . .] once blossomed out of man's primitive perceptive capacities as a necessary expression of feeling"; *OD*, 282/143). No difference of substance interferes here with the principle of (self-)explicatory unity: the entities in question become more precisely and more perfectly themselves by assuming ever higher forms which, bodily speaking, are all the same—and in this case, are all sound. The distance between origin and end is a function not of supplementary being but of organizational degree, and thus of the *intensification* of sounding structure; it takes place in Wagner's text as a temporal process of solidification or *Verdichtung*, an implicatory "folding in" at the level of tonal-emotional substance that has for its consequence an explicatory unfolding at the level of tonal-emotional form: "From an infinitely flowing faculty of feeling, man's sensations gradually concentrated themselves into more and more definite contents, expressing themselves in the *Ur*-melody in such a way that its naturally-necessary progress finally intensified into the formation of a pure word-language" (*OD*, 281/142).[10] Musical condensation appears here, in essence, as the flip side of language growth.

It is the task of the poet to continue this process by actualizing the potential of ordinary language: the *Dichter* must perform the *Verdichtung*[11] that transforms language substance into poetic form. In order to do so, however, he must first rediscover the material roots of everyday speech, since, long covered over by conventional categories and thus lost to consciousness, these roots are nevertheless the sounding condition of all linguistic possibility. Poetry, for Wagner, occurs only as an etymological return to the state of linguistic nature, and to the generative *potential* of its most basic, signifying forms: "Pin-ing for salvation, the poet now stands in the winter frost of language, and looks longingly across the surface of the pragmatic prosaic snow [. . .]. But here and there, wherever his painfully hot breath pours out, the frozen snow begins to melt; and see there!—from out of the bosom of the earth fresh green buds are sprouting toward him, shoot-ing forth all new and lush from the ancient roots he took for dead" (*OD*, 265/128). In harmony with the linguistic tradition, Wagner believes he can presume for these still-living roots—and thus for the origins of language as such—a very particular structure. The vowel, "this sonorous sound [*dieser tönende Laut*], whose fullness, when fully enunciated, becomes of itself a musical tone" (*OD*, 266/129), is for him the immediate expression of human sensation and therefore of sense more generally; it is the sounding substance of all meaning and the ultimate source or *Quelle* of all linguistic power. It cannot, however, communicate anything *in particular* without a consonant to circumscribe its infinite possibilities: "The first activity of the conso-nant resides therein, that it raises the sonorous sound of the root to a definite characteristic, by firmly delimiting its infinitely fluid ele-ment" (*OD*, 267/129). The consonant bestows the contours of con-centrated form on the liquid life force of language; it acts, in Wagner's own words, as an articulatory "boundary post" (*Grenzpfahl*; ibid.) plunged into the vocalic flow, and in its unbending rigor plays fertil-izing phallus to the potentially fruit-bearing womb of sound. (A sec-ond metaphor, developed at some length, involves a fusion of blood and breath *coagulating* at the consonantal boundary, in a grotesquely carnal version of the formal *Verdichtung* [*OD*, 272–73/134–35].[12]) The marriage, or "Vermählung,"[13] of these two radically disparate phonetic categories quite literally gives voice, at the etymological ori-gin of language, to the moment when musical substance first assumes signifying form.

In doing so, it provides both model and material for the poet, whose job is to distill the substance of everyday language into the higher structure of poetry, and who does so by exploiting—and assuming—the delineating function of the *Stab*. The alliterative line, alone among compositional modes, does justice to the substance-form interaction out of which language emerges, because, alone among compositional modes, it focuses on the substance-form interaction that unfolds at the level of the roots themselves. Behind the spring-bringing, root-revealing breath of Wagner's pining poet stands therefore the structure of the *Stabreim* technique, whose apparently arbitrary rules turn out to relate organically to the particularities of the poetic task.[14] The alliterating letters bestow contours on a sounding substance, like their consonantal cousins at the foundations of language, distinguishing main words from marginal ones in a way that makes *sense* of the poetic line. Their proximity to the origin, however, rests on more than mere analogy, since, according to Wagner, the sounding substance to which the *Stäbe* give shape is in essence *identical* to the etymological root. In a move that places him definitively outside the scholarly tradition, whose basic assumptions he here drives well beyond their commonly accepted limits, Wagner insists that the identity of initial letters *necessarily implies* an identity of etymological meaning.[15] This etymological identity of first letters means, in turn, that the alliterative line unavoidably develops as a poetic *explication* of the etymological root—or at least, of the root's meaning-bestowing consonantal component. Wagner's own example of *"Die Liebe giebt Lust zum Leben"* (Love gives pleasure to life; *OD*, 292/152) becomes in this context a reflection on the etymological relationship of three key concepts, whose common link to the sounding L-matrix the line rescues from everyday obscurity, but whose mutually determining differences it simultaneously, syntactically, reflects.[16]

If Wagner had ended his discussion of the *Stabreim* with this "insight" into the etymological force of its alliterating consonants, his theory would have remained merely a particularly radical variation on the standard philological model, which sees in those consonants the living letters of linguistic fantasy, and in the *Stabreim* itself a kind of protolinguistic analysis. He goes on, however, to ask about the vowels, and in doing so transforms the philological understanding to make room for a deeper and different mechanics of animation. The real source of life in language is for him not alliterative poetry but alliterative poetry unfolded into a poetics of musical sound: "Science has laid bare to us the organism of language; but what it showed us was a *dead* organism, which only the greatest poetic urgency [*nur die höchste Dichternoth*] can bring back to life, precisely by closing up the wounds with which the anatomic scalpel had gashed the body of language, and by breathing into it the breath that could ensoul it with self-movement. *This breath, however, is—music*" (*OD*, 265/127). Wagner's plaidoyer for a poetry that takes seriously the sounding elements of language intersects, here, with his critique of language science, which in his view does no such thing. Linguists such as Jacob

Grimm, who focus all too exclusively on the possibilities of a consonantal writing—as evidenced, Wagner thinks, by their willingness to
postpone the question of specifically *vocalic* sound laws—end up dissecting the organism of language at the expense of the very phonetic
essence they had hoped thereby to reveal. While their cuts may be
true, the same cannot be said for their assumption that a purely consonantal analysis prepares the way for a future writing of the vocalic
substrate ("[S]houldn't it be possible, supported by the relation among
the consonants, to trace the points of contact among the vowels as
well?"[17]), since, according to Wagner, pure sound must be allowed to
obey its *own* rules.[18]

Wagner's proposed alternative acquires more precise contours as a
theory of poetico-musical *modulation*, which is also, at least implicitly, a surprisingly far-reaching "solution" to the philological conundrum of vocalic alliteration.[19] While (other) scholars of the *Stabreim*
struggle to explain the fact that vowels alliterate differently from consonants (that *a*, for instance, rhymes with *i* and *e* while *l* rhymes only
with other *l*'s), Wagner accounts for the discrepancy by embracing
the principle of vocalic uniqueness—and demanding its even more
radical expression in the technique of *total rhyme*. All vowels, regardless of their position in the word or their relationship to verse accentuation, "speak" primarily of their similarity to one another: "An
understanding of the vowel, however, is not based upon its superficial
kinship with a rhyming vowel from another root, but rather—since
all vowels are originally related to one another—on the disclosure of
this *Ur*-kinship through the full realization [*Geltendmachung*] of its
emotional content, by means of the musical tone" (OD, 275/137). The
practice of alliterating first vowels, which acknowledges the vocalic
similarity only at the beginnings of accented syllables rather than
everywhere it occurs, performs, from this perspective, a relatively per-
functory explication of the "meaning" of vowels in verse.

The poet who wants to do real justice to the sounding and feeling foundations of language has therefore no choice but to become a
musician, integrating the vowels, which are actually notes ("the vowel
is itself nothing but condensed [*verdichtete*] tone"; OD, 275/137), into
a single melodic-harmonic network of emotional and tonal kinship:
"Now the tone-poet must determine the tones of the verse in accordance with the familial affinities of their expressive capacities, such
that they do not only announce [*kundgeben*] the emotional content
of this or that vowel, as a *particular* vowel, but rather represent this

content, at the same time, as one familially related to *all* tones of
the verse, and thus as *a particular member of the Ur-family of all
tones*" (*OD*, 278/140). Isolated vowels need to be linked "horizontally" to all the other vowels within the line and beyond. Thus, the
i's and *u*'s of "*Die Liebe giebt Lust zum Leben*" rhyme not only with
each other but also with the *e*'s and *o*'s and *a*'s of the lines that precede and follow. They do so, however, without relinquishing their
particular character, which resides in the emotional resonance of the
vowel as *specified* by the consonant within a given syllabic configuration. *Leben* (life), for instance, has a different emotional valence than
Weh (woe), despite the fact that the two words share a vowel, while
Lust (pleasure) and *Wonnen* (bliss) clearly have much in common,
despite the fact that their vowels differ. The technique of total vocalic
rhyme, according to Wagner, manages to maintain this precarious
balance between sameness and difference, oneness and multiplicity,
because it is in fact identical to the musical technique of modulation,
which offers a method for moving systematically among different but
genetically related tonal "families" or keys (*Tonarten*). The horizontal connection of vowels corresponds, in its temporal succession of
tones, to the musical category of melody, and melody, when composed in accordance with these rules of harmonic relation, can progress through innumerable transformations of tonal "mood" without
ever endangering the more originary unity of the whole: "[M]elody
has acquired the most astonishingly varied ability, by means of harmonic modulation, to set its initial tonic key into relationship with
the most distant tonal families, so that in a larger composition the
Ur-family of all keys is presented to us, as it were, in the light of one
particular tonic key" (*OD*, 288/149). Genetic relation, in the context of tone, means harmonic proximity, and harmonic proximity,
in the context of the nineteenth-century theories with which Wagner is familiar, refers to the ability of particular tones to "sound well
together" (the word "consonant," from Latin *consonare*, means literally "to sound together") within the vertical constellations known
as chords: "Harmony grows upwards from below like a perpendicular pillar, out of the joining together and overlaying of familially
related tonal material."[20] A tonal family or key is thus the set of tones
required to form an enclosed system of consonant triads, centering
around a given primary tone or "tonic." And the degree of genetic-
harmonic intimacy among the various tonal families is determined
by the number of connecting links required to transition from one to

another. (The triad *g-b-d* , for instance belongs to the key of G major as the chord of the tonic, and to the key of C major as the chord of the dominant; these two keys are therefore separated from one another by a single degree of harmonic relation.) The resulting network of potential sound relations—Wagner's "*Ur*-kinship" of tones, which is simultaneously a kinship of vowels and feelings—receives several different graphic representations over the course of the nineteenth century (see figure 7 for an early example, with which Wagner may well have been familiar), and it is against the backdrop of such musical "family trees" that vocalic rhyme, according to Wagner, derives its true meaning. The technique traces the sounding elements of language back to a common source more primitive than all purely linguistic beginnings, and in doing so, takes the etymological thrust of alliterative verse to a new and necessarily extrapoetic extreme. Only where the particular mood, or *Stimmung*, of the poetic line stands in audible relation to a more general tonal-emotional *capacity* for musical combination (attunedness or *Gestimmtheit*), as systematized in the theory of harmony, can the alliterating poet claim to have fully exploited the originary significance of his vowels. As a "knower of the unconscious" (*Wissender des Unbewußten*; OD, 265/128), he calls to consciousness the conditions of poetic possibility—where sound laws of tonal kinship take precedence over sound laws of phonetic change—by transcending the disciplinary boundaries of his trade: "This *Ur*-kinship, which is preserved in the language of words as an unconscious moment of feeling, is brought to feeling's unmistakable consciousness via the full language of tone" (*OD*, 276/138).

Vocalic rhyme, then, acts in every respect to intensify the etymological power of the consonantally alliterating line. It allows the poet to explore deep-structure emotional relationships that would otherwise go unexpressed, and so to provide a more comprehensive analysis of the conceptual relationships established by the *Stab*. "*Die Liebe bringt Lust und Lied*" (love brings pleasure and pain), for instance, shares a common L-root with "*die Liebe giebt Lust zum Leben*" (love gives pleasure to life), yet the precise nature of the bond tying pleasure to pain clearly differs from the bond that ties pleasure to life. The modulating musician makes this disparity audible by transitioning to a different but related key on the way from *Lust* (pleasure) to *Leid* (pain).[21] In a further gesture of supplementation, he can then go on to link the *l*'s of "die Liebe bringt Lust und Leid" to the *w*'s of "*doch in ihr Weh auch webt sie Wonnen*" (but into her woe she also weaves

TABELLE

der Tonartenverwandltschaften.

C — a — A — fis — Fis — dis — Dis — his — His — gisis

F — d — D — h — H — gis — Gis — eis — Eis — eisis

B — g — G — e — E — cis — Cis — ais — Ais — fisis

Es — c — C — a — A — fis — Fis — dis — Dis — his

As — f — F — d — D — h — H — gis — Gis — eis

Des — b — B — g — G — e — E — cis — Cis — ais

Ges — es — Es — c — C — a — A — fis — Fis — dis

Ces — as — As — f — F — d — D — h — H — gis

Fes — des — Des — b — B — g — G — e — E — cis

Bes — ges — Ges — es — Es — c — C — a — A — fis

Eses — ces — Ces — as — As — f — F — d — D — h

Ases — fes — Fes — des — Des — b — B — g — G — e

Deses — bes — Bes — ges — Ges — es — Es — c — C — a

Figure 7. "Table of the Key Relationships," Gottfried Weber, *Attempt at an Ordered Theory of Tonal Composition* (*Versuch einer geordneten Theorie der Tonsetzkunst*), 3rd ed., vol. 2 (Mainz: B. Schott, 1830–32), 86.

bliss)—thereby definitively transcending the domain of the traditional *Stabreim*, which has no technical means of articulating the relationship between individual lines of verse—by simply reversing the order of the pleasure-pain key change: "Thus 'webt' [weaves] would become a tone leading back again to the first key [*ein Leitton in die erste Tonart*], just as, from here, the second emotion returns, now enriched, to the first" (*OD*, 292–93/153). Pleasure, after passing through the foreign territory of pain, reemerges on the other side of the two-line couplet as the deeper, fuller emotion of bliss, and the transition from tonic to new key back to tonic makes this relationship of intensificatory fulfillment clear. A more expansive version of the same dynamic binds *all* lines together at the level of the epic-operatic whole (it is in this context that Wagner introduces his notion of melodic "motives," later termed *leitmotifs*, which bestow unity by recurring in various stages of transformation throughout the work), thereby exponentially increasing the articulatory reach of the traditional *Stabreim*.

Wagner's revolutionary revaluation of vocalic rhyme does not imply, of course, that consonantal alliteration has been rendered superfluous, since the space of pure sounding potential can never, on its own, acquire the contours of an actual work. Music acts as metalanguage when bearing audible witness to a poetic motivation, for which purely poetic means are not sufficient. When sounding alone, without the justification of meaning, it *acts* not at all: "The musician thus receives the justification for his procedure, which if unconstrained would seem to us arbitrary and incomprehensible, from the intention of the poet—from an intention toward which the latter could only gesture or, at the most, realize approximately and fragmentarily (i.e., in *Stabreim*)" (*OD*, 293/153). The *Ur*-kinship of vowels, itself unlimited by any signifying tendency ("The tone poet must have at his disposal a kinship context that reaches to infinity"; *OD*, 279/131), can make itself heard in a verse melody only where a verse exists to provide it with teleological direction. Truly animatory, therefore, is neither music nor meaning, but the womb of harmony as penetrated by the "procreative power of the word" (*OD*, 117), and the intentional-intensity of language as "aroused" to action by the sensual beauty of sound. The desire-driven synthesis triggers an orgy of mutual determination that culminates in a revolution of compositional technique, and a new art, or *technê*, of fusing two arts into one.[22]

The resulting energy figures in *Opera and Drama* as a gale with the power to wake the dead—or move the world.

Through the redemptive love-kiss of that melody, the poet is now
initiated into the deep, infinite secrets of the feminine nature: he
sees with other eyes, and feels with other senses. To him, the bottomless sea of harmony, from which that beatific vision rose to meet
him, is no longer an object of dread, of fear, of terror, as it earlier
appeared when it was an unknown, foreign element; not only can he
float upon the waves of this ocean, but—gifted with new senses—he
now dives down to its deepest ground. [. . .] His rational sense pen-
etrates everything clearly and tranquilly all the way to the ocean's
primal font [*Urquell*], from which he orders the wave columns that
are to rise to the sunlight, so as to run in ripples under its radiance, to
softly plash with the sighing west-wind, or to rear their crests man-
fully with the storms of the north; for the poet now commands even
the breath of the wind—for this breath is nothing other than the sigh
of infinite love, of the love in whose rapture the poet is redeemed, in
whose power he becomes the lord of nature [*Walter der Natur*]. (*OD*,
285–86/147–48)

Like its later Saussurean counterpart, this Wagnerian parable of
"*l'eau et l'air*" performs a simultaneous rereading of both the Biblical and the nature-philosophical creation narratives, which is to say
also of the etymological origins of rhythm. (The traditional derivation traces the Greek *rhuthmos* back to *rhein*, "to flow," across the
mediation of repetitively flowing *waves*.[23]) Wagner's coupling of two
masses, however, serves the purpose of intensification rather than
parody, since his substrates fuse where Saussure's evaporate—thereby
subjecting the agency of *(Sprach)Geist* to an amplificatory, because
bidirectional, intermingling of breath.[24] Two developmentally distinct
but substantively identical flows, resonating together, create here a
sounding structure that far exceeds the vibrations of its parts. The
vertical columns of harmony rise up toward the sunlight from out of
the depths of the sounding sea, independently of any external influence exerted by wind or heavenly bodies, in order to punctuate the
surface in the form of melodic waves. In doing so, they do not yet
establish a rhythm: "The feeling of necessary care for the beauty of
this horizontal motion [*Bewegung nach der Breite*] is foreign to the
nature of absolute harmony; she knows only the beauty of her columns' changing play of colored light, not the charm [*Anmut*] of their
temporally perceptible ordering—for this is the work of rhythm."[25]
It falls to the propulsive force of the poetic breath, in other words,
to "set" the watery *Stäbe* by arranging them in the forward-tending
order of an alliterative line. This teleological, temporal ordering, however, will remain a mere ornament—"arbitrarily atop the peaks of

those harmonic pillars" (*OD*, 116/87)—so long as it does not emerge directly out of the poet's intimate familiarity with the *Ur*-font, conceived here as the universal alphabet of tones and feelings. The rhythm of the total artwork unfolds, for Wagner, not as the imposition of objectively counted time from above but as the future-oriented intensification of emotional fluctuations (*Affektwechsel*) already inherent to its sounding medium.

The language-scientific priority of the *Stabreim*, which alone among poetic modes gives (further) shape to language spirit without subjecting it to alien forms, is thereby assured precisely at the moment when it opens itself most radically to the possibilities of an a-linguistic ground.

WOTAN'S STAFF

Against conservative contemporaries who see in the metric freedom of the Wagnerian melodies only a decadent refusal to keep accurate time, Wagner therefore can and does insist on the historical, etymological, and national rootedness of his technique.[26] In the notion of an essentially linguistic music, as defined by the "sense-filled and sensual" (*sinnig-sinnliche*) rhythms of the poet's organically ordered wave pillars, he has a theory that ties Liliencron's Pythagorean sound figures to the problem of *Stabreim* meter by tying both to the innovations of his own alliterative solution. Pillars of poetico-harmonic intention, which erupt into waves not of sound but of sounding sense, distill the traditional, wooden materiality of the runes into the energy of rhythmic direction, and in doing so restore the power of a specifically Germanic poetry at the level of compositional praxis. The point, however, is less the future of German verse than of the universe, as evidenced by the extraordinary proliferation of *w*'s that punctuates the account of rhythm in the German version of the passage cited above: *Weib*, *Weihen*, *Wogen*, *Wonne*, *Wellen*, *Wallen*, *West*, and *Walter* (the uncited lines contain *weit* and *Wundern*) intertwine here with "*w*ind" and "*w*ater" to link the activity of the Wagnerian poet alphabetically—which for Wagner always also means, etymologically—to the world-structuring force of the Germanic god Wotan. Wagner the German composer is thus quite literally inseparable, according to the logic of his own alliterative self-presentation, from Wagner the formative principle of the All.

The chapter "Wuotan" in Jacob Grimm's *Germanic Mythology* (1844), which Wagner knew well, includes the following reflection on the god and his name:

> It can scarcely be doubted that the word is immediately derived from the verb Old High German *watan, wuot*, Old Norse *vada, ôd*, corresponding literally to the Latin *vadere* [to go, to walk] and meaning *meare, transmeare, cum impetu ferri* [to go, to wander, to flow, to go through, to be carried along]. From here derives the noun *wuot* [*wut*, fury], just as μένος [anger, energy, force] and *animus* [wrath, passion] actually mean *mens, ingenium* [mind, spirit], and then also by extension impetuosity, wildness; in the Old Norse *ôdr* the meaning of *mens* or *sensus* is still entirely retained. Accordingly *Wuotan, Odinn* would appear to refer to the all-powerful, all-penetrating being, *qui omnia permeat* [which permeates the all]; as Lucan says of Jupiter: *Est quodcunque vides, quodcunque moveris* [he is whatever you see, wherever you go], the spirit-god [*die geistige Gottheit*]. In the Bavarian dialect, *wueteln* means to arise and move [*sich regen und bewegen*], to seethe [*wimmeln*], to grow rampantly [*üppig wachsen*] and thrive.[27]

To the root word *watan*, here identified with the Latin for "to go" or "to move," Grimm adds his own full complement of explicating *w*-verbs, of which the most significant is clearly the one that serves him as common denominator: Bavarian *wueteln*, modern German *wimmeln* and *wachsen*, Old High German *watan*, and the by now familiar constellation of spirit, breath, wind, and life all come together at the very end of the passage in the crucial concept of self-propelled movement, or *Bewegung*.[28] *Bewegen*, the transitive modus of *wiegen*, Gothic *vigan*, has, according to the Grimms' *German Dictionary*, the originary, "concrete" meaning "to move or make vibrate."[29] And it is in this capacity to move or make vibrate that Wotan/Odinn haunts the margins of the texts on the *Stab*, setting nature in motion from behind the scenes and inspiring—or rather *motivating*—the writing of a runic poetry whose living, and thus *motivated*, letters he himself is said to have invented. Grimm can therefore conclude his investigation into the etymology of the divine name with a summation that makes explicit the identity between German god and cosmic *Bildungsprinzip*: "If we are to briefly sum up the attributes of this god, he is the *all-penetrating, creating,* and *form-bestowing* power [*die alldurchdringende, schaffend und bildende kraft*], who endows humans and things with both shape and beauty."[30]

It is in this same capacity, as moving force of both history and poetry, that Wotan reemerges, rune-knowing, onto the center of the

Wagnerian stage, where his wielding of a spear, which is simultaneously a runic staff ("Binding runes / are carved into its shaft: / the
one who wields the spear / holds in his hand / control of the world"
[*Treue-Runen / sind in den Schaft geschnitten: / den Haft der Welt/
hält in der Hand, / wer den Speer führt*]),[31] drives the plot of the *Ring*
toward the downfall of the gods. The notion that thereby arises of a
relationship between fate and the *Stab*, and of a specifically *runic* wisdom pertaining to the mysteries of cosmic time, explores at the level
of epic plot the implications of a natural Germanic writing like the
one Liliencron and the Grimms propose. It does so, however, by rei-
magining the whole idea of natural writing, and so also the idea of a
time-telling *Stab*, as a viable contemporary goal for all humankind, to
be accomplished via the universalizing redeployment of the Germanic
people's oldest-known origin story.

The story in question is the medieval *Poetic Edda*'s enigmatic
depiction of the god Wotan's rune-inventing travails. The alliterative
rendition of Karl Simrock—one of the two translations of the *Poetic
Edda* with which Wagner himself would have been familiar[32]—gives
the pertinent lines of the Old Norse poem as follows:

Odhins Runenlied

Ich weiß, daß ich hing am windigen Baum
Neun lange Nächte,
Vom Sper verwundet, dem Odhin geweiht,
Mir selber ich selbst,
Am Ast des Baums, dem man nicht ansehn kann,
Aus welcher Wurzel er sproß.

Sie boten mir nicht Brot noch Meth;
Da neigt' ich mich nieder
Auf Runen sinnend, lernte sie seufzend:
Endlich fiel ich zur Erde.

Hauptlieder neun lernt ich von dem weißen Sohn
Bölthorns, des Vaters Bestlas,
Und trank einen Trunk des theuern Meths
Aus Odhrörir geschöpft.

Zu gedeihen begann ich und begann zu denken,
Wuchs und fühlte mich wohl.
Wort aus dem Wort verlieh mir das Wort,
Werk aus dem Werk verlieh mir das Werk.

Runen wirst du finden und Rathstäbe,
Sehr starke Stäbe,

Sehr mächtige Stäbe.
Erzredner ersann sie, Götter schufen sie,
Sie ritzte der hehrste der Herscher.

Odin's Rune-Song

I know that I hung on a windswept tree
For nine long nights,
Wounded by a spear and consecrated to Ódinn,
Myself to myself,
On a branch of the tree about which it cannot be known
From which roots it emerged.

They offered me neither bread nor mead.
I then bent downwards
Pondering the runes, I learned them, sighing,
Finally I fell to earth.

Nine mighty spells I learned from the white son of
Bölborr, Bestla's father,
And drank a drink of the precious mead,
Scooped out of the Ódrerir.

I began to thrive and to think,
I grew and felt well.
Word coming from word led me to the [next] word,
Work coming from work led me to the [next] work.

Runes you will find, and staffs of council,
Very strong staffs,
Very powerful staffs.
Arch-orators devised them, gods created them,
They were carved by the noblest of the rulers.

Simrock, who here adds *Wort* (word) and *Werk* (work) to the list
of Wuotan-related *w*-words, points out in his accompanying notes
that this oldest of *Stab* texts in fact tells the story of several different
staffs, or *Stäbe*. First comes the tree itself, identified by the notes as
the "world tree" (*Weltesche* Yggdrasill) of Nordic mythology, under
whose second root the sacred spring of spiritual energy (Odhrörir)
flows.[33] Next comes the spear with which the god claims to have been
wounded, and finally the *starke Stäbe*, or "strong staffs," to be identified either with the runes themselves, or with the "staffs of council" into which they were carved. Behind and beneath all three of
these *Stäbe* stands, however, the vertical shadow of a unifying and
thus potentiating fourth. The *Stab* of the god's all-powerful phallus,
which for Simrock plays the role of rhetorical transition from tree

trunk to weapon, functions also as a generative origin on par with the life-giving liquidity of the sacred spring: "As the fruit of the world-tree, from which he must first still detach himself, the god hangs on the stem [*Stiel*], and this stem, *or that which corresponds to it in the human fruit*, can here be compared with the piercing spear."[34] Together, the phallic seed and the watery womb yield the fruit of the enlightened god, which is also, not coincidentally, the fruit of the living letters, and which therefore, as a passage from *Stab* to *Stab*, must be said to engender *itself* ("*mir selber ich selbst*").[35] Remarking in this context on the common use of runes in midwifery, and on the parallel of nine days on the tree to nine months in the womb, Simrock concludes, "Through the invention of the runes, Odin here assists in his own birth."[36]

The *Ring* cycle pays tribute to this potentially infinite regress of self-reflexion with a proliferation of phallic staffs matched only by the proliferation of staff-wielding, rune-seeking scenes. From Wotan's originary, rune-producing "rape" of the "spring, whispering wisdom" (*Quelle, Weisheit raunend*),[37] to Siegfried's subordination of (and to) the rune-knowing Brünnhilde;[38] from the tree to the spear to the sword and back again to the tree, this time in fragments[39]—the entire opera undermines the traditional understanding of dramatic direction by unfolding as a series of punctual rebirths. Separated less by a distance in clock time than by a difference in the intensity of (self-)consciousness, these rebirths redefine narrative chronology as a stepwise progression of articulatory levels, with the result that the experience of temporal flow comes to depend for its effect on the chronology-disrupting structure of repetition. This counterintuitive, yet utterly typical, nineteenth-century dynamic finds its most emblematic expression in the prologue to the cycle-concluding *Götterdämmerung*, where the cosmic time of the weaving fates can be seen to depend for its very continuity on the interruptive violence of Wotan's time-stopping *Stab*: it is the youngest fate's vision of the shattered runic spear that causes the thread of the epic narrative, quite literally, to snap, leaving the fates to scream their famous lament ("*Es riß! Es riß! Es riß!*"[40]) as much for the god as for themselves.

An earlier and more explicit reflection on the nature of Wotanic movement—the context is the essay "The Wibelungen," which treats the relationship of epic to history and, more particularly, of epic to the possibility of its modern reactualization—makes clear that the paradox of a cyclical momentum does not confine itself to the epic realm.

The ancient battle [as described in the *Nibelungen* material] is therefore now carried on by us, and its ever-changing result is precisely the
same as that constantly recurring change [*wiederkehrende Wechsel*]
of day to night, summer to winter, and, finally, of the human race
itself, which moves onward and onward [*sich fort und fort bewegt*]
from life to death, from victory to defeat, from joy to sorrow, and
which, through this continual process of rejuvenation, brings the eternal essence of man and of nature, in itself and through itself actively
to consciousness [*an sich und durch sich thatvoll sich zum Bewußt-
sein bringt*]. The quintessence of this eternal movement, i.e., of life,
found its ultimate expression in "Wotan" (Zeus), as the highest god,
the father and all-penetrator of the All.[41]

The passage proceeds via a series of identifications, which perform
a repetitive unfolding very similar to that of the explicatory structure they describe: in the ups and downs of epic struggle Wagner
sees the cyclical change of days and seasons, the birth and rebirth
of generations, and the oscillations of fate; in the oscillations of fate
he sees the gradual coming to consciousness of the all-penetrating,
all-infusing, and all-inspiring god who is simultaneously synonymous
with the essence of movement conceived as the energy of life. The
teleo-logic of iteration—the peculiar, stuttering trajectory of Wagner's "onward and onward" *(fort und fort)*—drives here the history
of the world in addition to the plot of the *Ring*, in perfect accord
with the Schellingian *Schweben* (to oscillate, to hover) that conditions, nature-philosophically speaking, all forward-surging power.
Whether in epic or in opera, world history or the artist's own personal trajectory,[42] every significant temporality must unfold according to the interruptive rhythm of a "constantly recurring change," or
rather, of a *wiederkehrende Wechsel*.

It is this rhythm that the artwork, as microcosm, so emphatically
reflects: "Rhythm is [. . .] the very soul of these necessary movements, rendered conscious to the artist."[43] Yet for the Wagner of
Opera and Drama, who presages the even more radical Wagner of
the *Ring*, art has already ceased merely to mirror the rhythms of the
cosmos, and become instead the site where these rhythms find their
modern, which is to say, *ultimate* fulfillment. Just as, in the beginning, time and writing originate as one—such that Wotan, the driving force of history, can birth *himself* together with the runes—so,
too, toward the teleological end, can they be expected to reconverge
in an orgy of self-reflexive entanglement, yielding an artwork capacious enough to actually *contain* the entirety of the existing universe.

Wagner's historic rewriting of the *Edda* poem, which dares to replace the mythical spring-drinking god of a primordial Germanic past with the deep-diving poet-composer of a fast-approaching universal future, makes clear that he understands his intensified *Stabreim* as a tool for articulating the All. The goal is nothing less than a poetico-musical realization of the nineteenth-century linguists' metalinguistic fantasies, and so also of the world spirit's gradual trajectory toward the *telos* of total self-explication. The resulting "infinite melody" will run continuously, articulated but unbroken, from the beginning of the world to the end, without the need for traditional fragmentation into components (such as aria and recitativo) or categories of being (such as nature and art).

In a final "summary" of the theoretical position worked out at such length in *Opera and Drama*—a summary first published in French translation under the title "The Music of the Future" ("La musique de l'avenir," 1860)—Wagner envisions the auditory experience of this comprehensive melody as an encounter with the sounds of a primeval forest:

> I will have recourse to metaphor once more in order to delineate for
> you what is most characteristic about the kind of melody I intend,
> which encompasses the entire dramatic tone-piece; and for this pur-
> pose I will focus on the impression that such a melody must engender.
> Its infinite wealth of divergent details should in no way reveal itself
> only to the connoisseur, but also to the most naïve layman, if only
> he has first arrived at the appropriate level of concentration [*sobald
> er nur zur gehörigen Sammlung gekommen ist*]. At first the melody
> should exercise an effect on his mood [*Stimmung*] like the one engen-
> dered by a beautiful forest on a summer evening in a solitary visitor
> who has just recently left the noise of the city behind; the peculiar
> stamp of this impression—which I leave the reader to elaborate in all
> its psychological effects—is that of a silence growing more and more
> alive [. . .]. But when, overwhelmed by this first general impression,
> the forest's visitor sits down to ponder; when, the last burden of the
> city's hubbub cast aside, he girds the forces of his soul to a new power
> of observing; when, as if hearing with new senses, he listens more and
> more intently—he perceives ever more distinctly the infinite diversity
> of voices waking in the wood. Ever new and different voices join in,
> voices he does not believe he has ever heard before; as they wax in
> number, they grow in peculiar strength; the woods ring louder and
> louder; and many though the voices be, whose individual strains he
> hears, the overwhelmingly bright, swelling sound nevertheless seems
> to him just one great forest-melody: that very melody which from the
> outset had so bound him to devotion [*zur Andacht fesselte*], like the

deep-blue firmament of night binds the gaze, allowing the countless multitude of stars [*zahllosen Sternenheere*] to appear ever more
distinctly, brightly, and clearly, the longer one remains sunk in the
spectacle.[44]

The wave pillars of a future Wagnerian art return, here, after a detour
through the arrhythmic cacophony of modern, urban life, to the runic
trees of their Germanic roots, with the difference that the structure of
the system is now in principle available to everyone. Where previously
only German priests and language scientists could gather—"*Temple*
also means *forest* [. . .] the godhead dwells there and conceals its
image in the rustling leaves of the boughs"[45]—in order to commune
with the spirit of a Germanically inflected cosmic gathering (*logos*,
legein, from proto–Indo-European **leg*, "to collect, to gather"), the
denationalized layman can suddenly be counted upon to penetrate,
even unto infinity, if only he has first sufficiently *gathered* his own
spirit (*sobald er nur zur gehörigen Sammlung gekommen ist*). The
source of the shift lies, for Wagner, in the technical transformation
undergone by the artwork itself, which is here imagined to culminate
in an *audible* apotheosis of unlimited self-delimitation. The musical poetry of tomorrow, for those who have ears to hear, will render the "standing together" of its *Stäbe* simultaneously as distinct as
trees—or stars—and as indivisible as the moving air—or *spiritus*—
without which systemic distinction could never, acoustically speaking, take place.

CHAPTER 5

Pythagoras in the Laboratory

THE WAGNERIAN SOUND OF SENSE

Richard Wagner's insistence on a music *of* language as the only legitimate medium of future art, combined with his willingness to anticipate this fusion in the form of a present-tense, public spectacle, set the stage for more than half a century of music-inflected encounters with the conundrum of expressive sound. The result was a gradual transformation in the nineteenth-century understanding of language-as-system, over the course of which language spirit—for nature-philosophical linguists still a *qualitas occulta*, inaccessible to direct sensory perception—took on the contours of an audibly vibrating, measurable mass. The story begins, unsurprisingly, with the famously intensive reception of Wagnerian tenets among the Symbolist poets of France.[1] I will argue here that the same story culminates, somewhat less intuitively, in a new, natural scientific theory of the mind-matter interaction, associated primarily with the names of early experimental psychologists such as Gustav Fechner and Wilhelm Wundt. Wagner's musical amplification of language spiritual principles receives in the wave-based models of a Wundtian "psychophysiology," which borrows its fundamentally vibratory structure from contemporaneous acoustic theory, an ostensibly "measurable" corroboration. The Romantic hypothesis of *Sprachgeist* reemerges toward the end of the nineteenth century, in consequence, as a viable, language-scientific alternative to the system-denying materialism of the Neogrammarians and their positivist allies.

Charles Baudelaire plays the role of forerunner in this context, as in so many others, when he celebrates Wagnerian opera for demonstrating that musical sound can *mean* independently of all conventional reference. His early, pre-*Ring* essay, "Richard Wagner and *Tannhäuser* in Paris" (1861), which was almost certainly written under the influence of Wagner's "La musique de l'avenir" from the same year, puts forward the following thesis: "The reader knows the aim we are pursuing, namely to show that true music suggests analogous ideas in different brains."[2] Citing lines from one of his own most famous *Fleurs du mal* poems, "Correspondences," Baudelaire goes on to imply the existence of a deep affinity, or rather analogy, between the sound-sense correspondences of Wagner's harmonizing, alliterating *Stäbe*, and the musical, but also visual and olfactory, potential of his sonnet's "living pillars": "Nature is a temple in which the living pillars / Sometimes yield confused words; / Man passes there through forests of symbols / which observe him with knowing eyes."[3] Both modes of writing seek to express, according to him, an originary experience of the world "as [*comme*] a complex indivisible totality," where sounds suggest colors, colors melodies, and sense always occurs, "by a reciprocal analogy," in tandem with sensory transport.[4] And both therefore also assume, despite the radical modernity of the techniques they presuppose, the timeless shape of primeval trees, through which the winds of an undivided totality, or infinite melody, can be assumed to perpetually whisper.

By the time of the first performance of the complete *Ring* cycle, Baudelaire's portrayal of Wagner as a patron saint of synesthesia and sound symbolism had become a central tenet of orthodox Wagner reception. In 1876, the Wagner scholar and acolyte Hans von Wolzogen—himself an amateur poet, contributor to Symbolist periodicals, and, later, the editor of Wagner's house journal, *Bayreuther Blätter*—published a Wagner-approved treatise titled *Poetic Sound Symbolism: Psychological Effects of Language Sounds in the Alliterative Verse of R. Wagner's "Ring of the Nibelungen"* (*Poetische Lautsymbolik: Psychische Wirkungen der Sprachlaute im Stabreime aus R. Wagner's "Ring des Nibelungen"*), in which he claimed to deduce the "most essential symbolic meanings" of fifty consonants and consonant clusters directly from the sensory experience of the alliterating *Ring* text.[5] Under the cluster *st*, for instance, with its hard, *sh*-stopping *t*, he discovered "a movement that reaches a position and remains there, arrested, thus everything still-standing, static, stiff, stony, and also standing [*daher alles Stillstehende, Stätige, Steife, Starre und auch Stehende*]."[6] In the "forward-pressed breath" of Wagner's *w*, he discerned a "wafting (waving) movement [*eine wehende (wallende) Bewegung*]."[7] Wolzogen's methods were explicitly antiphilological, in line with the Wagnerian theory of an *unmediated* tonal-emotional relation; his treatise purposely eschewed all historical speculations that could potentially conflict with the psychological impression presumed to communicate itself, immediately, through the sounds ("With certain consonants, we clearly sense a symbolic message conveying certain ideas [*Vorstellungen*]. It does not matter what this is based upon: the effect is there"[8]). Letters operate within the alliterative context of the *Ring*, so the argument goes, according to the same rules as the sound symbols of Wagner's musical melodies, for which Wolzogen would later propose the term *Leitmotiv.*[9] In doing so, they prepare the way for an etymological practice that prioritizes universal rules of natural sound-sense affinities over particular laws (like Grimm's sound shift) of historical phonetic change.

The Baudelaire-Wolzogen interpretation of Wagner's alliterating letters had particular relevance for the early sound experiments of the French Symbolist poets, whose attempt to discover the sensory sense of sound drove them beyond and beneath the conventional linguistic sign. Rimbaud's "Vowels" ("Voyelles," 1871), which assigns a color to each vowel, grew out of this context, as did Mallarmé protégé René Ghil's *Treatise on the Verb* (*Traité du verbe*, 1866), which follows a paean to Wagner's "poetic instrumentation" with three lists of letters keyed to musical instruments and a set of instructions for "orchestrating" linguistic music.[10] Both works insist, in a way that seeks to fulfill the avowedly Wagnerian promise of Baudelaire's living pillars, on the possibility of a meaning somehow inherent to the alphabetic elements themselves, irrespective of their participation in words, and thus also irrespective of any demonstrable philological connection to these words' "actual" historical trajectories. The Symbolist-affiliated journal the *Wagnerian Review* (*La Revue Wagnérienne*), founded in 1885 with the stated goal of promoting Wagner as the creator of a new kind of linguistic art (Stéphane Mallarmé, Paul Verlaine, René Ghil, and Wolzogen himself were all among the contributors) provided a forum for the consolidation of this perspective. And by the end of the nineteenth century, the fascination had spread to the Symbolist movements in Moscow and Saint Petersburg, where musicians such as Alexander Scriabin, artists such as Wassily Kandinsky, and poets such as Andrei Bely paid explicit tribute, in the course of their respective synesthetic experiments, to the significance of Wagnerian principles regarding the unity of sense and sound.[11]

The poetic radicalization of the Wagnerian *Stabreim* at the hands of the French and Russian Symbolists presupposed an understanding of language that may well have remained a quasimystical footnote in the history of language science had it not been for the simultaneous emergence, from within the academe itself, of a theory that could claim to stand guarantor over the sound-sense correspondences of modernist poetic experiments. Most prominently developed by the psychophysiologist Wilhelm Wundt in his laboratory at the University of Leipzig—under the influence of his predecessor and fellow Leipzig scholar, Gustav Fechner—the principle of "psychophysical parallelism" posited that changes in the physical substrate of the body could be shown to correspond, verifiably and predictably, to changes in the psychic condition of the organism. The capacity for speech could then be interpreted, in turn, as one particularly significant modification of a more fundamental, psychophysiological analogy. Wundt's perspective offered a counterpoint to the materialism of his colleagues in the Leipzig linguistics department, since while Neogrammarians such as Brugmann and Osthoff strove to subtract the "metaphor" of spirit from the study of language sound, Wundt and his students worked to enfold sound *back* into an all-encompassing science of *Geist*. The hypotheses of psychophysiology were designed to be laboratory-testable, its object of investigation measurable, its claims falsifiable, and the result was a respiritualization of language science capable of rivaling the Neogrammarian "literalization" for the allegiance of a profoundly positivist age. By the beginning of the twentieth century, Wundt's one-room laboratory at the University of Leipzig had become the international center for a new field of study that sought to measure stimulus-response interactions, with the aim of establishing the relational laws governing all facets of mental life. His monumental *Principles of Physiological Psychology* (*Grundzüge der physiologischen Psychologie*) had gone through six separate editions.[12] And the six-hundred-odd students to whom he lectured every semester had carried his central ideas well beyond the confines of German academia, with the result that avant-garde artists all over Europe, from Symbolists to Futurists to Dadaists, could call on the principle of psychophysical parallelism to stand guarantor for their Wagnerian innovations.[13]

WAVE SYSTEMS (ACOUSTICS)

Wundt's theories, then, came to count for his contemporaries as science, in a way that the poetic experiments of the avant-garde Wagnerians, or of Wagner himself, clearly could not. They did so, however, not simply because of Wundt's laboratory, as commonly assumed, but rather because his theories managed to adapt for psychological use—so I will argue here—the enormously powerful physical concept of *naturally occurring wave systems*, particularly as formulated by Wundt's teacher, the great physicist (and Wagner admirer) Hermann von Helmholtz.[14] Helmholtz himself had treated this concept most thoroughly in the context of his midcentury studies on acoustic harmony. Any attempt to render readable the true stakes of the Wundtian perspective for a rethinking of the sound-sense bond must therefore begin by taking into account this surprisingly intimate relationship between a new, nineteenth-century science of spirit and an existing nineteenth-century science of sound, between a science of air waves in motion and a science of (self-)moving *psukhē*.

Helmholtz first turns to the question of acoustic harmony in the 1850s, on the heels of his groundbreaking investigations into light waves and the perception of color, in an attempt to do for the domain of the ear what he had already done for the eye. Beginning with his 1857 lecture "On the Physiological Causes of Harmony in Music" ("Über die physiologischen Ursachen der musikalischen Harmonie")[15] and concluding with his book-length study *On the Sensations of Tone as a Physiological Basis for the Theory of Music* (*Die Lehre von den Tonempfindungen als physiologische Grundlage für die Theorie der Musik*, 1863),[16] Helmholtz treats the age-old problem of musical pleasure primarily as a kind of case study, to be interrogated for its ability to shed light on the relationship of stimuli to sense perception. By studying the correlation between acoustic phenomena, in the form of sound waves, and the perception of sound, in the form of auditory impressions, he hopes to arrive at an account of what it means, physically and physiologically, to hear.

Helmholtz is perfectly aware, of course, that the "case study" of harmony enjoys a particular philosophical cachet. Ever since the ancient discovery, traditionally attributed to Pythagoras himself, of the relationship between consonant musical intervals and rational numerical ratios, harmonic principles had been understood to model cosmic laws. A string stretched between two points (Pythagorean tradition calls this a monochord) produces one tone when plucked in its full length, a tone an octave higher when held down precisely at its midpoint, such that the sounding length stands to the original length in a ratio of 1:2, a tone a fifth higher at a ratio of 2:3, and one a fourth higher at a ratio of 3:4. These three most consonant intervals, together with the permutations they make possible (which yield the somewhat less consonant intervals of major and minor third, second, sixth, and seventh), form the basis of what the Greeks called *harmounia*, from *harmos*, "joint," proto–Indo-European **ar*, "to fit together." The "joints" in question are the nodes that define the various string lengths, and these nodes "fit together" into an articulated system of sounds (*articulare*, also from proto–Indo-European **ar*, "to fit together"), insofar as they interact with reference to a single, structuring principle or standard of measure. Only lengths that share a common unit can relate to one another in the manner of whole number ratios, which is to say, *rationally*—every "rational" number, mathematically speaking, can be remainderlessly expressed as a ratio of two whole numbers, for which the technical Greek term is "logos"; every "irrational" number fails this test—and only such rational string-length ratios yield compound sounds that please. It is this ability to render aesthetically manifest the configuration of a purely numerical rationality that turns musical tones, together with bodily organs, the letters of the alphabet, and the rational numbers themselves, into time-honored figures for the privileged "joints," or foundational units, of Being and thought.[17]

When, therefore, nineteenth-century Germany's best-known proponent of an exclusively mechanist physics undertakes to provide a purely physiological account of harmony—why, Helmholtz wants to know in his 1857 lecture, do we *hear* ratios of small whole numbers as pleasurable, whether or not we are actually *aware* that we are hearing a numerically "rational" interval?—his attempt cannot help but intervene in a much larger debate regarding the "naturalness" of conceptual structure. Helmholtz accepts the burden of the problem's long history ("this is an old riddle, propounded already by Pythagoras, and hitherto unsolved"; *PC*, 47/58), while remaining true to his own quasi-Kantian standpoint, which forbids the presumption of an inherently rational universe, and insists instead on scientific experiments to establish mechanistic causal chains. ("Let us see whether the means at the command of modern science will furnish an answer"; ibid.). His methods allow him to demonstrate conclusively, for the first time, the extent of the ear's capabilities: even the least "mathematical" of listeners turns out to analyze compound sounds according to the mathematical compatibility of their components, and in this sense, the privileged *logoi* of Pythagoras's consonant intervals do indeed appear to originate in the objective reality of nature. These same experimental methods, however, also allow Helmholtz to propose an interpretation of the ear's abilities that depends solely on the physiological makeup of the human organism, without reference to the hypothesis of an ostensibly cosmic "harmony of the spheres."

Since at least the time of the mathematicians Leonhard Euler, Jean d'Alembert, and Daniel Bernoulli, with their respective contributions to the eighteenth-century conversation known as the "vibrating string controversy,"[18] it has been common knowledge among theorists of harmony that the Pythagorean ratios of string lengths correspond to ratios of air vibrations, which have the mathematical structure of periodic functions. Such functions describe continuous yet repetitive movements, like those of a pendulum or a beating heart, and they consequently have continuous yet repetitive, which is to say *wave*-shaped, graphs. Helmholtz agrees that the mathematics of periodic functions must form the foundation of any modern science of harmony, but he takes issue with the idea, proposed by Euler, that an innate psychological preference for the orderly elegance of rational ratios can explain the audible phenomenon of consonance. Euler's understanding, he thinks, fails to account for the way in which the rational order of harmonious sound waves becomes accessible to the human psyche. "We must keep in mind," says Helmholtz in *On the Sensations of Tone*, "that man in his natural state is scarcely aware that tone rests upon vibrations. There exists, moreover, no cognitive means whatsoever for immediately and consciously perceiving the different numbers of vibrations, or for discerning with the senses the fact that these numbers are larger for higher tones than for lower and that they stand in particular relationships for particular intervals."[19] It is difficult to see, in other words, how the pleasure of consonance could be based on the pleasure of simple, rational relationships, if these relationships are never, in any meaningful way, *perceived*.

Helmholtz finds the prototype of *perceptible* wave relations in the "instructive spectacle" (*lehrreiches Schauspiel*) of the sea, from which, as he points out, the word "wave" enters the vocabulary of physics in the first place:

> I have often spent hours in contemplation on the steep, richly forested coast of Samland, where the sea takes the place of the Alps for us

inhabitants of East Prussia. It is rare not to see there, in incalculable numbers [*in unabsehbarer Zahl*], wave systems of different lengths, propagating themselves in different directions. The longest tend to come from the deep sea and dash against the shore. Shorter ones arise where the larger ones, upon breaking, burst apart, and then run back again out to sea. Perhaps a bird of prey darts after a fish and sets in motion a system of circular waves [. . .]. Thus there unfolds before the spectator—from the distant horizon, where white lines of foam on the steel-blue surface first betray the approaching procession of waves, down to the sand beneath our feet, where the waves draw their arcs in the sand—a sublime image [*ein erhabenes Bild*] of immeasurable power [*unermesslicher Kraft*] and constantly changing variety, which, since the eye easily recognizes therein order and law, captivates [*fesselt*] and exalts [*erhebt*] without confusing the mind [*den Geist*]. (PC, 57/70–71)

Beneath the philosophically loaded image of a lone observer standing in rapt contemplation of the sea—the passage recalls a famous Leibnizian rumination on the relationship between waves, perception, and the harmony of the All,[20] but it also restages the Romantic painter Caspar David Friedrich's most famous, Kant-inspired tableau of an encounter with the dynamic sublime (see figure 8)[21]—Helmholtz provides an almost comically prosaic catalogue of possible wave behaviors. The wave system generated by the lunar tides differs from the wave system generated by the wind, which differs, in turn, from the system set in motion by a diving bird or a passing ship. The waves in each of these systems occur with their own particular frequency or speed (measured in terms of the number of waves-per-time-interval passing through a given point); they have their own particular shapes, and their own particular heights or "strengths" (measured in terms of the distance between crest and trough). Where these different wave types intersect, however, they combine to form a new kind of wave, whose frequency, height, and shape are the sum of the frequencies, heights, and shapes of its components.

By scanning the horizon and taking account of the paths of the various wave systems in play, the interested observer of the ocean surface can actively follow the formation of compound waves with a precision that would be unthinkable, for instance, in the otherwise wholly parallel case of sound waves combining in a dance hall. Sound waves operate by periodically "compressing" the air particles through which they flow, and, like their watery counterparts, they vary in frequency, strength, and shape according to the nature of their sounding source: the ear hears the particular number of air condensations

Figure 8. Caspar David Friedrich, *The Monk at Sea* (*Der Mönch am Meer*), 1810, oil on canvas. Photograph by Andres Kilger. Reproduction courtesy of bpk, Nationalgalerie, Staatliche Museen zu Berlin..

per second as pitch (*Klanghöhe*), the intensity of the condensation as amplitude or volume (*Klangstärke*), and the "shape" of the condensation as timbre (*Klangfarbe*). Where multiple such oscillations come together, they combine to form compound systems (a waltz played by the orchestra, the rhythmic clacking of the dancers' feet, the simultaneous buzzing of multiple conversations), but the combinatory process itself fails to register consciously, since the ear hears the new systems only as already formed. The uniqueness of the oceanic spectacle thus resides, for Helmholtz, in the fact that the naked "bodily eye" can see happening on the surface of the water "what otherwise could only be recognized by the mind's eye [*das geistige Auge*] of the mathematical thinker, with respect to the air traversed by sound waves" (*PC*, 57/70), namely, the interactions out of which complex wave systems turn out to *emerge*. The play of waves renders phenomenal the "rhythmic" dynamic of vibratory synthesis—the rational patterns of periodicity underlying the apparent lawlessness of liquid flux—and in doing so provides the scientifically educated observer with a physicist's version of aesthetic pleasure ("a certain physicalist delight"; ibid.).

The particular mathematics of wave combination that Helmholtz has in mind was discovered in the first decades of the nineteenth century by the French physicist Joseph Fourier, in the course of his groundbreaking work on the behavior of heat waves. Fourier's *Analytic Theory of Heat* (*Théorie analytique de la chaleur*), published in 1822, contains the statement of a revolutionary new law pertaining to the structure of compound waves, which Helmholtz paraphrases for the lay audience of his harmony lecture: "*Any arbitrary wave-form can be composed out of a definite number of simple waves of different lengths. The longest of these simple waves has the same length as that of the given wave form, the others have lengths one-half, one-third, one-fourth, etc. of this length. One can produce, through different kinds of overlap among the crests and hollows of the simple waves, an infinite variety of forms*" (*PC*, 62/77).

The term "simple or pure wave form" refers here to the perfectly symmetrical, classically wave-shaped oscillations described by the so-called sinusoidal function, of which Helmholtz provides the following graphic illustration (see figure 9). The majority of audible sound waves, including those generally considered to be particularly "musical," do not, as it turns out, share a shape with this most "regular" of all periodic graphs. A bowed violin string, for instance, vibrates periodically according to a function that has a saw-toothed form (see figure 10). And a bowed violin string observed from the perspective of certain "nonnodal" points on the string acquires further "puckers" (*Kräuselungen*; see figure 11). Yet despite the jagged asymmetry of the latter two periodic graphs—and herein lies the power of Fourier's mathematical paradigm—each can be represented as the sum of a certain number of larger and smaller simple sine curves, all of which have the smooth, symmetrical form of the elementary wave (figure 12 illustrates this process for the case of two such "sharp-edged" functions). No matter how irregularly shaped the periods of a given oscillatory graph, in other words, the law guarantees the mathematical possibility of analyzing these periods uniquely and without remainder, into a multiplicity of paradigmatically wave-shaped units. The method thus allows the mathematician to rigorously relate every conceivable kind of oscillation to every other, across the "universal equivalent," or common measure, of the simple sine function.

Figure 9. Hermann von Helmholtz, "On the Physiological Causes of Harmony in Music" ("Die physiologischen Ursachen der musikalischen Harmonie"), 55/68.

Figure 10. Hermann von Helmholtz, "On the Physiological Causes of Harmony in Music" ("Die physiologischen Ursachen der musikalischen Harmonie"), 57/70.

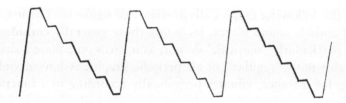

Figure 11. Hermann von Helmholtz, *On the Sensations of Tone as a Physiological Basis for the Theory of Music* (*Die Lehre von den Tonempfindungen als physiologische Grundlage für die Theorie der Musik*), 84/145.

Helmholtz's primary contribution to the science of acoustic wave combination resides in the series of ingenious experiments and instruments he designs to demonstrate, beyond all possible doubt, that Fourier's mathematical analyses have real-world correlates at the level of actual sound waves, both in the air and in the ear. The largest curve in the graph in figure 12, for instance, when interpreted in terms of audible tones, corresponds to what Helmholtz calls a "fundamental tone" (*Grundton*), while the smaller curves with higher frequencies, which "shift" the shape of the base curve in the direction of the final, compound curve, represent "upper partial tones" or "overtones" (*Oberpartialtöne*). Together, the fundamental with its accompanying

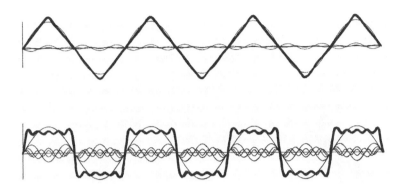

Figure 12. Fourier analyses for two different discontinuously oscillating functions. From Ian Johnston, *Measured Tones: The Interplay of Physics and Music*, 3rd ed. (Boca Raton: CRC Press, 2009), 98.

partials, whose frequencies are always some whole number multiple of the frequency of the fundamental, make up the compound tone, or *Klang*, which corresponds to the sound the ear actually hears. Nearly every acoustic phenomenon is such a compound, according to Helmholtz (exceptions include the simple sinusoidal vibrations produced by tuning forks and pipe organs), and under the right experimental circumstances, the various components can also be made to *sound* by themselves. Helmholtz's principle method of causing partial tones to appear—which is to say, of *analyzing* or *dissecting* the so-called *corps sonore*[22]—exploits the long-known but little understood phenomenon of sympathetic resonance: a body with the potential to vibrate at a particular frequency or pitch, like an undamped piano string, will begin to oscillate spontaneously, without itself being struck, in the presence of a second vibrating body of corresponding frequency. The discovery that sympathetic resonance occurs even when the frequency in question is only an upper *partial* of the sound being generated (an undamped G string on the piano will sound, for instance, when a C is played on the violin, since in this case G is one of the overtones of C) makes clear that partials can "act" on their environment independently of the whole they help compose.

Having demonstrated that most sounds are in fact wave *systems*, and that these systems can be analyzed with the help of the phenomenon of sympathetic resonance, Helmholtz proposes that the ability to hear consists in the largely unconscious performance of precisely

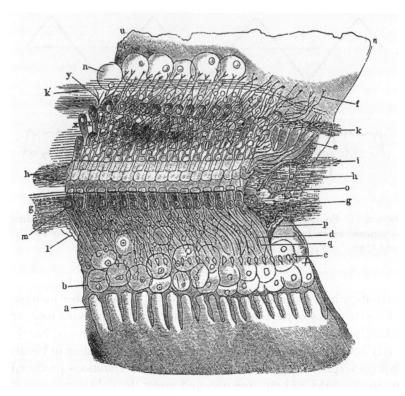

Figure 13. Hermann von Helmholtz, "On the Physiological Causes of Harmony in Music" ("Die physiologischen Ursachen der musikalischen Harmonie"), 61/75.

such analyses. Ordinary untrained laboratory subjects, with the help of certain amplicatory techniques, can be brought to perceive partials at the expense of the compound; it therefore seems likely that the ear itself operates much like an undamped piano, with nerve "strings" set to vibrate sympathetically at various simple, sinusoidal frequencies (see figure 13).

This physical-physiological account of what it means to hear provides the basis, in turn, for Helmholtz's physical-physiological theory of harmony, in which the human ear is presumed to do the same "work" as the wave-analyzing physicist, and in consequence to display the same "preferences" for the most easily accomplished wave analyses. Compound sounds register as consonant when their components relate to one another in simple, elegant ways (as when the pitch frequencies of the partials are low-number multiples of the pitch frequency of the fundamental, or when two different, simultaneously sounding fundamental tones share at least one partial). Compound sounds register as dissonant noise, on the other hand, when their components have little in common.

Since, however, nearly all such wave analyses take place beneath the level of ordinary aural awareness, Helmholtz must go on to explain how this unconscious distinction between structure and chaos, consonance and dissonance, elicits the *conscious* emotions of aesthetic pleasure and displeasure. He finds his answer in the phenomenon of "beats," which occur whenever component wave curves intersect. Defined as alternations in the amplitude, or volume, of the compound sound, beats function as a perceptible indicator of how often and how haphazardly component wave curves "interrupt" each other's course. Whereas component waves of a consonant compound will tend to intersect with each other only at relatively infrequent and "natural" intervals, the partials of a dissonant compound will intersect with each other many times over the course of a single, fundamental period. The result, in the former case, will be relatively infrequent variations in the *volume* of the compound sound (given at every point by the sum of the heights of each component curve), while in the latter, such variations will be all but incessant: "While every individual musical tone produces in the auditory nerve a uniformly sustained sensation, two tones of different pitches mutually disturb one another and chop each other up [*sich zerschneiden*] into individual tone jolts [*Tonstösse*], which produce in the auditory nerve a discontinuous stimulation, and which are as disagreeable to the ear as similar, intermittent but rapidly repeated stimuli are to other sensitive organs, for example, flickering and glittering light to the eye, scratching with a brush to the skin. This roughness of tone is the essential character of dissonance" (*PC*, 71–72/87). Graphically, then, the concept of dissonance describes a scenario in which the component waves fail to "fit together" into the harmonious *corps sonore* of an organic whole. In the absence of a clear common divisor or principle of relation, they cut into each other every which way (*zerschneiden*) rather than dissecting each other cleanly at well-defined, natural "joints" (*zergliedern*). Acoustically, this mathematical disorder makes itself manifest in the perceptual phenomenon of "too many" jolts: the human ear cannot make sense of alternations in amplitude that come too fast to be registered or counted, and the result is a conscious experience of sensory displeasure.[23]

Helmholtz's major work on acoustics, *On the Sensation of Sounds*, uses this physical-physiological theory of harmony to undergird a teleological account of musical development that stretches from ancient Greece to modern Germany. The account follows the lines laid out by Wagner himself, with whom Helmholtz would later become personally close, in portraying Germanic music from the eighteenth century onward as the fulfillment of sound's material potential for increasingly complex harmonic combination.[24] Like Wagner, Helmholtz locates the turning point of this trajectory in the discovery of a "vertical" dimension to musical composition—the dimension of chords, with their simultaneous sound combinations—and like Wagner, he assumes that this discovery has found its culmination in the innovative Wagnerian techniques of harmonic transition, which bind manifold, seemingly unrelated chords together into an intricate, infinitely melodic whole: "In this way one achieves modulations that lead in a single step to comparatively distant keys [. . .]. These means are often employed by modern composers (in particular *Richard Wagner*)."[25]

Unlike Wagner, however, Helmholtz retells this triumphalist story in pronouncedly physicalist terms, and in doing so claims its consequences for the domain of natural science. Wagner's image of chords as wave columns reappears, here, literalized, as the foundational structure of all sound—since even single notes can be shown to consist of multiple, superimposed tones—and the introduction of chordal harmonies becomes legible as a technique for rendering conscious, via an act of explicatory redoubling, the demonstrably *physiological* conditions of musical possibility. (The C-major triad of C-E-G, for instance, turns out simply to "highlight" the first and second partials of the fundamental tone C—by converting them into individual notes.) At this new, higher level of harmonic organization, the tonic root of the chord plays the role of the fundamental tone, but the structuring principle remains otherwise the same. A consonant chord is one where only consonant intervals are found, and a consonant interval is one in which two notes share at least a single partial tone.

The gradual accumulation and systematization of knowledge regarding the principles of this second-order capacity for combination provides new insight into the relationships among individual notes, since two notes that have little in common at the most basic level of the interval could turn out to relate to one another in significant, useable ways at the level of the various chords in which they participate: "We shall consider two compound tones [*Klänge*] to be related in the first degree when they have two identical partial tones, and related in the second degree when they are both related in the first degree to the same third compound tone [*Klang*]."[26] This new insight, in turn, prepares the way for a transformation in the structure of the scale. Where once there had been many different options from which the composer could feel free to choose (Ionian, Dorian, Phyrgian, Lydian, Mixolydian, Aeolian, Locrian), the requirements of rational chord relation distill the abundance down to two privileged types— the "genders" (*Geschlechter*) of major and minor—whose scale steps, or *Stufen*, carve up the sound continuum according to the criteria of a higher-level sound kinship, or *Klangverwandtschaft*.

The resulting musical analysis of the acoustic given is not, in itself, "natural," in the sense that it could be said to exist, independently of cultural preference, within the realm of artistically unmanipulated sound; Helmholtz makes very clear that the winnowing process he describes occurs exclusively at the hands of European artists, on the basis of an idiosyncratically European prioritization of chordal harmony. If it is true, however, that all musical systems are necessarily the product of culturally contingent, aesthetic selection, it does not thereby follow that certain choices are not *objectively* more compelling than others, in the sense that they manage to more comprehensively actualize the possibilities available within nature: the music of eighteenth- and nineteenth-century Austro-German composers, culminating in Wagner, represents for Helmholtz a particularly emphatic realization of the physical-physiological principles that govern all tonal relationships, and the possession of a more sophisticated system for rendering sound relationships comprehensible to the listener allows, in turn, for greater freedom in the use of combinatory means. A composer who need not worry that more distant relations will register only as noise can afford to employ more dissonance, and to modulate more radically, in the process of unfolding a musical thought.[27] The result will be a compositional "concept" capable of subsuming many different kinds of relation under a single principle of development: "It is obvious that the great breadth and wealth of expressive gradations, with which modern compositions can be endowed without endangering their artistic unity, rests essentially on this state of affairs."[28]

The question of what, exactly, such a compositional concept conceptualizes—and hence also the question of what cognitive status should be assigned to the unprecedented integrative power of

Wagnerian music—is the topic to which Helmholtz turns at the conclusion of his public harmony lecture. (The same reflections reappear,
slightly reformulated, at the end of the later monograph, under the
heading "Relationship to Aesthetics" ["Beziehungen zur Ästhetik"].[29])

> For the higher, spiritual beauty of music, harmony and disharmony
> are only means, but they are essential and powerful means. In dishar-
> mony the auditory nerve feels tormented by the beats of incompatible
> tones. It longs for the pure outpouring [*Abfluss*] of tones in harmony,
> and pushes toward this state in order to linger there, appeased. Thus
> both harmony and disharmony alternatively drive and subdue the
> flow [*Fluss*] of tones, in whose incorporeal movement the mind [*das
> Gemüth*] beholds an image of its own stream [*Strömung*] of ideas and
> moods. As with the undulating [*wogende*] sea, the mind is here cap-
> tivated [*fesselt*] and carried along by the rhythmically repeated and
> yet ever-changing mode of movement. But whereas in the case of the
> sea only blind, mechanical powers of nature are at work [*während
> dort nur mechanische Naturkräfte blind walten*], and the impres-
> sion of chaos [*der Eindruck des Wüsten*] thus, finally, prevails over
> the spectator's mood, in the musical artwork the movement in ques-
> tion follows the currents [*Strömungen*] of the artist's agitated soul
> [*der erregten Seele des Künstlers*]. Now gently flowing along [*sanft
> dahin fliessend*], now gracefully springing about, now intensely
> excited [*heftig aufgeregt*], punctuated by or working violently upon
> the natural sounds of passion [*Naturlauten der Leidenschaft*], the
> flow of tones transmits to the soul of the listener, in their originary
> vitality [*in ursprünglicher Lebendigkeit*], unsuspected moods that the
> artist has overheard within his own soul—in order, finally, to carry
> the listener upward [*emportragen*] into the peace of eternal beauty,
> of which God has allowed but few of his elect favourites to be the
> heralds [*zu dessen Verkündern unter den Menschen die Gottheit nur
> wenige ihrer erwählten Lieblinge geweiht hat*].
>
> Here, however, lie the boundaries of natural science, which force
> me to halt. (PC, 75/91)

In this remarkable passage, with its uncharacteristically flowery prose
and reckless profusion of speculative inferences, lies the kernel of a
tentative, Helmholtzian account of psycho-physical interaction, which
definitively distinguishes the cultural experience of musical wave systems from the natural spectacle (*Schauspiel*) provided by the surface
of the sea. The fact that every musical development is experienced as
an expressive trajectory has to do, for Helmholtz, with the universal human tendency to (falsely) interpret *all* movement, regardless of
origin, as the "expression" of a teleological moving force. Where the
source of the movement is known to be human, however, as in the

"moving tones" of music or language, greater familiarity with the mental (and thus truly teleological) forces in play renders legitimate this otherwise dubious interpretive tendency.[30] The composer's carefully calibrated alternation between the acoustic phenomena of consonance and dissonance—between the physiological experiences of pleasure and pain—sets in motion for the listener a temporal dynamic of desire and fulfillment, of conflict and resolution, that in this case can be *reasonably* presumed to mirror the movements of the composer's own soul.

Such a mirroring effect possesses no referential, representational content, according to Helmholtz, since the sounds of music, unlike those of language, never transcend themselves to become symbols of particular emotions or things. Rather, musical sounds relate to one another as moments in a continuous, undulating flux ("the rhythmically repeated and yet ever changing mode of movement"), reflecting in the course of their stuttering progression those fluctuations of mood (*Stimmung*) that form the necessary backdrop for all specific, conceptual determinations (*Bestimmungen*). Musical *rhythm*, in other words—for Helmholtz as for Wagner—has less to do with some externally imposed and in this sense ultimately arbitrary "tact," like the naturally occurring rhythm of ocean waves, than it does with the irregular periodicity of the composer's own tonal-emotional oscillations. And this latter ordering principle has more power to structure experiential time for the composer's audience precisely because the rhythm of affect need not conform, as waves do, to the blind governance of mechanical forces, which, for all their metrical predictability, still leave behind an impression of chaos in the mind of the human observer.[31] The rhythm of affect shapes time according to the trajectory of a purposive consciousness, with the goal of transmitting the experience of the artist's specifically psychological inspiration ("unsuspected moods, which the artist has overheard within his own soul") from the soul of the divinely favored source to the souls of the less emotionally-acoustically attuned.

THE UNDULATING ALL (PSYCHOPHYSICS)

Helmholtz himself, true to his fundamentally Kantian commitments, refuses to speculate further about the actual, scientific foundations of the motion-emotion analogy he postulates ("Here lie the boundaries of natural science"). The nature of the link between sound

waves and affect fluctuations remains vague throughout all of his
writings on the topic of harmony, as does the nature of the relationship between mechanical and teleological modes of movement, matter and mind, physical law and volition. It therefore falls to others to
explicitly explore these relationships—and, in the process, to reject
the Helmholtzian-Kantian schism between *ratio* and world—by turning Helmholtz's submerged Pythagorean promise of an empirical-
harmonic parallelism (back) into the foundation for a physics of the
human psyche.

Wilhelm Wundt is preceded in this attempt by his older colleague,
Gustav Fechner, who began his career as an avowed disciple of
Schelling's nature-philosophy,[32] and whose monumental *Elements of
Psychophysics* (*Elemente der Psychophysik*, 1860)[33] cites Helmholtz's
nearly contemporaneous work on harmony in support of a profoundly
un-Helmholtzian hypothesis: *All* cosmic movement, including that
of the psyche, possesses for Fechner a periodic form, and *every* cosmic complex therefore arises for him as a superimposition of simple
waves, in the manner of a compound sound. Fourier's discovery that
any movement, no matter how arbitrary and irregular, can be represented in terms of sinusoidal curves, merges here with Helmholtz's
demonstration of the acoustic reality of such representations, to yield
the basis for Fechner's combining-wave model of the universe.[34] He
begins his elaboration of the theory by noting that the intensity of
human consciousness can be conceived as periodic, due to the daily
rhythm of sleep and waking. He calls this overarching oscillation the
"total wave," or *Gesamtwelle*, and proceeds to offer a representation
of mental life that interprets this total as the sum of numerous smaller
vibrations—from the underlying oscillations of conscious attention
and distraction (this is the fundamental wave, or *Unterwelle*, which
corresponds to the fundamental tone of a compound sound), to the
vibrations caused by individual sensations or mental events (these are
the harmonic overtones, or *Oberwellen*).[35] He compares the wave
patterns of consciousness to the wave patterns that can be observed
on the surface of the sea ("The sea presents us with the reality of our
scheme, as it were *in itself*"[36]) before going on to ascribe to all Being
the character of rhythmic undulation: "[F]or the whole activity of
the earth's system can be represented under the schema of one large
wave, to which the systems of activity of the individual organic beings
belong as small surface ripples [*Oberwellen*]."[37] The waves of mental attention play the role of upper-level overtones to the more basic

waves of organic life, which in turn provide harmonic embellishment
for the fundamental tone of Nature ("of the universal system of all the
movements of nature"[38]) in all its vibrating, pulsating entirety.

Though Fechner seldom makes the connection to music explicit,
his theory thus quite clearly reformulates the Pythagorean principle
of cosmic harmony, using terms intended to attract and persuade a
generation of Helmholtz-reading scientists. In the process, he repeatedly draws the kind of conclusions that Helmholtz himself had explicitly refused to consider, but he does so—and herein lies the appeal of
his work for more cautious successors, such as Helmholtz's student
Wilhelm Wundt—without jettisoning a commitment to natural scientific modes of investigation. Citing the laboratory experiments of his
own teacher and fellow Leipzig physiologist, Ernst Weber, Fechner
argues that the psychological experience of sensation can be shown
to vary proportionally with the physical phenomenon of the stimulus,
in a mathematical correlation now frequently referred to as Fechner's
Law.[39] Such a correlation implies that the psyche moves *in measurable
tandem* with its more accessible, physiological substrate. (The algebraic equation, on which Fechner founds his entire project of a quantitative psychophysics, dictates a geometric increase in the intensity
of the sensation for every arithmetic increase in the intensity of the
stimulus.) Whereas Helmholtz, therefore, feels compelled to remind
his audience that the psychological meaning of musical pleasure must
forever remain a question of subjective aesthetics rather than objective science, despite its demonstrable foundation in the nonnegotiable
physiology of the ear, Fechner happily interprets the discovery of harmony's bodily basis as (further) proof that both the subject and the
aesthetic can in fact be objectively quantified.

True knowledge of nature, Fechner then goes on to claim, is
always and only possible on the basis of an underlying unity that
binds the structure of the mind *analogically* to the structure of the
world: "The consequence of this view leads to the belief in a conscious
God, omnipresent in nature, in which all minds/spirits [*Geister*] live,
move, and exist [*leben, weben, und sind*], as God does in them, with
the individual spiritual levels [*geistigen Zwischenstufen*] inherent to
the astronomical bodies standing between him and us [. . .]. This
belief can be further developed and supported on the basis of the
analogies and connections provided to us by the hierarchical, steplike
structure [*Stufenbau*] already present in humans themselves."[40] The
notion of analogy, however—and with it the notion of the human

as microcosm—has acquired a new precision in light of Helmholtz's acoustic experiments, since an organism built to resonate *sympathetically* with the universe can now be expected to mirror nature's vibratory combinations in a manner at least potentially accessible to future laboratories. Fechner, for whom the "brain" is no isolated organ but rather a network of nerves extending throughout the entire somatic apparatus, views the Helmholtzian ear as a model for the physiological substrate of sense production per se. The body as a whole thus comes to operate, on his account, like Helmholtz's undamped piano, its various nerve strings constructed to vibrate in tune with particular external phenomena, and to combine—or rather, to harmonize—in chordal concepts of unlimited scope and variety:

> A piano, despite its relatively small number of fixed keys, allows for the possibility of executing the widest variety of melodies and harmonies, and no matter how various or how elevated the thoughts a human might conceive, 25 letters suffice to express them; in both cases success depends only on the connections and on the order in which one runs through the keys or the letters. The brain, with its countless fibers, all of which operate in different ways, contains in this sense incomparably richer resources, so that there cannot be any barrier to presuming it capable of accomplishments, internally, at least as great as those we execute externally by means of it.[41]

The mysterious phenomenon of concept formation becomes legible here, in Fechner's Helmholtzian adaptation of the Platonic letter-note analogy, as the product of a vibratory *brain writing*, its combinatory power a microcosmic reflection of nature's own boundless capacity to self-articulate. Fechner's entire theory thus reveals itself to be, in turn, an extension of the nineteenth century's various Pythagorean, world-spiritual fantasies: Schelling's notion of an oscillatory *Logos*, the Grimm brothers' "living, trembling movement" of sticks and letters, Liliencron's world-moving incantations, and Wagner's alliterative wave pillars all find their (ostensibly) laboratory-ready realization in the sinusoidal swinging of Fechner's human psyche.

Wilhelm Wundt's primary contribution to the science of psychic harmony—a science in which he tellingly nowhere explicitly admits to being involved—is to render Fechner's model acceptable to a scientific mainstream concerned with observable facts. By disavowing the quasimystical, esoteric language of world souls and communion, on the one hand, and proposing techniques for actual testing, on the other, the *Principles of Physiological Psychology* paves the way for a respectable, experimentally reproducible investigation into the structure of the psychic substrate. What for Fechner had remained an "image" with the potential for future concretization, thus becomes, for Wundt, a working hypothesis to be affirmed or discredited in the present, with the help of experiments designed to *measure* fluctuations in the intensity of mental attention. If these fluctuations turn out to possess a periodic structure, then mental life as such, and with it, all phenomenona of consciousness, from the most elemental emotions to the most abstract concepts, will assume the nonmetaphorical shape of a Helmholtzian compound wave.

Wundt quietly unfolds the consequences of this hypothesis—via a series of punctual reflections scattered seemingly at random throughout the thousands of pages of the later editions of the *Principles*—into a full-blown theory of brain rhythm. While comparing, for instance, the patterns of attentional intensity associated with adding single-digit numbers and memorizing meaningless syllables, he allows himself the following general remarks regarding the oscillatory structure of thought:

> Since the fluctuations that appear in both cases recur in all curves, no matter what the type of task may be, it becomes permissible to relate them back to those oscillations of apperception that can already be observed whenever the attention is directed toward a single, homogeneous impression, and which then recur in a similar way in all possible forms of apperceptive functions. [. . .] These apperception waves then likely also correlate with the tendency to *rhythmically* analyze [*gliedern*] mental as well as physical performance [*Arbeitsleistungen*], a tendency that becomes more pronounced the more that physical and psychological performance coincide: namely, everywhere that the articulatory movements of the language organs, or other expressive movements, accompany mental labor, as in reading, writing, memorizing, adding, and the like.[42]

The graphs, or "work curves," of the two laboratory-friendly activities in question have markedly different shapes (see figure 14). Both oscillate, however, at regular temporal intervals between maxima of concentration and minima of distraction, and Wundt claims in this passage that they do so necessarily rather than contingently. Consciousness consists, for him, in an energy of attention that can also be characterized as the energy of a *tension*: a concentration, or *Verdichtung*, of powers by which the mind gathers itself together around a singular point of focus, and without which no experience of any kind, whether sensory or conceptual, could occur ("The

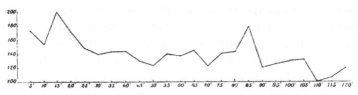

Fig. 381. Arbeitscurve bei der Addition einstelliger Zahlen, nach KRAEPELIN.

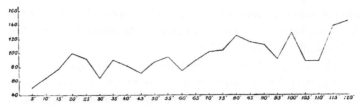

Fig. 382. Arbeitscurve beim Erlernen sinnloser Silben, nach KRAEPELIN.

Figure 14. Work curves for the addition of single-digit numbers and the memorization of meaningless syllables. From Wilhelm Wundt, *Principles of Psychological Psychology* (*Grundzüge der physiologischen Psychologie*), vol. 3, 5th ed., 616.

fundamental phenomenon underlying all intellectual performance is the so-called *concentration of attention*"[43]). Such energy comes in pulses, or waves, meaning that mental activity itself must always unfold according to an underlying, all-encompassing *rhythm*, beneath and before the various patterns imposed by any particular cognitive tasks.

This rhythm can be experimentally discerned, Wundt thinks, at every conceivable level of consciousness. And it can therefore be presumed to *precede* all encounters with the ostensibly unstructured chaos of a mechanistic universe. An experience of the world as a collection of randomly colliding parts, and of world time as a seamless, arhythmic flow of moments, presupposes on the part of the experiencing subject the ability to strip away—and thus to retroactively ab-stract from—an always prior experience of the All as periodically *articulated* into waves.[44] Concepts form and develop as an outgrowth of the "gathering" power exerted by the intensificatory energy of attention, a force Wundt will explicitly, if cautiously, go on to identify with the directional striving of an underlying "will."[45] This volitional tendency toward *Verdichtung* drives the mind of humankind ever further away from mere associative inferences and ever closer to the ideal of total synthesis, through a trajectory of increasingly complex and capacious *self*-articulations, whose task is effectively to explicate the potential of the most originary psychic rhythms. Conceptual collections therefore operate, for Wundt as for Fechner, according to the logic of Helmholtz's Austro-German chords, which is to say they parse and, in parsing, render "audible" the diverse capacities for combination that would otherwise remain hidden within the simple shapes of the foundational brain waves.

What this means is that the mechanist worldview so energetically promulgated by prominent nineteenth-century scientists such as Helmholtz can now be seen, from Wundt's unspoken perspective, to stand in conflict with the fundamental tone of all scientific thought. And just as artificially imposed "task rhythms" provide demonstrably suboptimal conditions for the intellectual labor of adding or memorizing—"in general, mental work is qualitatively most supported when the individual follows a self-chosen rhythm, while this same work [. . .] displays errors and imprecision when the rhythm is externally dictated"[46]—so, too, can mental production more generally be expected to suffer in cases where the conceptual superstructure fails to harmonize with the tonic. If the essence of rationality consists in a teleological striving for ever more encompassing "harmonic" arrangements, then any nonteleological theory of the universe must necessarily falsify the activity of thought: the mechanist perspective eliminates the scientist's ability to reflect on his or her own teleological, because rational, methods of inquiry.

At his most daringly un-Helmholtzian, Wundt will then go on to suggest that the conflict between a mechanist science and its own essentially teleological conditions of possibility bears witness to a fundamental inadequacy in the mechanist conception of matter. Since, in the case of minded beings, psychic changes can be shown to correspond to physical ones, as mathematically expressed by the Fechner-Weber law, a properly scientific concept of matter must be able to accommodate the *potential* for psycho-physical, teleological-mechanical interaction, even in cases where this potential has not (yet) been realized: "Since, in the end, we must presuppose that the life-expressions [*Lebensäußerungen*] developed by complex substances of organic nature have their preconditions in the simpler configurations of inanimate nature, so we also cannot escape the assumption

that in the simplest element of substance, the atom, the most elemental drive-forms [*elementarste Triebformen*] are already preformed
[*vorgebildet*]."⁴⁷ Every atom of physical substance, in other words,
must actually bear within itself, in the form of a built-in proto-telos,
or *Trieb*, the conditions of possibility for its eventual participation
in the rhythmically articulated, volitional flow of the psyche. Wundt
elaborates on the large-scale implications of such a teleological physics in the final lines of the fourth edition of the *Principles*—one of
the only places in his entire oeuvre where he makes (almost) explicit
the fundamentally nature-philosophical premises on which his oscillatory theory of the psyche depends:

> This view of the problem of interrelation [between physiology and
> psychology] unavoidably leads, moreover, to the presupposition that
> mental Being [*das geistige Sein*] is the reality of things [*die Wirklich-
> keit der Dinge*], and that the most essential feature of this reality is
> development [*Entwicklung*]. Human consciousness is for us the apex
> of this development: it forms the node in the flow of nature [*bildet
> den Knotenpunkt im Naturlauf*], through which the world reflects
> upon itself [*sich auf sich selber besinnt*]. Not as simple Being, but as
> the developed product [*Erzeugnis*] of countless elements, does the
> human soul become what Leibniz called it: *a mirror of the world.*⁴⁸

At stake, for Wundt, in his experimental investigation of vibrating
brains, is thus never a science of human Being in contradistinction
to the science of its natural counterpart, as dictated by the mechanist perspective of his teacher, Helmholtz, but rather a science of
human Becoming that reflects the true essence of the physical universe—by microcosmically rearticulating a truly *universal* propensity
to progress.

A PHILOLOGY OF THE EAR (POETICS)

Language plays a familiar role in the context of this new paradigm
("It is the responsibility of physiological psychology to investigate the
outer and inner conditions under which language, as the highest form
of human life-expression, emerges"⁴⁹), with the crucial difference that
Wundt no longer feels bound by the ban against primal sound-sense
speculations. Where Bopp and Grimm had viewed the question of
ultimate origins, and of the nature of the bond tying first sounds to
first senses, as forbidden territory beyond the purview of a rigorous
language science, Wundt believes the principle of psycho-physical

correspondence provides him with the beginnings of a testable hypothesis. This principle, which effectively turns *every* physical movement into the "expression" of a psychic shift, and thus into a representative of the category Wundt calls *Ausdrucksbewegung*, or "expressive movement," paves the way for a theory about the emergence of language from gesture. Just as the originary humans instinctively moved their limbs in response to various forms of psychic stimulation, so they also moved their larynx in tune with the interior vibrations of their emotions: "The sounds of language [*der Sprachlaut*], like the gesture, arise out of the innate human drive to accompany feelings and affects with movements, which stand in immediate relation to emotion-arousing impressions."[50] The result is the *Klanggebärde*, or sound-gesture, a primitive, immediately comprehensible form of articulation that precedes the separation of Language into languages and in doing so resists all attempts at traditional, etymological explanation. Wundt's later, more expansive work on language origins offers as the paradigmatic example of such sound gestures-turned-words the near-universal pairing of "mama" and "papa," which—so the argument goes—aligns an inherently "weak" consonant with the inherently "weak" sex across a wide swath of historically unrelated idioms.[51]

Only the methods of a physiological psychology, according to Wundt, can give access to these most original of all roots, and to the true meaning of the further developments they presage, since only the methods of a physiological psychology can subject the actual, expressive *dynamics* of such sound-sense parallels to the scrutiny of laboratory testing. Wundt himself does not devote a great deal of experimental energy to the project of charting particular sound-sense relationships. He does, however, make explicit the direction that any such future project must take:

> The development of language is intimately bound up with the development of musical sound-expressions [*Lautäußerungen*]. [. . .] The deeper psychological reason behind this connection lies, however, in the universally human disposition toward rhythmic analysis [*Gliederung*] of impressions and harmonious sound sequences, which reveals itself in numerous guises beyond the realm of specifically musical production. The linguistic expression of thought and song thus likely have a common point of origin in a song-like form of speech, which was neither singing nor regular speech in our contemporary sense, and which for precisely this reason could develop both of these latter forms out of itself.[52]

The sounds of language, in other words, represent a form of music that runs parallel to the underlying "music" of the intellect ("The movement of rhythm thus corresponds to the progression of the affects, the feeling of harmony to the resolution [*Lösung*] of this progression"[53]). For where the conundrum of the sound-sense bond crystallizes into a question of vibratory correspondence—Wundt refers the reader in a footnote to his various discussions of mental periodicity—language roots becomes audible, or rather, experimentally *realizable*, as the manifestation of originary mental rhythms. The details of these rhythms, which are governed by the oscillations of a subjective mood, or *Stimmung*, and not by the objective time of tact, can therefore be recovered, despite their lack of a fixed standard of measure, by an interpreter sufficiently versed in the psychophysical principles of expression. And the developmental trajectory they turn out to follow, which mirrors the developmental trajectory of the mind itself, can be explained in the terms of a teleological psychology. Sound shifts occur, according to Wundt, whenever a group of people requires easier and more efficient forms of phonetic articulation to correspond to their faster and more complex processes of mental analysis. Similarly, poetic meter progresses from the stringency of syllable counts to the flexibility of free verse, with an intermediary stop at the *Stabreim*, in response to the ever-increasing expressive power of the modern poet's structuring mind.[54]

Wundt's own rather tentative attempts to draw specific, language-scientific consequences from his theory of mind-matter correspondence—as formulated, for instance, in the volumes titled *Language* (*Die Sprache*), from his monumental study of collective consciousness or *Völkerpsychologie*—sufficed to provoke several exasperated responses from his Neogrammarian colleagues in Leipzig, whose mechanist-materialist perspective prohibited all such explanatory recourse to the "telos" of a progressively developing collective psyche.[55] The Wundtian principles found perhaps their most ambitious linguistic application, however, at the hands of the one-time Neogrammarian phoneticist and *Stabreim* analyst Eduard Sievers, who did not shrink back, as Wundt had done, from putting psychophysiological principles into actual philological practice (or from proclaiming the obsolescence of the mechanistic Neogrammarian alternative). By the time Sievers arrived at the Wundt-inspired methods of his *Rhythmic-Melodic Studies* (*Rhythmisch-Melodische Studien*, 1912),[56] he had directly confronted both the philological dilemma of a definitive *Stabreim* meter (in the volume *Old High German Metrics* [*Altgermanische Metrik*], 1893, as discussed in Chapter 3), and the phonetic impossibility of an adequate alphabetic transcription (in his magnum opus *Principles of Phonetics* [*Grundzüge der Phonetik*], 1876, as discussed in Chapter 1). From Sievers's perspective, therefore, the notion of experimentally recoverable sound-sense vibrations held out the promise of a much-needed new approach to language science. Sievers dubbed this approach "ear philology" (*Ohrenphilologie*), to distinguish it from his predecessors' exclusively text-based "philology of the eye," and he conceived of it as a quest for the original tonal-emotional "substance" of transmitted works. More important for the practice of philological critique than the genealogical rules of textual transformation—because more immediately connected to the originary *movement* of meaning that animates every text—became for him the honing of a specifically physiological sensibility to the phenomenon of psychological resonance.[57] The only truly legitimate interpretation of a poetic text, he argued, must take the form of an archeological "sound analysis" (*Schallanalyse*), which requires the reader to submit as fully as possible, in accordance with a series of teachable "listening" techniques, to the particular vibratory rhythms of an originary "tone writing" buried beneath and before all conventional modes of transcription.[58]

Sievers's decision to realize this fantasy of total recovery on the thorniest of all nineteenth-century philological problems—by *aurally* reconstructing the genetic history of the *Nibelungenlied* manuscripts, together with the most "original" text of the *Edda*[59]—testifies to the radicality of his psychophysiological conversion. It also, of course, links his proposal for the philological future to the musical language of its Wagnerian past, and, by extension, to the "deployment" of Wagnerian-Wundtian principles already long since under way in the domain of avant-garde poetics. For Sievers, as for Wundt, the expressive "will" that courses through poetic compositions, giving rhythmic direction to the harmonic whole and turning collections of psyche-sound waves into microcosmic articulations of an undulating, developing cosmos, becomes newly accessible to direct, natural scientific observation in the context of the psychophysiological laboratory. Privileged insights into the tonal-emotional "stuff" of language spirit, which Wagner and Grimm and even Schelling had attributed to the mysteries of an artistic or linguistic inspiration, can therefore now be produced at will with the help of proper experimental conditions and techniques ("Experience shows that these kinds of observations can only be set up, carried out, and controlled with sufficient dependability inside the quiet little chamber of one's own office, or in the most intimate and trusted community of a laboratory"[60]). The result is a nature-philosophical equation of language system with language spirit, and an etymological project of deducing spiritual-systemic roots, that have been updated to reflect early twentieth-century concerns. Sievers's psychophysiological theory of poetry as a vibratory *corps sonore*, and of reading as an exercise in sympathetic resonance, holds out to his contemporaries the promise of a realized "scientific poetry" (Ghil), and of a corresponding "science of sound symbolism" (Wolzogen).[61] It can therefore claim to herald the beginning of that infinitely melodic future toward which Wagner himself was presumed to have so clairvoyantly gestured.[62]

CHAPTER 6

Jakobson's Zeros

When, in 1949, Roman Jakobson concluded a paper titled "The Identification of Phonemic Entities" with a comparison between language science and physics—"Linguistic analysis, with its concept of ultimate phonemic entities, signally converges with modern physics, which has revealed the granular structure of matter as composed of elementary particles"[1]—there could be little question that the wave theories of Wundt and Sievers belonged among the positions he intended to reject. Having come of age amid the Russian Futurists, whose intimate if troubled relationship to both Wagner and "ear philology" so powerfully influenced, in turn, the perspective of their Russian Formalist contemporaries, Jakobson was well-versed from the outset in the seductions of the psychophysiological paradigm. His first book, the polemical *Newest Russian Poetry: First Attempt: Viktor Khlebnikov (Novejsaja russkaja poezija,* 1921), tackles the question of the language-music relation as it pertains to the phenomenon of sound poetry; his second, *On Czech Verse (O cheshskom stikhe,* 1923), explores the problem of establishing poetic meter and, by extension, of measuring language time.[2] Both books, in their respective ways, treat traditional psychophysiological concerns in the service of a decidedly antipsychophysiological hypothesis: poetic language, according to Jakobson, owes its appeal not to the deployment of universal, natural correspondences between psychology and physiology but to the deformation of particular historical conventions of sociolinguistic behavior. A general theory of naturally expressive sound must therefore necessarily fail to account for the specificity of poetry's culturally determined effects.[3]

The early projects, then, testify to the significance of a polemic against psychophysiological principles for the development of Jakobson's own approach. By the late 1920s, however, as Jakobson began to turn his attention away from poetry and toward the fledgling theory of phonological relations, references to psychophysiologically inclined interlocutors such as Eduard Sievers largely disappeared from his writing. The primary antagonist in the battle for well-defined minimal units of speech—and thus also for writeable letters—became instead the mechanist-materialist model associated with the Neogrammarians and their successors, which denies the very possibility of discovering real boundaries—and thus also real structure—in the continuum of language sound. Jakobson began his public phonological career with an antimechanist manifesto titled "The Sound Law and the Teleological Criterion" (1928), in which he argues that the true laws of sound change can be identified only if languages are presumed to diachronically *develop* toward a *telos* of systemic self-preservation ("[P]honetic changes must be analyzed in relation to the phonemic system which undergoes these mutations. For instance, if the order within a linguistic system is disturbed, there follows a cycle of sound changes aimed at its renewed stabilization"[4]). In a series of later texts, he borrows the terms associated with Wundt's teleological harmony theory of the psyche to describe the "sounding together" of a phonological system, comparing language to music, and phonemes to chords.[5] His final monograph portrays the identification of phonological units as the culmination of an analytic trajectory that runs from the Sanskrit priests, through Plato's *Philebus*, to the Norse poet-magicians with their alliterating runes.[6] And a late, brief rumination, "Structuralism and Teleology" ("Structuralisme et Téléologie"), bookends a career spent combatting the mechanistic viewpoint by harnessing two avowedly antiteleological forebears (Jakobson names Saussure together with the Polish linguist Jan Baudouin de Courtenay) to his longstanding teleological cause: "Even today, there are people for whom teleology is synonymous with theology. And yet it must be noted that intuition obliged these two great precursors of modern linguistics to distance themselves from this dogma in their research."[7]

The problem, of course, with such explicit declarations of teleological allegiance—for an interpretation that seeks the *specificity* of

the new phonological contribution to the thinking of language structure, rather than the site of its continuity with older models—is that
Jakobson's all-encompassing polemic against mechanism leaves little
rhetorical room for a sustained reflection on the problematic of teleology itself.[8] Jakobson clearly knows, in other words, something he
nowhere actually says, which is that he cannot simply implement the
notion of goal-oriented agency as he inherits it from its nineteenth-
century German proponents, and expect it to yield something other
than a "theology" of German *(Sprach)Geist*. In order to remain true
to the deflationary thrust of his own fundamentally untheological
project, he must therefore somehow ensure that the phonological fulfillment of his predecessors' teleological fantasies unfolds exclusively
as an *emptying* of their respective animating energies. The divinity
in whose service the Vedic priests perform their analyses; the "god
or divine man" who taught the Greeks, according to Plato, to carve
up the sound continuum; the Wotanic world-spirit whose alphabetic
power the *Stabreim* attempts to harness; the Wundtian-Wagnerian
will to which the music of language ostensibly gives voice—all must
be made to disappear, here, into the remainderless purity of a jointure
defined only by difference.

   Jakobson began to work through the linguistic ramifications of a
purely differential teleology in a series of short studies from the 1920s
and 1930s, and the resulting theory of language analysis finds its
first general exposition in a pair of lectures from 1939.[9] The lectures,
which were held in Copenhagen—Jakobson's first point of refuge
after his flight from Nazi-occupied Czechoslovakia—were published
under the title "On the Structure of the Phoneme" ("Zur Struktur des
Phonems"). Together they lay out a unified theory of the phonological
unit.[10] The first lecture opens with the dilemma of continuous, infinitely divisible sound:

> More than a hundred years ago, a Russian Romantic writer depicted
> a knowledge-hungry hero who strove toward the complete elabo-
> ration of external experience. An evil witch granted his wish, and
> suddenly the speech of his lover and the music of poetry broke
> down [*sich zerlegte*] for him into countless [*zahlose*] articulatory
> movements and uncountable [*unzählige*] sound impressions, which
> remained utterly meaningless [*sinnlos*] and unalluring [*reizlos*].
> The triumph of a naive, dismembering [*zerstückelnden*] naturalism
> could not have been foretold with more prescient insight than in this
> uncanny dream image of the wise Romantic. The sound-content of
> language disintegrated for the naturalistically inclined researcher into

a vast number [*Unmenge*] of fluctuating motoric or acoustic atoms, which he painstakingly measured, but whose purpose [*Zweck*] and meaning [*Sinn*] he consciously ignored [. . .]. This path threatened to lead to the hopeless devastation of verse theory [*Verslehre*], and indeed of sound theory [*Lautlehre*] per se, as well as to the remainderless elimination of sound theory from a linguistic—i.e., from a sign-related [*zeichenartige*], semiotic—problematic.[11]

The problem of properly carving up the sound continuum, Jakobson here suggests, can be solved only by taking into account the common *purpose*, or *telos,* that binds and gives shape to the whole. Such a unity of purpose provides the sole criterion for distinguishing between a formless infinity of possible sound variations and a countable number of linguistically significant units. The passage unfolds as a cautionary tale pertaining to the late nineteenth-century mechanization of language science, but the perspective it advocates is clearly in many ways an ancient one, as the Platonic dictum first cited in chapter 1 can attest ("And once we have grasped [this unity], we must look for two, as the case would have it, or if not, for three or some other number. And we must treat every one of those further unities in the same way, until it is not only established of the original unit that it is one, many and unlimited, but also how many kinds it is"[12]). When early twentieth-century innovators begin to (implicitly) champion teleological reasoning as the only possible means of escape from a Neogrammarian morass of unwritable noise—"[T]he terrifying image of chaotic multiplicity required the antithetical principle of ordering unity. Two ingenious linguists, Baudouin de Courtenay and Ferdinand de Saussure, initiated the inquiry into the purpose of language sounds" (*SP*, 280–81)—they would therefore appear to be performing a revolution by way of return.

The notion of *telos,* however, almost immediately begins to undergo a metamorphosis more radical than any mere erasure. Drawing on Saussure, Jakobson proposes to locate the defining essence of the most elementary linguistic sounds in their capacity to differentiate semantic "chunks" (we hear "bat" rather than "pat" or "cat," for instance, because of a difference in the initial sound segments of the respective utterances). Alone among linguistic phenomena, he then claims, such sounds possess no *positive* semantic function. The linguistic purpose of the phoneme, and so also the *telos* that distinguishes this acoustic fundament of language from all linguistically irrelevant noise, resides exclusively in *the fact of its distinguishability*—"the mere fact of 'being other' (*altérité*, according to the excellent French philosophical term)" (*SP*, 304)—for the speakers of a particular language:

> The formula *aliquid stat pro aliquo* obtains in equal degree, as we see, for all grammatical and lexical units of language [. . .]. To the difference between two morphemes corresponds a definite and constant difference in meaning. To the difference between a questioning tone and an exclamatory tone corresponds a definite and constant difference in the realm of what is expressed. [. . .] What, however, corresponds to the difference between two phonemes? To this difference corresponds solely the *fact* of a difference in meaning, whereas the *content* of this difference in meaning is neither definite nor constant. (*SP*, 291–92; emphasis mine)

The system of phonemes, as a set of differential incisions (*encoches*) that run absolutely no risk of being reified into symbols, thus turns out to provide precisely the kind of empty yet writeable linguistic foundations that Saussure had despaired of ever truly defining. Against the enspirited runes of Grimm, Pott, and the *Stabreim* theorists, Jakobson here holds up his strictly oppositional *articuli* as the new "foundational pillars" (*Grundpfeiler*) of a reconceptualized language science (*SP*, 301).[13] "The thesis of the *Cours de linguistique*— '*The phonemes are above all oppositional, relative, and negative entities*'—has become the departure point of phonology" (*SP*, 294).

The pressing question still remains, however, of how exactly this transformation in the concept of *telos* should actually be understood to work, for the dynamic of differentiation so triumphantly proclaimed above is clearly (and viciously) circular. The foundational units of language, having been emptied of their relationship to a definition-bestowing language spirit or psyche, are now simply being redefined in terms of their own—still utterly mysterious—*capacity to differentially define*. Before the theory of phonological interaction can legitimately claim to have realized the Saussurean ideal of despiritualized letters, it must therefore first provide a nonparadoxical account of the new, emptied *telos* in action, which is to say, in the *process* of calling the *unity* of a system into being.

Jakobson begins the labor of explicating this process with an analogy that involves three hypothetical scenes of reading:

> By way of illustration, let us transpose the problem into the optical realm! Let us say, for example, that we want to acquire facility with an unfamiliar script, for instance Coptic. The task is tremendously difficult if this script is for us a mere agglomeration [*Anhäufung*] of meaningless arabesques. The task is easy if every letter possesses

for us a constant and unified positive value. There is an in-between case: the positive value of the letters remains unknown to us, but we know the meanings of all the words in the given Coptic texts, and the letters function thereby as purely differential signs [*reine Unterscheidungszeichen*]. The acquisition of the alphabet is certainly more manageable than in the first case, but significantly more difficult than in the second. The fewer and more orderly *the external differences* appear, to which the variety of letters can be reduced, the more soluble the task becomes. (*SP*, 302)

The first of Jakobson's three scenarios in this passage recalls the predicament of the mechanist phoneticians, who steadfastly refuse to presume any teleological function at all ("a mere agglomeration of meaningless arabesques"). The second parallels the situation of a competent language speaker in the process of learning to read: it involves "matching" various symbols to a set of preexisting, already familiar referents for which the symbols can be presumed to stably stand (*aliquid stat pro aliquo*). The third and final scenario corresponds, as Jakobson will go on to make clear, to the case of a deaf child faced with deciphering a phonetic script, and it is this third case, according to him, that most resembles the task of an ordinary language learner or phonologist who seeks the building blocks of spoken language. Neither the deaf child nor the language-learner/linguist possesses any kind of "key" for assigning positive definitions to the various bits and pieces acoustically perceived. Both must therefore proceed purely negatively, piecing together an awareness of unit identity from the totality of observable oppositions.[14] The project of discovering the true joints of a linguistic system, in other words, is equivalent to the project of learning to read in the absence of a glossary, guided only by an innately rational ability to perform the simple and singular operation of distinguishing "this" from "that."[15]

Jakobson's extremely condensed account of the actual analytic procedure, which moves from ratios, to ratios of ratios, to units without ever passing through the forbidden territory of positive content, requires substantial unpacking to become clear. What follows is a reconstruction based on the brief outline offered in the lectures but also on the scattered hints to be found in other accounts dating from the same general period.[16]

The search for ultimate units begins with a process of trial and error, the goal being to locate, within the parameters of a given language, the smallest differences in sound that correspond to a difference in meaning. By somewhat arbitrarily choosing to compare the word "most" with the word "mint," for instance, we can isolate the sound segments /in/ and /os/ as capable of corresponding to such a difference. By then continuing on to compare "most" with "mist," we can isolate the even smaller sense-differentiating sound segments /o/ and /i/. Finally, by *failing* to discover any (English) words capable of differing from "most" by a sound segment smaller than /o/, we can hypothesize that "most" and "mist" represent what the phonologists call a "minimal pair." "Most" and "mist" could not differ any less, in other words, without becoming variants of the same utterance.

It is important to note that the isolation of such a minimal pair does not, in itself, result in the identification of a phoneme, since there are as yet no grounds for assuming that the particular sound segment by which "most" differs from "mist" appears anywhere *else* in the English language. Once enough minimal pairs have been collected, however, it becomes possible to form minimal pair *classes*—Jakobson calls them "paradigms," and, later, "paradigm sets"[17]—in which every word differs minimally from every other. The grouping beet : bit : bait : bet : bat is one such class; pill : till : kill : bill, and nap : gnat : knack : nab are others.[18] Phonetic analysis provides a consistent conceptual terminology for describing the various sound differences that distinguish the two members of each binary pair, such that every member can then be defined, within the limited context of a particular class, by the collection of purely differential characteristics—Jakobson calls these "distinctive features"—required to set it apart from all other members. Just as a given bodily organ depends for its definition on the role it turns out to play within the context of the organism, rather than on any available catalogue of stable, physical characteristics, so, too, does the linguistic meaning of a phonetic attribute such as "voiced" depend entirely on its functional relationship to the alternative "unvoiced," rather than on any objectively measurable threshold for voicedness. The /b/ of "bill" is voiced insofar as it can be distinguished, in English, from the /p/ of "pill," and the /p/ of "pill" is bilabial (which is to say, an articulatory product of the lips) insofar as it differs audibly, for a native English speaker, from the dental /t/ of "till" and the guttural /k/ of "kill." Where the need to distinguish /p/- from /b/- or /t/-words disappears, so too do the audible phenomena of voice or labial articulation ("Breadth cannot be conceived without narrowness, nor posterior formation without anterior, roundedness without unroundedness, etc."; *SP*, 303[19]).

Within the limited context of the bill-pill-till-kill class of minimal
pairs, then, the meaning-differentiating sound segment of "pill" can
be distinguished from all other meaning-differentiating sound segments by the short feature list "voiceless, bilabial." Once such lists
are present for every member of a given class, this class can in turn
be compared with *other* minimal classes, thereby paving the way for
the following definition of a phoneme: given two or more classes of
minimal pairs in which all members relate to one another in precisely
the same way, a "phoneme" just *is* any list of features held in common
across the respective classes. Where "pill" relates to "till," "kill," and
"bill" just as "nap" relates to "gnat," "knack," and "nab," a formal
equivalence relation can be established between the structural positions of "pill" and "nap." ("pill" is to Class 1 as "nap" is to Class 2.)
And the content of this equivalence relation—namely, those properties in virtue of which the two entities turn out to hold *structurally
analogous* positions within the context of the English language—can
be located in the list of their shared differential features, which can
in turn be designated, somewhat arbitrarily, as /p/. Each of the shared
feature lists in the table below comes to correspond in this way to a
particular phonological element.

Class 1	Class 2	Shared Feature List
pill	nap	bilabial, voiceless = /p/
till	gnat	alveolar, voiceless = /t/
kill	knack	palato-velar, voiceless = /k/
bill	nab	bilabial, voiced = /b/

Saussure's intuition of the linguistic unit as a "point of intersection
of several series <of> analogues" (*Notebooks* I 68a / *CLG*, 289:2),[20]
which caused him so much trouble at the level of the word (where
"analogue" always turns out to imply some unavoidably substantive
sameness of meaning), finds here, in the analogies of the minimal pair
paradigms, an unexpectedly precise confirmation: to define a phoneme is simply to locate the "place" where multiple series of oppositional ratios intersect.

The procedure described above represents, in its broad outlines, a
method for isolating phonemes on which most of the early contributors to the phonological movement would presumably have agreed.
Jakobson's central innovation in the 1939 lectures is to suggest that

this method does not, in fact, go far enough. Rather than stopping with the analogies that yield the phonemes—conceived here as relationships of positional equivalence between members of different but structurally identical word classes—language analysis should continue, he argues, until the possibility of further analogy has been exhausted. In the sample case of Class 1, this would mean going on to note, for instance, that "till," "kill," and "pill" all share a distinguishing feature with respect to "bill" (namely, the feature of not being voiced), such that it becomes possible to write "till" : "bill" = "kill" : "bill" = "pill" : "bill" (or, in the case of Class 2, "gnat" : "nab" = "knack" : "nab" = "nap" : "nab"). Analogies of this kind, which rest on the foundation of a single feature rather than a list, allow access to a common ground even *deeper* than the one carved out by the phoneme, since single distinguishing features generally participate in the definition of several different phonemes. Jakobson's friend and cofounder of phonology, Nikolai Trubetzkoy, had proposed for such identities—without ever assigning them, as Jakobson does here, the status of most basic phonological *unit*[21]—the French term *correlation* (from Latin *co*, "together" + *relatio*, "relation"), which translates the most technical sense of the Greek *analogia* as a metarelation among relations. A report to the First International Congress of Linguists in 1927, delivered by Jakobson and cosigned by both Trubetzkoy and their fellow Russian linguist Sergei Karcevski, provides the following preliminary definition of the new technical term: "A phonological correlation is constituted by a series of binary oppositions defined by a common principle, which can be considered independently of each pair of opposing terms."[22]

Jakobson illustrates the formative influence of this "common principle" of correlation by analyzing the Ottoman Turkish vowel system into three different equivalence chains, each of which represents the set of binary relations called into being by a different, single-featured opposition:

The phonemes /o/, /a/, /ö/, /e/ stand opposed to the phonemes /u/, /ɯ/, /ü/, /i/ as broad against narrow, the phonemes /o/, /u/, /a/, /ɯ/ to the phonemes /ö/, /ü/, /e/, /i/ as posterior to anterior, and the phonemes /o/, /u/, /ö/, /ü/ to the phonemes /a/, /ɯ/, /e/, /i/ as rounded to unrounded.

  1) o : u = a : ɯ = ö : ü = e: i
  2) o : ö = u : ü = a : e = ɯ : i
  3) o : a = u : ɯ = ö : e = u : i[23]

Taken individually, the eight Ottoman Turkish vowels *o*, *a*, *ö*, *e*, *u*, *ɯ*, *ü*, and *i* can combine with one another in a total of twenty-eight different binary ways. By classifying these various differential possibilities analogically, however, it becomes possible to reduce the number of relevant relationships to three: the "fundamental oppositions" of broad/narrow, back/front, and rounded/unrounded *underlie* the oppositions of individual vowels, like the common factors of an arithmetic equation, and in doing so, they greatly simplify the decoding of the phonological system. The mysterious process of learning to "read" phonologically, whether in the role of linguist or babbling infant, becomes comprehensible, in consequence, as an extension of the basic human capacity for comparing things to things—and comparisons to other comparisons ("[W]e easily detect a series of ratios"[24])—*until the limit of commensurability has been attained.* Jakobson concludes,

> Not only the differences among Ottoman Turkish vowel phonemes, but, indeed, the differences among all phonemes in all languages break down remainderlessly [*zerlegen sich restlos*] into simple binary oppositions, and accordingly all phonemes of all languages— consonants as well as vowels—dissolve remainderlessly into further indivisible [*unzerlegbare*] distinctive qualities. Not the phonemes, but rather the distinctive qualities appear as the primary elements of word-phonology. (*SP*, 302)

The linguist who identifies the limits of commensurability with the phoneme, in other words, has simply stopped comparing things to things too soon. The phoneme divides, and when it does it yields new ratios—broad : narrow, unvoiced : voiced, unrounded : rounded— which cannot themselves be further simplified.

Beneath these fundamental phonological *logoi* lies, for Jakobson, an acoustic version of the Platonic *apeiron*. The irrational realm of linguistically unstructured sound to which, on the one hand, the phonologist must refer in order to uncover polarities such as lax and tense, must, on the other hand, be kept rigorously separate from the domain of differentiated structure, if phonology is not to regress into mere phonetics: "The more phonetic matter [*Stoff*] phonology manages to examine and process, the better—but these givens must actually *be* phonologically processed; it is not permissible to simply carry over raw phonetic material into phonology, so to speak with hide and hair [*mit Haut und Haaren*]" (*SP*, 281; emphasis mine). The terminology makes clear that Jakobson quite consciously situates his language/noise distinction within the context of the ancient form/matter dynamic, and that he sees in his "primary elements" the result of a successful, *apeiron*-delimiting dieresis. Traditional phonetics, from this perspective, counts as a "science of matter" (*Stoffwissenschaft*) because it deals directly with the chaos of the continuous sound stream, while phonology merits the title "science of form" (*Formwissenschaft*) because it studies the system of relationships that *organize* this stream (*SP*, 281).

On the one hand, then, Jakobson's entire theory exemplifies the familiar teleological idea that language structures matter according to the demands of a single unifying principle, which in turn serves to actualize matter's inherent potential to take shape.[25] Language operates, according to Jakobson, by intensifying a continuum of degree into an opposition of kind, such that the seemingly indecipherable infinitude of "more" and "less" becomes comprehensible, instead, as the binary foundation for a countable collection of relational units. Sound "strives," of its own accord, toward the expressive fulfillment of its material "phonic essence" in the articulated forms of a harmonious whole, where function first gives contour and solidity to substance.[26]

On the other hand, something odd has clearly happened here to the twin notions of striving and fulfillment, for though the concept of purposive action remains all-powerful, it has been *deprived of all conceivable content*. Not only do common properties such as voicedness lack any objectively measurable or observable identity; not only do they owe their existence solely to the ratios that bind them to their opposites (these things, after all, were already true of the elements or "joints" in the nineteenth-century organicist models); rather, they also, despite serving the "common" purpose of differentiation, share absolutely nothing *in common* with one another. No analogy links the opposition of voiced/unvoiced to those of round/unround, narrowed/unnarrowed, and so on, which means, in the terms of both the ancient and the Helmholtzian harmony theories, that the units of language possess no common measure. Jakobson can write without difficulty such analogical series as o : u = a : ɯ = ö : ü = e: i, o : ö = u : ü = a : e = ɯ : i, and o : a = u : ɯ = ö : e = u : i. But he has no means at his disposal for going on to compare, in turn, these three different ratio chains. The language system, despite its functional unity, remains internally incommensurable, devoid of a singular substrate from which its individual parts could be understood to collectively emerge. The Jakobsonian distinctive feature differs radically, for this reason, from the Aristotelian precursor whose name it only almost shares, since the *differentia specifica* of scholastic philosophy always carves its definitional distinctions into the preexisting foundation of a more encompassing genre of being. *Logos*, conceived as the specific difference whereby certain entities distinguish themselves as human, requires the categorial backdrop of "animal" in order to operate as a conceptual "cut," whereas the slash of the voiced/unvoiced opposition can define—in the sense of sufficiently distinguishing meaning—without reference to any underlying, pan-systemic domain. Distinctive features do combine, like the properties in more traditional systems, to form higher-level complexes or "concepts." These concepts, however, remain every bit as bare of positive, conceptual essence as their purely differential constituents. Phonemes are aggregates, not syntheses, characterized by lists of irreducible, because incommensurable, ratios. They have the structure of musical chords that have been drained of their harmony-ensuring relationship to an anchoring root, or of organisms that have relinquished all reference to a unifying life force, and Jakobson will therefore speak of "bundles" instead of bodies when referring to the jointure of the whole.[27] The wave-shaped flow of sound, blood, will, and time gives way, here, to a *telos* that precedes all liquid spirit, as the quantifiable condition of the latter's mysterious, meaning-bestowing energy.

ZERO DEGREE RHYME

Or rather, flow *would* give way if not for the conundrum of the vowel-consonant relation, which threatens to destabilize the differential purity of the system. Plato's notion of vowels and consonants as qualitatively different *kinds* of letters, and of vowels as the fluid "bond" underlying the disjunctures of consonantal articulation ("More than the other letters the vowels run through all of them like a bond, linking them together, so that without a vowel no one of the others can fit with another"[28]), reappears in every later linguistic paradigm, from Aristotle onward, as the foundational dialectic of language matter and language form. Sound must first exist, it would seem, in order to be articulated, and it must first be articulated, in turn, in order to actualize its inherently linguistic potential. The vowel *a*, with its unobstructed, continuous "fullness of sound" (Sievers), is thus almost universally assumed to represent the origin of the alphabet but also, and for precisely this reason, the letter most likely to regress, when consonantally unbounded, into the semantic indeterminacy of a pre-linguistic cry.

Taken by itself, of course, the notion of a privileged First Vowel poses little danger to phonology, since the unique content of the *a* can be remainderlessly evacuated into a binary system of vocalic relations like the one Jakobson offers as an example in his lectures "On the Structure of the Phoneme." Where *a* is defined solely by the various oppositions and correlations in which it participates, and so ultimately by its relational position vis-à-vis *o, ö, e, u, w, ü,* and *i*, the sounding fullness of a fully open larynx loses all claim to positive priority. Not so, however, the content of the vowels more generally, for while *a* may prove integrable with respect to the other vowels, the vowels as a class continue to differ from consonants according to criteria that resist differential subsumption. The "feature" that distinguishes the two categories cannot be written as a correlation of empty, oppositional relations—according to the analogical formula "$vowel_1$: $consonant_1$ = $vowel_2$: $consonant_2$ = $vowel_3$: $consonant_3$, etc."—for the deceptively simple reason that vowels and consonants cannot alternate with one another at the level of minimal pairs. Paradigm sets such as "till," "kill," "pill," "bill" or "pat," "pit," "pet," "pot," in which the variable sound chunks are all either consonantal or vocalic, have no mixed vowel-consonant equivalents, with the result that the linguist can nowhere directly compare, in the mode of binary ratios, the positional structure of the two kinds.

Jakobson's campaign to "dry up" the sound-spirit continuum can only truly succeed, therefore, to the extent that he manages to render this vowel-consonant relationship susceptible to phonological formalization. And it is presumably no accident that one of the most radical in his long line of oblique attempts to do just that unfolds as a confrontation with the Germanic tradition of the *Stabreim*.

The brief essay "On the So-Called Vowel Alliteration in Germanic Verse" (1963) tackles anew the metrical problem of panvocalic alliteration to which Wagner, in 1851, had proposed his influential musical answer. Jakobson's initial question formulates the constitutive terms for his version of the debate, which revolves around the issue of poetic and phonological identity: "[I]f in the so-called vowel alliteration the vowels are merely variables, and even preferably dissimilar, what does remain as invariant?"[29] Where, he wants to know, is the elemental constant ("the phonemic unit"; *VA*, 189) that governs the vocalic variance in "rhymes" such as *alls—eini—odauðlegi*, setting it apart from both the consonantal *Stabreim* and the absence of all alliteration? What differential fact about the sound structure of the Germanic languages—about the relationship of Germanic vowels to Germanic consonants—does the peculiarity of the alliterating vowels presuppose and, thereby, poetically express?

The two prevailing nineteenth-century explanations fail to answer this all-important question. The "sonority theory," of which Wagner's technique of musical modulation and Wundt's philology of poetic melody provide such powerful teleological interpretations, sees in the fullness of vocalic sound a common denominator universal enough to justify treating all vowels as variants of a single letter. This hypothesis offends Jakobson's sense of symmetry, since nowhere else does the alliterating poet "rhyme" disparate letters on the basis of a single, shared feature (*VA*, 192–93). The real problem, however, lies less in the inconsistency of the proposed rules than in the sheer phonological impossibility of the proposed, panvocalic correlation. The notion that all vowels are somehow the "same" for the poet in way that they could never be for ordinary speakers of a language—the vowel-consonant distinction differentiates no meanings at the level of the minimal pair, and only meaning-differentiating features have acoustic reality for the speaker—threatens to turn poetry into a frivolous preoccupation with linguistically nonexistent entities. The alternative theory, which looks for the *Stab* in an unwritten consonant known as the *Knacklaut*, or glottal stop, at the head of all apparently vocalic syllables—the view held, for instance, by Eduard Sievers prior to his Wundtian-Wagnerian "conversion"—manages to avoid the attribution of a common vocalic essence, and is in this sense phonologically preferable. It errs, however, by assuming the existence of a sound for which no independent evidence exists, since while modern German makes widespread use of the glottal stop, modern Icelandic, with its living tradition of alliterative poetry, does not. There is therefore no reason to suppose that the German of the *Stabreim* poets necessarily contained such a consonant (*VA*, 193).

Jakobson introduces his own solution by providing a partial phonological analysis of Icelandic, the only living language for which the *Stabreim* still remains a viable compositional mode. Having identified several pairs of phonemes (/k/:/g/, /p/:/b/, /f/:/v/, /t/:/d/, /s/:/þ/) where the determining phonetic opposition is the presence or absence of vocal cord *tension*,[30] he goes on to note that all the participants of these pairs can be rearranged into other pairings as well, in accordance with other oppositions (the phonemes /f/, /v/, /s/, and /þ/, for instance, can be distinguished from the phonemes /p/, /b/, /t/, and /d/ by the presence of sound *continuity*, since, unlike those in the second group, they at no point completely block air flow[31]). What this means is that all the phonemes listed above possess more than one distinguishing characteristic within the Icelandic system, whereby "possession" can of course imply *either* the presence or absence of a particular feature. So long, for instance, as the tensionless /b/ occurs in a position where the tensed /p/ *would have been a possibility*—so long, in other words, as /b/ can alternate with /p/ at the level of the minimal pair—its lack of tension will remain every bit as capable of differentiating meaning, and thus every bit as meaningful, as the presence of tension in /p/.

Only one opposition, in Icelandic but also in the Germanic languages more generally, involves phonemes that do *not* possess multiple features. The *spiritus asper*, defined exclusively by the single characteristic of vocal cord tension, and phonologically symbolized as /h/, has for its binary counterpart a phoneme distinguished from all others by its lack of *any distinguishing quality whatsoever.* Jakobson calls the /h/ and its counterpart "glides" because they involve none of the air stoppage normally associated with consonants: "In motor terms, we may say that while in the tense stops aspiration is only one of the features, the tense glide has aspiration alone (and correspondingly a heightened *Durchschnittsluftvolumenverbrauch* [. . .]) whereas the lax glide carries no mark whatever" (*VA*, 190). The *spiritus asper*, then, in line with its literal Latin meaning of "rough wind," finds phonetic realization in the intensity of uninterrupted breath. The lax glide, however, which exists for Greek writing in the diacritic of the *spiritus lenis*, or "smooth breathing," occurs linguistically only as an *absence* of all occurrence. For this reason, it receives from Jakobson the designation "zero phoneme" (*VA*, 190), and is as such neither necessarily phonetically silent—modern German speakers, for instance, realize it as a glottal stop—nor ever, even when silent, inaudible: "The phonemic status continues unchanged, whatever glottal activity has been used to introduce the vowel; thus both the smooth onset of Modern Icelandic or English and the German *fester Ansatz* present the unmarked lax glide in opposition to the marked tense glide symbolized by /h/, or, in the spelling of Ancient Greek, by *spiritus asper*" (*VA*, 194). In languages for which /h/ plays a phonological role, the lack of aspiration remains, in all its possible phonetic manifestations, "perfectly audible" as *non-*/h/ (*VA*, 194).

	t	d	s	þ	p	b	f	v	k	g	h	#
Consonantal	+	+	+	+	+	+	+	+	+	+	−	−
Tense	+	−	+	−	+	−	+	−	+	−	+	−
Compact	−	−	−	−	−	−	−	−	+	+		
Grave	−	−	−	−	+	+	+	+				
Continuant	−	−	+	+	−	−	+	+				

Figure 15. Roman Jakobson, "On the So-Called Vowel Alliteration in Germanic Verse," in Jakobson, *Selected Writings*, 5:191.

Though Jakobson only hints at the larger significance of this phenomenon in the context of his work on the *Stabreim*, there can be no question that the Icelandic glides possess a unique phonological status, as represented by the seemingly paradoxical composition of the chart seen in figure 15. The analogical correlations uncovered by Jakobson's analysis of Icelandic provide definitions for the various distinctive features that organize the class of phonemes normally referred to as "consonants." In the chart in figure 15, the "+" and "−" signs indicate the presence or absence, respectively, of the feature in question, while the lack of either sign indicates nonparticipation. Within this consonantal class, every phoneme participates in at least two of the listed oppositions except for the phoneme pair h/zero (the zero phoneme is represented by the symbol "#"). Thus the phonemes /h/ and /#/ differ from one another according to the opposition tense/lax, but they also differ from all other phonemes in the class by virtue of *not* differing in any *other* ways. And it is this distinguishing *in*difference that gets expressed in the very first horizontal line of the table, where the two glides receive "−" signs with respect to the somewhat enigmatic feature of *being consonants*. The loaded label "consonantal," which Jakobson nowhere explicitly defines, turns out to refer, here, to a metafeature with the phonological content "participates in more than one of the oppositions below."

Like all phonological features, this opposition has a phonetic corollary: the presence of multiple features, within the context of the system charted above, registers acoustically as an interruption of air flow, since compactness, gravity, and continuity are all various ways of modulating the closure of the vocal tract. Only tension can be achieved without closure, which means that only the glides, alone among phonemes that alternate with what have traditionally been called consonants, can be heard as "nonconsonantal." Yet the question of the precise phonetic relationship between tension without closure, tension with closure, and the sounding substrate they both presumably serve to delimit—which is to say, the question of the phonetic relationship that obtains among consonants, glides, and vowels—does not and need not arise within the context of the chart. The anomalous presence of the glides at the boundary of a system in which they participate precisely *by not participating* provides Jakobson, instead, with a wholly negative way of characterizing the otherwise positive experience of articulatory noise. The phonetics of sound stoppage, long considered indispensable for the differentiation of consonants from vowels, disappears here into the emptiness of a second-order phonological opposition, whose definition depends not at all on the notion of a sonorous substrate to be stopped. To count as a consonant in the context of modern Icelandic means simply to differ in a formalizable, binary fashion from a more elemental mode of formalizable, binary differing—and in doing so, to sound *other* than a glide ("Besides consonants, i.e., phonemes with consonantal and without vocalic feature, and vowels, i.e., phonemes with vocalic and without consonantal feature, in Modern Icelandic there are [. . .] glides—nonvocalic and nonconsonantal one-feature phonemes"; *VA*, 190).

The glides, then, turn out to constitute a kind of vanishing point for the category of consonance, an absence that delimits from within rather than without, such that the border need share no substance with the vocalic alternative it nevertheless manages to hold at bay. It would be hard to imagine a more thorough inversion of the dynamic traditionally associated with the phenomenon of aspiration among Germanic proponents of an enspirited language. When, for instance, theologically inclined thinkers such as Jakob Boehme and Johann Georg Hamann argue for the priority of the letter *h* as the representative of God's animating breath, or *Hauch*—and thus also as the orthographic manifestation of linguistic spirit par excellence—they do so under the implicit assumption that articulation originates in the *intensificatory* gesture of adding *spiritus*, or air.[32] The first syllable emerges, for them, as a primordial union between the form-giving energy of breath and the form-taking matter of sound ("and the wind of God hovered over the face of the waters"), whereby the act of giving form coincides, both literally and figuratively, with an event of exhalatory emphasis: the protovowel of the prelinguistic, open-mouthed cry requires the punctuation of an initial gust in order to qualify for the status of true word.

For Jakobson, on the other hand, the privilege of an aspiration both *asper* and *lenis* lies not in its relationship to some mysterious, animating spirit but in its radical lack thereof, which is to say, in the proximity of this most minimally featured of oppositions to the foundational *emptiness* of language. Non-/h/, as the nothingness that first renders meaningful the articulating breath of the /h/—and thus also, one level higher up in the hierarchy, the articulating breath of speech per se—possesses a special, boundary-defining status for the phonological systems within which it occurs, because it marks the meaningless, soundless limit in reference to which all language unfolds. And the *Stabreim* poets, according to Jakobson, bear witness to this all-important placeholding function not by indiscriminately alliterating their sonorous vowels or disingenuously presupposing a fictitious glottal stop but by rhyming the empty spaces where a nonzero phoneme *could have been*. Such a poetics of potential expresses better than any purely consonantal play the phonological essence of what Jakobson here calls "the basic, universal model of the syllable" (*VA*, 192)—precisely insofar as it expresses *nothing substantive whatsoever* about the relationship between consonants and vowels: "Let us add that a more appropriate and realistic label for the alleged *vowel alliteration* would be *lax glide* (or *zero glide*) alliteration or simply *zero alliteration*" (*VA*, 196).

THE SILENT *E*

In the late 1940s, in the context of his work on the phonological system of French, Jakobson appears to suggest that the dynamic of the zero phoneme offers a potentially generalizable model for delineating the boundary between system and environment, language and nonsense, meaning and noise. The article "Notes on the French Phonemic Pattern" (1949), which he cowrote with the Hungarian linguist János (John) Lotz, has for this reason drawn considerable attention from French structuralist and poststructuralist thinkers such as Claude Lévi-Strauss, Jacques Derrida, and Gilles Deleuze, all three of whom see in the paragraph cited below the kernel of a provocative new theory of structure's relationship to its foundations:

> A ZERO-PHONEME, which may be symbolized by ə, or, in an analytical transcription, by #, is opposed to all other French phonemes by the absence both of distinctive features and of a constant sound characteristic. On the other hand, the zero-phoneme ə is opposed to the

absence of any phoneme whatsoever. In the initial prevocalic position this phoneme is known under the name "h aspiré"; although under emphasis it can be implemented as an aspiration, usually it is a lack of sound, which acts in the given sequence as do the French consonants. The vocalic variant of the zero-phoneme, which appears in other positions, is called "e caduc" and alternates between the presence and absence of a vowel.[33]

The famous French schwa (/ə/), also called *e muet* ("mute e") or *e caduc* ("dropped e"), presents a challenge to phonological analysis because of its extraordinary phonetic instability. French speakers may opt in many contexts not to enunciate it at all, as for instance in the phrase *je t'aime*, which is often pronounced /ʒtɛm/ rather than /ʒətɛm/; or they may enunciate it in the form of a generic, minimally articulated vocalic noise ("central, more or less rounded, lax"[34]), which is "both quantitatively and qualitatively indefinite."[35] In the passage above, Jakobson proposes to solve the puzzle of a phoneme that participates in no discernible oppositions by interpreting it as the vocalic corollary of the equally protean *h aspiré*. Since both the schwa and the aspirated *h* can often simply vanish, in French, without affecting the meaning of the utterance, and since the latter can only appear where the former cannot, namely, in the word-initial position, it makes sense, he thinks, to view them as two sides of the same zero-valued coin. Their sole phonological function would then reside in the fact that their nondistinctiveness distinguishes them from all other phonemes within the French system, on the one hand, while their potential pronounceability stands opposed to the absence of all phonemes, on the other.

For Claude Lévi-Strauss, whose presentation in 1950 profoundly influenced the later Derridean and Deleuzean treatments of the same material, the anomalous zero-degree phoneme of the "Notes" represents Jakobson's implicit acknowledgment of an unavoidable compensatory mechanism at the foundation of the language system. Such a "floating signifier" would be distinguished from the absence of all signification by its participation within a signifying order, and from all other signifiers by its lack of any particular, significatory content. Its function would thus be comparable to that of an algebraic variable, which allows the system of mathematical equations to accommodate new material, and thus to remain open to the undomesticated dynamism of empirical experience, without sacrificing the principle of structural closure. Lévi-Strauss sees a primitive instantiation of this compensatory vacancy in the Polynesian concept of *mana*, which he interprets as the ubiquitous but amorphous "stuff" or "energy" that keeps the system of magical correspondence supple:

> I believe that notions of the *mana* type, however diverse they may be, and viewed in terms of their most general function (which, as we have seen, has not vanished from our mentality and our form of society) represent nothing more or less than that *floating signifier* which is the disability of all finite thought (but also the surety of all art, all poetry, every mythic and aesthetic invention), even though scientific knowledge is capable, if not of staunching it, at least of controlling it partially. [. . .] That explains the apparently insoluble antinomies attaching to the notion of *mana*, which struck ethnographers so forcibly, and on which Mauss shed light: force and action; quality and state; substantive, adjective and verb all at once; abstract and concrete; omnipresent and localized. And, indeed, *mana* is all those things together; but is that not precisely because it is none of those things, but a simple form, or to be more accurate, a symbol in its pure state, therefore liable to take on any symbolic content whatever? In the system of symbols which makes up any cosmology, it would just be a *zero symbolic value*, that is, a sign marking the necessity of a supplementary symbolic content over and above that which the signified already contains, which can be any value at all, provided it is still part of the available reserve, and is not already, as the phonologists say, a term in a set.[36]

This interpretation of the linguistic boundary as an infinitely flexible catch-all capable of mediating between the *peras* and the *apeiron*, between the delimitations of finite structure and the chaotic profusion of an uncountable continuum—an interpretation that reappears in Derrida's idea of the supplement, as well as in Deleuze's idea of the empty square, or *spatium*[37]—places Jakobson's zero in the position of systemic substrate. From a French poststructuralist perspective, the unactualized potential for signification, once accepted into the sanctum of an otherwise fully articulated system, disrupts the conceptual-acoustic arrangement just enough to make room for the otherwise static joints to *move*. The zero thus becomes comprehensible as the wellspring of a system-binding, closure-thwarting flow, the source of a *hyle* that destabilizes and in doing so redynamizes all structure under the name of a Derridean "freeplay" or a Deleuzean "circulation."[38]

The problem, of course, from the perspective of a reading of Jakobson—and so also from the perspective of a specifically phonological contribution to the modern understanding of system—is that such a poststructuralist interpretation of the zero flouts the most fundamental requirement of the phonological method. The two-pronged opposition "zero phoneme vs. no phoneme" and "zero phoneme vs. all phonemes" has none of the legitimacy of the similar-sounding opposition that governs the Icelandic glides, since the distinction between glides and all consonants rests on the distinction between one glide and another, while the distinction between the schwa and all phonemes does not. The binary structure of a +/– relationship, with respect to the feature of tension, can be compared one by one with other binary structures, and the result is a second-order +/– relationship with respect to the "feature" of possessing multiple features. The zero phoneme of Icelandic can therefore be analyzed, just like any other phoneme, into distinctive featural components—one "–" sign for "tense," another for "consonantal"—which find their place within the table of Icelandic correlations reproduced above.

The French zero as defined in the "Notes," however, allows for no such binary analysis: neither "all" nor "none" are entities to which the *h aspiré* or the schwa could be said to stand phonologically opposed, for the simple reason that neither "all" nor "none" qualify as true phonological categories. No rigorous procedure of binary comparison yields an analogical correspondence capacious enough to define the expression "all phonemes," and in the absence of such a procedure both the "all" and its opposite necessarily remain profoundly *substantive* concepts, which is to say, concepts whose meanings resist remainderless translation into the terms of the system they ostensibly undergird. This resistance finds adequate, if implicit, expression in the symbolic representations provided by the "Notes" themselves, all of which fail in various ways to formalize the French zeros. Jakobson's tree-shaped diagrams of French vowel and consonant relations, for instance, simply omit any reference at all to the troublesome boundary phonemes (see figure 16). And the phonological transcription of a sample French sentence, with which the analysis of the "Notes" is intended to culminate, decomposes each consecutive phoneme into a distinctive featural bundle of "+" and "–" signs but leaves the schwa, alone among letters, unanalyzed (see figure 17). The nonbinary symbol "#," assigned to represent the *e muet* of *maître*, *permettre*, and so forth, lingers on undissected in the midst of an otherwise wholly binary rewriting of the article's opening lines, calling into question both the purity of the proposed notation and the coherence of the system concept therein quite literally spelled out.

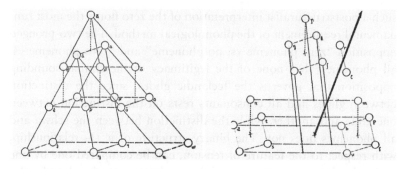

Figure 16. Roman Jakobson, "Notes on the French Phonemic Pattern," in Jakobson, *Selected Writings*, 1:433. The different kinds of lines represent different kinds of binary oppositional relationships, as indicated by the following key: Nasal/Oral –––––––––––; Saturated/Diluted –––––––––; Grave/Acute – – – – – – –; Tense/Lax -·-·-·-·-· ; Continuous/Intercepted · · · · · · · · · .

Jakobson returns to the seemingly intractable problem of the French zero exactly thirty years later, in his final, full-length work, *The Sound Shape of Language* (1979), in order to rectify his earlier failure of phonological formalization with an account that definitively, if retroactively, invalidates the Lévi-Straussian, Derridean, and Deleuzean interpretations of his "Notes." The pertinent passage appears under the section heading "Glides":

> Both *h aspiré* and *e muet* may be viewed as contextual variants of one and the same phoneme, a latent glide in opposition with a compulsory zero: *il hait* with an optional glottal catch vs. *il est* without a glottal catch, or *ferais* with an optional schwa vs. *frais* with no schwa permitted. [. . .] If one were right in interpreting the difference between the opposition of a potentially omissible *h aspiré* or *e muet*, on the one hand, and a compulsory zero, on the other hand, as the difference between greater strength and absence of effort, then the opposition in question could be identified with the tense ~ lax opposition which functions both in the vocalic and in the consonantal subsystems of French.[39]

Here, as in the "Notes," *h aspiré* and *e muet* appear as the two faces of a single, idiosyncratic phoneme, which can operate *either* as consonant *or* as vowel depending upon its position in the acoustic string. Crucially, however, in this later account, the phoneme in question no longer "alternates," paradoxically and illegitimately, with the

Thus we transcribe our epigraph:

		ʃ ɛ r	m ê t r ə	v u l ê	v u	n u
1.	Vocality vs. Consonantness . .	−+±	−+−±#	−+±+	−+	−+
2.	Nasality vs. Orality	−−	+−−	−− −	−−	+−
3.	Saturation vs. Diluteness	+±	−±−	−− ±	−−	−−
4.	Gravity vs. Acuteness	−	+−−	++ −	++	−+
5.	Tenseness vs. Laxness	+	++	− + −		
6.	Continuousness vs. Interception	+ −	−−	+ +	+	

	p ɛ r m e t r ə	d ə	v u	p r ɛ z Ã t ê	n o z-	o m a ʒ ə	ê	n o
1.	−+±−+−±#	−#	−+	−±+−+−+	−+−	+−+−#	+	−+
2.	−− +−−	−	−− −	−−+−−	+−−	−+−−	−	−+
3.	−± −±−	−	−− −	±−+−±	−±−	±−++	±	−±
4.	+− +−−	−	++ +	−− −−	−+−	++	−	−+
5.	+ −+	−	− +	− ++	−	−−	+	
6.	− − −−	−	+	−− + −	+.	+		

	m ɛ ɪ ø r	v ø	d ə	s Ã t ê	d ə	p ʌ r f e	b o n ø r	ê	d ə
1.	−++++±	−+	−#	−+−+	−#	−+±−+	−+−+±	+	−#
2.	+−−−	−− −	−+−−	−	−− −−	−−+−	− −		
3.	−±−±	−± −	−+−±	−	−+ −±	−±±±	± −		
4.	+−−±	+± −	− −−	−	+ +−	++−±	− −		
5.	− −	+ ++	−	+ +−	−−	+ −			
6.	− +	−	+ −	−	− −±	−	− −		

	t r Ã k ɪ l ɪ t ê	d - ä m ə
1.	−±+−+±+−+	− +−#
2.	− +−− −−−	− −+
3.	− ++− −−±	− +−
4.	− − −−−	− +
5.	+ + ++	− +
6.	−− − + −	−

Figure 17. Roman Jakobson, "Notes on the French Phonemic Pattern," in Jakobson, *Selected Writings*, 1:434.

pseudocategories of all language and none but rather with a well-defined binary counterpart that takes the shape of what Jakobson terms a "compulsory zero." The word *ferais* thus forms a perfectly unobjectionable minimal pair with the word *frais*, and the phrase *il hait* with the phrase *il est*, despite the fact that the members of each pair *can* be pronounced in a phonetically identical fashion. Since the French speaker *in principle* has the option of articulating a semantically relevant, phonological difference with respect to these two words—by pronouncing the optional schwa of *ferais*, in opposition to the compulsory zero of *frais* —such a difference must *in principle* still exist even where its potential for phonetic realization remains unactualized. And since this difference possesses the familiar oppositional structure of presence/absence, feature/no feature, tension/laxness, the French zero can now be unproblematically subsumed into the table of French phonological correlations. The hieroglyphic cipher "#" dissolves without remainder into the deeper, more foundational writing of the binary "+" and "–" signs—into the system of purely differential marks or cuts—without which no (other) letters could come into being.

Jakobson's late return to the question of the French zero makes clear, then, as his earlier analysis of the Germanic glides could not, that the dynamic of minimal differentiation operates simultaneously on *both* sides of the consonant-vowel divide. The opposition "optional schwa/compulsory zero" does for the vowels exactly what the opposition "h/zero" in Germanic (and the opposition "optional h/compulsory zero" in French) does for the consonants—with the result that the boundary between these two incommensurable categories becomes fixed internally *from either end*. Indirectly and oppositionally, therefore, the zero phonemes do indeed provide a means of approaching the problem of language foundations. As the minimally featured limit cases against which the twin notions of "all vowels" and "all consonants" first acquire phonological contour, the glides mark the contentless place where phonological analysis necessarily begins and ends. They do so, however, according to a logic quite different from the one traditionally attributed to systemic origins, since they precisely do *not* mark the place where inside and outside interpenetrate (regardless of whether this interpenetration is conceived nature-philosophically, as a fusion of structure and substrate, or poststructurally, as a paradoxical differing of structure from itself).

The article "Zero Sign" ("Signe Zéro," 1939), written in the same year as the "On the Structure of the Phoneme" lectures, represents Jakobson's earliest and most explicit attempt to theorize the relationship between the particular dynamic of minimal differentiation, as exemplified by single-featured phonological oppositions of the form "glide/nonglide," and the differential dynamic governing language in general:

> The organization of the grammatical system, as I have tried to show elsewhere, is based precisely on the "opposition of something to nothing," that is, on the *opposition of contradictories*, according to the terminology of formal logic. Thus, the nominal system and the verbal system can be decomposed into binary oppositions, where one of the terms of the opposition signified the presence of a certain quality and

the other (the uncharacterized or unmarked term, in brief, the zero term) *indicates neither its presence nor its absence.*[40]

The passage begins by announcing the most maximal of oppositions—between something and nothing, presence and absence, being and nonbeing, one and zero—as the foundation of grammatical structure. It ends, however, by emptying this foundational antagonism of all conceivable conflictual intensity. The clash of cosmos and chaos reemerges here, deenergized, as a purely formal disjunction between that which is marked and that which is not, whereby lack of mark turns out to imply *not* an absence of writing, delimitation, or order per se, but rather *an absence of decidability* relative to the existence of any defining feature. Such a feature might be present, and it also might not. So long as these two possibilities can peacefully coexist, the linguistic phenomenon in question will qualify as the logical contrary of an unambiguously marked counterpart, whether or not this binary finds expression in the force of a real, which is to say actualized, opposition.[41]

Jakobson's later work on the zero introduces the term "latent" as a synonym for "unmarked," and in doing so makes clear that the duality of absence/presence, nothing/something, nonbeing/being, zero/one corresponds for him, linguistically speaking, to the equally ancient dualities of *potentio/actio, dynamis/entelechia,* and *hyle/morphe.*[42] Investigations into the nature of the linguistic zero are thus always also investigations into the conditions of possibility of language, conceived as a domain of latent rather than actual marks. But the phonological interpretation of this latent domain definitively uncouples the notion of systemic potential from the notion of a formless, system-transcending matrix (*apeiron*), out of which formal divisions (*peras*) could somehow triumphantly and spontaneously emerge, like the biblical creation from chaos ("In the beginning the earth was unformed and void [. . .] and God divided the waters"). With the help of a seemingly minor modification in phrasing—from "unmarked" understood as the expression of a lack of mark to "unmarked" understood as the lack of an expression of mark—Jakobson drains the undifferentiated Nothing of its power to generate, destabilize, or otherwise exceed the orderly Something of structure. A single mark, sans substrate or "stuff" of incision, draws a distinction between itself and the rest without imputing to this "rest" any content whatsoever, *not even the content of being other than itself.* The ostensible paradox of a nonbeing that somehow *is,* to the extent that it could be presumed to differ *substantially* from being, gives way, here, to a difference that precedes both being and nonbeing, both substance and absence, in virtue of the very purity with which the notion of difference gets rethought. In the beginning, for Jakobson, only the distinction exists. Everything else gets bracketed, even the question of whether there actually *is* an "everything else" to bracket, such that the bracket itself, now empty, becomes the formal boundary against which the mark of difference can be sufficiently—and system-internally—defined.[43]

"Zero Sign" makes the case for the special relevance of the latency concept at every level of the language hierarchy, from morphological to lexical to syntactic to stylistic. The purest manifestation of the zero, however, remains phonological, since only at this most elemental of levels can the lack of a mark actually imply the lack of *any and all* demarcation.

> A correlation is formed by a series of pairs, where each pair contains, on the one hand, the opposition between one and the same quality and its absence and, on the other, a common substrate (for example the pair z'–z consists of an opposition between softening and its absence, and of a common substrate: sibilant, constrictive, voiced). But this common substrate could be absent from one of the pairs: in this case, the phoneme is reduced to the quality in question and is opposed quite simply to the absence of a phoneme (or to the *zero phoneme*). Thus, A. Martinet rightly insists, by virtue of structural analysis, on the fact that, in the correlation of aspiration which characterizes the consonants of Danish, one must recognize the opposition: initial aspirate /h/–initial vowel.[44]

Whereas the morphological zero, for instance, can only ever be unmarked at a level that supervenes on the "common substrate" of the phonologically marked, the phonological zero makes explicit the possibility of a substrate that disappears altogether. Jakobson's career-long interest in glides springs directly from the fact that, as single-featured phonemes, they necessarily instantiate this dynamic. The Nordic-Germanic /h/ and the French /h/-schwa share nothing in common with the various forms of nothingness to which they respectively stand opposed, and the correlations in which they participate have, correspondingly, no common measure.[45]

Only the French glides as reinterpreted in the late *Sound Shape*, however, provide an example of phonological latency radical enough to unfold the full potential of the theory first sketched in "Zero Sign." Unlike the Germanic-Nordic /h/, which alternates with an equally stable lack of /h/, the compulsory French zero alternates with an /h/-schwa that *may or may not* appear. It is therefore the French *lack* of a phoneme that deserves the title of mark, despite registering acoustically only as an absence, while the optional /h/-schwa functions as the unmarked, zero term. The phonological zero, in other words, is not always a nothing, and the nothing is not always a zero. Both *e muet* and *h aspiré* can exist or not, as the speaker chooses, and in this sense they are less than nothing, since to be nothing all the time, everywhere, is just another way of being something—namely, nothing—whereas to be nothing only sometimes, without discernible cause or consequence, is to elude the category of thinghood entirely. An entity that possesses no identifying features, including the feature of possessing no features, cannot fall under a conventional concept. And yet, for all its featureless indeterminacy, the French zero does not escape the grasp of phonological writing, as Jakobson's French structuralist and poststructuralist contemporaries contend, which means, in turn—and this conclusion is for Jakobson the real lesson of the French glides—that *the reach of phonological writing exceeds the reach of conventional conceptual thinking.*

More specifically, the reach of the phonological system exceeds the reach of its own conceptual definition. In "On the Structure of the Phoneme," Jakobson defines the phoneme as a purely differential sign. The purpose of a phoneme is to distinguish semantic units, and its conceptual "essence" is therefore oppositional in nature ("The only linguistically, i.e., semantically relevant content of a phoneme is its opposition to all other phonemes within a given system"; *SP*, 293). The French zero, however, need not distinguish anything in order to operate (the semantic unit *frais* pronounced with a compulsory zero sounds exactly the same as the semantic unit *ferais* pronounced without the optional schwa). It thus represents the absolute zero-point of phonological function, which is to say also the unmarked counterpart to the marked condition of being a phoneme, the contentless boundary of the phonological system, and—perversely but etymologically—the *telos* (Greek *telos* = end, border, extremity) of linguistic teleology. The notion of a system that includes the emptying of its own formative principle as its form-bestowing *finis* or limit, and in the process succeeds in defining or delimiting itself from within, casts an entirely new light on the poet Paul Valéry's famous characterization of the schwa, which Jakobson cites, in passing, during the discussion of French glides in *Sound Shape*: "The silent e [*L'E muet*], which sometimes exists and sometimes almost seems to erase itself entirely [*tantôt ne se fait presque sentir qu'il ne s'efface entièrement*], which produces so many subtle effects of elementary silences, which terminates and prolongs so many words by a kind of shadow."[46] Like the aspirated *h*, whose very name, in French, appears to reference the exhalations of a language-conditioning breath, or *spiritus*, mute *e* can be called upon to play the part of a letter ensouled. Its peculiarly protean (lack of) character can be understood to correspond to the supple fluidity of an underlying substrate, concealed yet omnipresent beneath the jointed structure of conventional articulation. Jakobson, of course, does not embrace this interpretation. But he does not refuse it either. Instead, he deflates it, in a gesture that precisely parallels the one he imputes to the form-bestowing fringes of language itself. The notion of language spirit, when viewed from a properly phonological perspective, reappears at the margins of linguistic formalization as a "kind of shadow" rendered indispensable by its very dispensability, a necessary yet superfluous ghost standing differentially opposed to the differential writing of function.

MAMA AND PAPA

Jakobson's insight into the system-founding power of minimally featured phonological oppositions has a curious antecedent in one of Saussure's earliest intuitions about the chronological origins of language sound. The latter's *Memoire on the Primitive Vowel System of Indoeuropean Languages* (*Mémoire sur le système primitif des voyelles dans les langues indo-européennes*, 1878), written when its author was just a twenty-one-year-old Leipzig student, dares to introduce into the alphabet of the reconstructed Indo-European proto-language a set of philologically unattested, phonetically indeterminate glides as a way of accounting for certain enigmatic etymological transformations in the attested development of modern Indo-European languages. Termed "sonic coefficients," or resonants, to distinguish them from both consonants and vowels, these hybrid, hypothetical letters found almost no proponents among Saussure's more conservative contemporaries, who continued to insist on a primal alphabet of unambiguous vowel-consonant distinctions, until the early twentieth-century discovery and decipherment of ancient Hittite unearthed evidence of their historical reality. Alone among known Indo-European descendants, Hittite turned out to possess written symbols for precisely the kind of barely articulated, vowel-lengthening half-consonants that Saussure had so presciently posited at the origin of all Indo-European languages everywhere.[47]

The idea of featureless phoneme categories that could be structurally rather than contingently prior, such that the necessity of their historical existence becomes deducible even in the absence of any philological confirmation, clearly has far-reaching consequences for the science of language systems in time. Saussure himself, however, did not go on to draw any wider implications from the transempirical "firstness" of his etymologically reconstructed zeros, and it therefore fell to Jakobson to formally explore the relationship between phonology and chronology, latency and beginning, linguistic zeros and language origins.[48] The central conundrum can be summarized as follows: On the one hand, the zero is clearly no origin in any conventional, chronological sense, such that, for instance, it would be appropriate to speak of zero phonemes as of historically earliest letters. Since all linguistic zeros require the prior presence of a system in order to meaningfully exist, they cannot be thought to represent a primal source that precedes (other) language, the way that potentiality has traditionally been presumed to precede realization. On the other hand, there is just as clearly some sense in which language *begins* with the marking of the entirely unmarked, since without this most basic of single-featured oppositions, no other differential relations are thinkable.

Jakobson tackles the question of what the proper sense of beginning might be in a brief monograph titled *Child Language, Aphasia, and Universal Sound Laws* (*Kindersprache, Aphasie, und Allgemeine Lautgesetze*, 1941), which appeared on the heels of his late 1930s writings about distinctive features and zero signs (an early draft of the material, presented in 1939 at the Fifth International Congress of Linguists, dates from exactly the same period as the lecture "On the Structure of the Phoneme" and the article "Zero Sign"[49]). The study analyzes data pertaining to language emergence among infants, on the premise that a better understanding of ontogenetic beginnings— the only level at which a language *in statu nascendi* offers itself to empirical observation—can shed light on the temporality of phonological foundations more generally. Written in German from the relative safety of Jakobson's Scandinavian exile, the book begins and ends as a not unpolitical plaidoyer for the concept of language universals, understood here as the structural constants that tie the languages of the world back to their common conditions of possibility. The first epigraph, from Edmund Husserl, makes the general philosophical point "Everything that truly unifies is a relationship of foundation" (*Alles wahrhaft Einigende sind die Verhältnisse der Fundierung*). And the second, from native Norwegian and fellow phonologist Alf Sommerfelt, to whom the book is also dedicated, contributes the linguistic specification: "There is no difference in principle among the phonetic systems of the world, although there are, naturally, many phonemes with relatively limited distribution."[50] Taken together, the two citations point the way toward a notion of ground more ecumenical than the Indo-European roots of an Ayran-German genealogy, and toward a notion of firstness less dependent on particular points in time. The irreversible, because logically hierarchical, development of phoneme structure, on which every national and dialectical system is founded and which Jakobson here intends to derive, does not occur only once, like Grimm's "historical, stepwise trajectory, which can neither be disrupted nor reversed."[51] Yet without the truly universal sound laws obeyed by every child who learns to speak—according to the contention announced already in the manuscript's mildly polemical title—historical pronunciation would never have undergone the regular shifts that made possible Grimm's formal definition of "German" and the notion of a law-abiding linguistics. Where traditional etymologists seek specific rules governing particular historical trajectories, the etymologizing phonologist, according to Jakobson, must therefore dive deeper, in search of the metahistorical laws that determine *all* historical unfolding: "Thus, the founding laws [*Fundierungsgesetze*] turn out to be *panchronic*: they retain their validity in every situation and through every transformation of all the various languages of the world" (*CL*, 59/367).

   The idea that such panchronic laws exist at all implies a fundamentally teleological approach to the history of language, since a mechanistic logic of accidental collision provides no basis for presuming that any two historical trajectories will necessarily unfold alike, in accordance with a set of identifiable rules. Like Kielmeyer, Schelling, Wagner, and Wundt, in other words, Jakobson sees each stage of language history as the momentary articulation of a much broader, developmental dynamic that governs all phonetic transformations (his word for the synchronic state, in this context, is *Entfaltetsein*,[52] which translates roughly as "the condition of having developed"). The dynamic he intends to locate, however, involves a deflationary emptying rather than an intensificatory upsurging: the phonological

study of first letters, in contradistinction to the psychophysical practice he elsewhere dubs "etymological phonetics,"[53] seeks to underpin
the chronological trajectories of language sound without reference to
an analysis-transcending principle such as language spirit or national
psyche.

The *Child Language* monograph correspondingly concerns itself
less with the question of originary mental "contents," in the sense
of a scientific sound symbolism, than with the essentially contentless
oppositions that establish, for Jakobson, the differential ground of
all rational being: "[I]n a child's mind the pair is anterior to isolated
objects. The binary opposition is a child's first logical operation."[54]
The investigation proceeds on the basis of existing data, which Jakobson culls from numerous psychological and phonetic case studies. In
the beginning, according to this data, there is babble, the so-called
Lallperiode, during which the child spontaneously and unintention-
ally produces all sounds that any documented language anywhere has
to offer, without assigning to them any particular function.[55] Eventually, however, this "aimless [*ziellos*], egocentric soliloquy" gives
way to a "communicative intention" (*Unterhaltungsintention*; *CL*,
24/337), and desire for dialogue directs the child's attention outward
toward the speakers in its immediate environment. Out of a "still
biologically-anchored 'tongue-delirium'" (ibid.) characterized primarily by a protean fullness of sound, emerge the acoustic and articulatory boundaries that define sound in relation to sense. It is at this
point that language, from the perspective of the phonologist, assumes
its most elementary empirical form, and it is therefore this stage that
attracts Jakobson's interest.

The first boundaries, as Wilhelm Wundt had claimed and others
after him had confirmed, are the ones separating the *p*, *a*, and *m* of
"papa" and "mama": "An *a* appears as the first vowel, and generally
a labial stop [*Verschlusslaut*] as the first consonant, in child language.
The first consonantal opposition to emerge is that of the oral and
nasal sounds (e.g., *papa-mama*)." (*CL*, 48/357). According to Jakobson, however, no psychophysical principle of expressive vibration or
gestural mimesis can be called upon to explain the fixed and universal order in which these proto-phonemes take shape for the child's
active vocabulary: as elements of a differentially organized language,
they require a differential interpretation to account for their emergence.[56] Such an interpretation, he argues, can be located in a law of
maximal acoustic contrast, which applies the principle of definition

by difference to the question of diachronic development. The phonetic expansion of a language progresses from the highest degree of audible opposition to the lowest, whereby the most fundamental distinctions acquire meaning for all peoples in all times, while the most subtle have the statistically smallest probability of being put to use.[57]

Under such conditions, it makes sense that the very first opposition of all would be the one separating language sound from language silence, as represented by the periodic alternation between fully closed and fully open mouths.[58] Citing August Böckh on "the pure *a*, the root and stem of the vowels" (*CL*, 75n16/380n103), Jakobson thus goes on to align Eduard Sievers's notion of vocalic "sound fullness" (*Schallfülle*) and the synesthetic principle of "the greatest colorfulness" (*die grösste Farbigkeit*; *CL*, 75/379) with the *a* as an "optimum of openness" (*Optimum der Öffnung*; *CL*, 69/375). This optimal openness, in turn, acquires linguistic significance from the contrast with "colorless," voiceless consonants such as *p*, "which obstruct the entire oral cavity" (ibid.). The notion of color plays in both cases a more than merely metaphorical role, for Jakobson devotes considerable attention to similarities in the development of optical and aural perceptions, even going so far as to take seriously the classic Symbolist topos of *audition colorée*.[59] Children who hear colors, he maintains, demonstrate "a clear inclination to link the more colorful vowels with the brighter colours, especially *a* with red." (*CL*, 82/386–87). The black-white row, on the other hand, is above all the province of the consonantal stops ("Incidentally, the relation between vowels and consonants is similar to the relation between the so-called tones or bright colors [*getönten oder bunten Farben*], on the one hand, and the color-tone-less [*farbtonlos*] grey series on the other hand"; *CL*, 75/380), since these distinguish themselves for the ear less by the quality of their particular sounds ("the consonant is generally voiceless in the beginning stage of child language"; *CL*, 70/376)[60] than by their particular modes of sound stoppage. As the most silent but also least colorful of language sounds, *p* represents the logical counterpart to the saturated intensity of the *a*. In the opposition of red and white-black, acoustic flow and articulatory interruption, vocalic substrate and consonantal mark, Jakobson thus discovers—like Grimm and Wagner before him—the most originary realization of a universal "model of the syllable" (*CL*, 71/377).

Within a phonological context, of course, this first syllable alone does not yet count as a word, since form acquires function only through opposition to other forms. *Pa*, in other words, is only really *pa* insofar as it is not *la*, *ka*, *ta,* or *sa*, and real language arises only at the point where one syllable can substitute for another. The opposition between *p* and *a* is thus followed by a second but nonetheless equally originary opposition between *p* and *m*, the latter a labial consonant produced with the supplemental assistance of an open nasal chamber. Phonetically speaking, *m* represents the acoustic *synthesis* of maximal opening and maximal closure, and is as such "a natural consequence of the opposition consonant ~ vowel" (*CL*, 71/377); side by side with *p* and in combination with *a*, it stands, phonologically speaking, at the origin of all language everywhere. The *ma/pa* opposition, as a pair of First Words, is thus simultaneously an empirically attested version of Saussure's imaginary *ba/la* language, the most fundamental expression of Jakobson's new, panchronic sound laws, and a phonological reinterpretation of the nature-philosophical coupling (between vowels and consonants, sound and structure, melody and rhythm, matter and form) from which the language of Bopp, Grimm, Schelling, and Wagner "descends."

This phonological reinterpretation, however, remains crucially insufficient so long as it fails to confront the dilemma of emotionally resonant sound. Within the framework of maximal acoustic contrast, *ma* represents nothing more than the purely differential negation of *pa*; yet the actual acoustic shape of the syllable derives, as Jakobson finds himself forced to admit, from the juxtaposition of two infantile noises, both of which turn out to correspond to instinctive expressions of desire: "The origin [*Quelle*] of these linguistic formations [*ma*] lies in the prelinguistic, affective sound utterances [*Lautäusserungen*]. Two ways of discharging displeasure [*zweierlei Unlustentladungen*] were familiar to the child from the very beginning—the cry [*der Schrei*] with a quasivocalic [*vokalartiger*], *a*-like oral opening, and the nasal murmur" (*CL*, 72/377). The apparent presence of such psychophysical, sound-sense correspondences at the very root of language threatens to undermine all arguments for the originarity of empty, binary opposition: if one pole of the nascent *ma/pa* system emerges out of a will to self-expression and emotional release rather than a dynamic of contentless self-differentiation, then the "matter" of psychic energy can no longer be legitimately excluded from the foundations of the phonological realm.

Jakobson's explicit response to the risk of contamination leaves the precise nature of the relationship between desire and language, derivation and system, unspecified, relying instead on a division of explanatory labor to keep the phonological and psychophysical levels apart. Phonology, he claims, can explain why *ma* appears as the second legitimately linguistic syllable of child language, which is to say, as the empty binary counterpart to *pa*. The science of affect, however, must explain why we say "mama" to the mama.[61] "Nasalization is particularly charged with affect for the child, whereas stoppage in itself [*der Verschluss an sich*] indicates rather an affective weakness, an appeasement [*Beruhigung*]. It is thus easily understandable that the nasal consonant, in opposition to the oral (pure) stop, acts as a carrier of affect at the threshold of child language, i.e., as a complaining, demanding, calling sound of pain, and finally as the name [*Rufname*] used to call those, 'who are above all called [*berufen*] to still [*stillen*, which also means "to nurse"] the affects of hunger and longing: mother and nurse.'" (*CL*, 73/378).[62] The science of affect must therefore also explain—in further conformity with a rhetorical strategy that commandeers the resources of the competition in order to ensure their ultimate exclusion from the domain of language science—why the *p* : *a* opposition of papa remains phonologically *prior* to the *m* : *a* of mama, even though the latter frequently first acquires the paradigmatically linguistic function of *naming*.[63] The child who says "ma-ma" before "pa-pa," according to Jakobson, speaks not a real name but a *Rufname*, or call of desire, an articulate, iterable cry that hovers somewhere between noise and language ("at the *threshold* of child language"), just as the hunger-stilling mama hovers somewhere between other and self. And the true transition from cry to word, the shift that anchors a self-sustaining system precisely by transcending the originary affect of need, must therefore await the boundary-drawing, difference-maintaining, identity-grounding naming of the papa:

> [T]he oral stop, on the other hand, appears on the scene as a formation that carries either less affect or no affect at all; it is not used for wailing but for "pointing, dismissing, refusing" [*Hinweisen, Fortweisen, Zurückweisen*; the repetition of the verb stem -*weisen*, "to show, to point," indicates that the common denominator of these three acts resides in the gesture of directing attention *away* from rather than *toward* the self] and for calmer, more neutral naming [*ruhigeren, gleichgültigeren Benennung*], and it thereby heralds the real transition from emotional expression [*Gefühlsausdruck*] to representational language [*darstellenden Sprache*]. (*CL*, 73/378)

Emotionally, expressively, substantially, and chronologically, then, the mama may very well come first. (Phono)logical priority, however, belongs exclusively to the papa, in all his rigid consonantal purity.

The explanatory bifurcation of the *ma* syllable into prelingual cry and phonological sign—two categorically different modes of expression that come together exceptionally, and without ever "actually" intermingling, at the boundary of child language—might legitimately be said to diffuse the danger of the affect-laden half-labial *m*, which can now be interpreted *either* as sound gesture *or* as phoneme. The quarantine approach accomplishes nothing at all, however, with regard to the equally affect-laden *Ur*-vowel. As one half of the first linguistic opposition and one half of both first words, the *a* plays an even more foundational role than the *m*, and it plays this role not despite but *because* of its radically antiphonological substantiality. The emotional-acoustic content it inherits from the unarticulated "cry with a quasivocalic, *a*-like oral opening" cannot be relegated to phonological irrelevance like the whine of the nasal *m*, since it is not an accidental byproduct of the transition from babble to speech. Rather, the sounding fullness of the *a* contrasts with the silent closure of the *p* as the sole condition of linguistic beginning.

Jakobson confronts this significantly more difficult challenge to the hypothesis of an originary emptiness by quietly performing on his primal vowel a version of the very deflation he holds more generally responsible for the emergence of language from *Lallen*. In doing so, he renders the case of the *a* paradigmatic for a phonological theory of linguistic beginning. Of the transition from babble to speech he writes, "In place of the phonetic abundance [*Fülle*] of babble, there emerges the phonemic barrenness of the first language stages, a kind of *deflation* which transforms the so-called 'wild sounds' of the babbling period into linguistic values" (*CL*, 25/338; emphasis mine). Substance, here represented by sound fullness and elsewhere by sound feeling—the two are inseparable for Jakobson—provides the matrix but not the ground for the boundary-drawing operations of language. It obeys no laws and bestows, of its own accord, no forms ("It is significant," says Jakobson, "that one cannot here ascertain any universal order to the acquisitions"; *CL*, 27/340). The trick, therefore, which both language and phonologist must perform, lies in allowing sound to enter the linguistic system, as the fundament of a vocable language, while simultaneously deactivating, or rather demotivating, its dangerously substantive abundance. And with respect to the particular abundance of the *a*, this maneuver turns out to involve a perversion of the traditional, vocalic *Verdichtung*: the phonological condensation distills *into nothingness* the expressive energy of the

infantile cry, such that the substance of sound becomes the basis for
an originary zero sign.

Jakobson accomplishes this crucial transformation in passing,
on his way to a discussion of the consonantal binary p/m. Having already dismissed the relation between voice and its absence as
a mere "accompanying manifestation" (*Begleiterscheinung*) of the
closed-open distinction (*CL*, 70/376), he pauses a few paragraphs
later to rephrase this quietly revolutionary claim in more formal phonological terms. The distinctive feature that governs the first vowel-
consonant opposition, according to the characteristic phonological
logic of presence and absence, finds its articulatory realization, for
him, in the mechanism of sound interruption rather than sound production. The defining characteristic of the *a* must therefore be sought
not in the positive phenomenon of sonorous fullness but in the negative phenomenon of unrealized closure of the vocal tract, which in
turn means that the consonant, rather than the vowel, will carry the
weight of the phonological mark: "[T]he vowel continues to be clearly
characterized by the absence of a closed cavity [*das Nichtvorhanden-
sein eines Verschlussrohrs*]" (*CL*, 71/377). As the unmarked half of
a single-featured opposition—in a context where one distinctive feature suffices to divide the language sea—the *a* is phonologically featureless and can assume any phonetic shape whatsoever, including,
at least theoretically, the silent shape that marks its counterpart. And
indeed, Jakobson appears to go so far as to suggest, based on the data
of language disappearance among aphasiacs, that total indeterminacy
is actually the *most* originary way of realizing this most originary
of zeros, since "in word-deafness, [. . .] the number of syllables in
a word is often grasped even when the vowels or consonants are no
longer distinguished" (*CL*, 64/372). At the beginning of language and
the fringes of language disturbance, where each "beat" signifies the
presence of a single distinguishing mark, the ability to count syllables
turns out to *precede* the ability to hear any difference between the
sounds of the vowels and the consonants.

Having thus reduced the sonorous fullness of the *a* to a mere latent
variant of the consonantal silence, Jakobson goes on to argue that this
variant does not, in fact, truly deserve the title of phoneme. Belonging
as it does to an entirely different *order* of language from the consonants, the *a* cannot take the place of either *m* or *p*, which means that,
so long as no other vowels exist with which it could alternate, the *a*
also cannot differentiate meaning:

In the child whose linguistic inventory consists of *papa* and *mama*, every phoneme contains a single distinctive feature: *p* = oral, *m* = nasal. As soon as *tata* (and eventually also *nana*) appears, there arises a coincidence of two features within a single phoneme (e.g., *p* = oral + labial). But here, too, the fundamental principle of "one phoneme, one word (or, indeed, one sentence)" still applies, so that one can speak of a *p*-word and an *m*-word (or a *p*-sentence and an *m*-sentence), and so on. The vowel is merely an accompanying manifestation [*Begleiterscheinung*]; at the origins of child language, it is by means of syllable reduplication that the linguistic value of the sound, which is to say, the closure [*Abgeschlossenheit*] of a linguistic unit, is signaled. (*CL*, 85/388)

The primal *a* participates in no minimal pairs, distinguishes no words, closes off no syllables, and as such belongs less to the arsenal of empty binary relation than to the unstructured profusion of emotionally saturated babble: "[T]he only vowel still functions solely as a support-vowel and as a carrier of expressive variations" (*CL*, 79/384). Since, however, the inability to differentiate semantic units is also what characterizes the *a* as an unmarked counterpart of *p*—the latter possesses, as its only defining feature, the ability to interrupt, to cut off, to delimit, to differentiate—the first vowel remains linguistically relevant even in the absence of all phonological function. On the one hand, it violates the differential principle of the emerging language system in virtue of its failure to make a difference. On the other hand, it fails to make a difference according to the profoundly differential logic of Jakobson's latent glides, which delimit by not delimiting and belong by not belonging. The same second-order mode of distinction between distinctiveness and its other, between a system and its limit, that had so fascinated Jakobson from a synchronic perspective in his work on the zero sign assumes here the role of diachronic origin for a language system *in statu nascendi*.

The result, I am arguing, is a new understanding of linguistic beginnings. Language develops, according to the binary logic espoused by a phonological account of origins, insofar as it empties rather than intensifies the acoustic materiality of its preconditions, whereby the "emptying" in question remains every bit as *formative* and *dynamic*—which is to say, every bit as *teleological*—as the intensificatory energy it replaces. In contrast to the more familiar notion of traditional abstraction, whose terms Jakobson frequently adopts in order to polemicize against a purely empiricist, Neogrammarian approach, the phonological model does *not* trace the emergence of language systems back to the exclusion of a recalcitrant, because category-resistant, materiality (*abstrahere* = to draw away). Instead, a confrontation between form and its zero-degree boundary is presumed to give rise to a system that continuously *includes* its own other as the very condition of its ongoing existence. Consonant and cry still establish together, for Jakobson as for his nature-philosophical and psychophysical predecessors, the chronological foundations of an articulated language, in the same way that black and blood-red lay the groundwork for a differentiated field of vision.[64] Where Grimm, Wagner, and Wundt, however, had seen a fertile marriage of stuff and structure, womb and phallus, flux and rigidity, blood and bone—"one can consider the colours white, red, and black the most primary and sensuous," writes Jacob Grimm in the *Old German Forests,* "in part because in them and their mixture the human body appears: white reigns in the skin, nerves, tendons, and bones; red in the blood, black in the hair and in the pupil of the eye"[65]—Jakobson insists counterintuitively on the generative power of a perpetually *un*consummated coupling. His originary union allows for no unity, in the sense of a common ground or measure, since it engenders a language precisely by eschewing all content that could possibly be held "in common" among its constituents. The one-over-many of the phonological system successfully reproduces itself in and over time, like the formative principle of the nature-philosophical organisms it resembles, but it does so by *evacuating* all principles of formation, and with them any tendency toward substantive sameness, from the definition of the phonological "self."

Jakobson has not forgotten, of course, that the dynamic of the *p/a* origin necessarily differs from the dynamic of the glides, since no system yet exists, in the former case, to guarantee the validity of a purely differential interpretation of differing. Both the nineteenth-century organicist and the twentieth-century phonological accounts of language essence possess equal merit at the level of the singular, systemless *pa*, for the simple reason that both rely primarily on claims about what the *pa*, once systematized, will *become*. The question of whether, at the very beginning of language, vocalic sound infuses the consonant or evaporates into its zero degree boundary, is one that can only be answered with reference to the infusive or evaporative character of the results. The results that count, however, are the glides themselves, whose absolutely minimal mode of distinction thus turns out to bear witness to a potential for emptiness that must always already have been inscribed within linguistic origins. Theories that fail to integrate the possibility of the phonological zero, with its undeniably significant function at the margins of (some) language systems, cannot reasonably be presumed, Jakobson believes, to accurately articulate the conditions under which sound acquires meaning for a community of speakers. The deflationary narrative of the *Child Language* monograph, despite the fact that it must remain mere hypothesis, offers therefore, in his view, the only credible interpretation of ontogenetic roots.

IN RETROSPECT: THE FUTURE

The glides do *not*, however, thereby assume the role of future goal for the trajectory of linguistic development. Unlike the labials *b* and *p*, which necessarily precede the dentals *d* and *t*, which in turn necessarily precede the gutturals *g* and *k*, and so on, zero phonemes seem free to appear, or not, at virtually any point along the path of language formation.[66] They can therefore play no role at all within the stepwise progression that Jakobson lays down in *Child Language* as his new, panchronic law of sound change. Rather, at the latent limit of the linguistic tendency toward differentiation, their existence forces a decision on the question of language origins precisely by bottoming out, without ever actually eliminating, the very notion of *telos*-as-goal upon which all stories of systemic development, including Jacobsen's, depend. The glides represent, for Jakobson, a future that must always be possible, but need never become actual, in order to definitively distinguish phonology not only from babble but also from *every historically prior theory of language systems*.

I will argue here, by way of conclusion, that this peculiar interpretation of futurity constitutes Jakobson's true contribution *both* to the modern rethinking of system *and* to the modern understanding of poetic verse. The phonological intervention, in other words, which on my reading derives its system-theoretical significance from a new approach to the relationship between systems and their preconditions, will counterintuitively turn out to entail—or perhaps even presuppose—a new approach to the relationship between language and the "telos" of lyric.

The role played by poetry in the development of a rigorous phonology, and, more specifically, the role played by Jakobson's early association with the Russian Futurist poetic movement in the development of his later linguistic theories, has long been a subject of contention among scholars of twentieth-century linguistics and poetics. Jakobson himself famously and repeatedly insisted that the one flowed directly out of the other. The first volume of his *Selected Writings*, titled *Phonological Studies*, includes as part of its conclusion, or "Retrospect," the following paean to his friend and fellow Futurist poet, Velimir Khlebnikov, about whom he wrote very first book: "[Khlebnikov's] search for the 'infinitesimals of the poetic word,' his paronomastic play with minimal pairs or, as he himself used to say, 'the internal declension of words' [. . .] prompted 'the intuitive grasp of an unknown entity', the anticipation of the ultimate phonemic units, as they were to be called some two decades later."[67] The slightly later "Retrospect" to the fifth volume of *Selected Writings*, titled *On Verse, Its Masters and Explorers*, opens with a survey of Russian Symbolist, Futurist, and early Formalist poetics, before stating succinctly, "[I]t was the analysis of verse which enabled me to descry the foundations of phonology."[68] Both declarations credit Russian avant-garde poetry in general, and Khlebnikov's Futurist verse in particular, with a kind of proto-phonological dissection of language that paves the way for Jakobson's discovery of a truly differential linguistic transcription.

In cases where the scholarly literature seeks to take such claims seriously, it has tended to respond by enumerating the ways in which Russian Futurist poetry could indeed be considered presciently "phonological" in its approach to language sound.[69] This line of argumentation, however, must necessarily fail to persuade, for there is no sense in which the phonetic analyses performed by the Russian Futurist poets can be said to deviate substantially from those of their more conventional contemporaries. Futurists such as Velimir Khlebnikov and his friend Aleksei Kruchenykh, who is often credited with the "invention" of sound poetry, do indeed display an unprecedented poetic interest in the phonic material "beneath" the word, even going so far as to coauthor a manifesto proclaiming the independence of "the letter as such."[70] Kruchenykh justifies his new philosophy of poem creation, to which he ascribes the Russian neologism *zaum*—from *za*, meaning "beyond" or "across," plus *um*, meaning "mind" or "rationality"—with an argument for the verse-constituting character of phonetic rather than semantic elements: "Verse unwittingly gives us a series of vowels and consonants. These series are inviolable. It is better to replace words with something else close not in sense

but in sound (*lyki-myki-kika*)."[71] Kruchenykh goes on to write poems
that play seriously with the notion of a "transrational" sound series,
and that thwart all attempts to rediscover conventional meaning by
thwarting all expectations about the proper *shape* of linguistic sound:
"Dyr byr schul," the first and most famous of the *zaum* poems, is
composed entirely of consonants, while later experiments include
purely vocalic poems, poems based on numbers and foreign (pseudo)
languages, and poems that recall the monosyllabic babble of infants
(a young Roman Jakobson, in a letter to Khlebnikov, suggests adding
poems that require the simultaneous pronunciation of multiple phonemes, according to the model of a musical chord[72]). None of these
experiments, however, presupposes a specifically phonological understanding of language on the part of either the poet or the reader. In
an era where letters had long since begun to signify without the help
of existing words—"to 'stehen' belong 'steif', 'starr', 'Stock', 'Stamm',
'steil', 'stopfen', 'stauen', 'Stab', 'stützen', 'stemmen',", the comparative linguist Georg von der Gabelentz could maintain by 1896, "*irrespective of whether and how much they have to do with the root
sthā*"[73]—the Futurist foray into a poetry of linguistic sound proceeds
rather in perfect accordance with preexisting, Wagnerian-Wundtian
paradigms.

The complicated proximity of the Russian Futurist radicals to the
psychophysical models of their Symbolist predecessors acquires particularly sophisticated articulation in the work of Jakobson's friend
and poetic paragon Velimir Khlebnikov. Khlebnikov's notebooks
contain multiple phonetic glossaries in which he assigns to individual Russian letters the abstract, movement-based meanings he claims
to have distilled, Wagner-style, from examining clusters of alliterating words. Both the definitional technique and the substance of the
definitions themselves bear a striking similarity to those of Wagner's
house "linguist," Hans von Wolzogen. Like Wolzogen, Khlebnikov
explicitly refuses to ground his derivations in the rigor of a traditional etymological method ("Comparative linguists may have fits?
Let them"[74]), since, like Wolzogen, he assumes that all homophones
are *necessarily* related in virtue of the inherently meaningful sounds
they share.[75] The Russian letter Л (*l*) turns out to mean, for instance,
"the diffusion of the smallest possible waves on the widest possible
surface perpendicular to a moving point" as deduced from words
such as *lob* (forehead), *laty* (armor), *lyzhi* (skis), *lodka* (boat), and
luzha (pond). The letter B (*v*, or German *w*) means "in all languages

[. . .] the turning of one point around another," but also "a wave-like movement," as in *veter* (wind), *vit'* (to twist), *volny* (waves), and *vir* (whirlpool). And the 3 (*z*) of Kruchenykh's neologism *zaum* means "the reflection of a moving point from the surface of a mirror" or "the impact of a ray upon a solid surface," as demonstrated by *zerkalo* (mirror), *zoi* (echo), *zvezdy* (stars), and *zaria* (dawn)—such that the name of Futurist poetry comes to be "etymologically" associated, via the power of alliteration, with both the doubling activity of reflection and the temporality of new beginnings.[76]

The notion of a natural language built up out of universally expressive articulatory movements, in the tradition of Wundt's psychophysical sound gestures and Bely's Symbolist dance of the tongue,[77] establishes the theoretical foundation from which Khlebnikov then goes on to create *new* sense. His favorite *zaum* technique, legitimated by the Wagnerian premise that "the initial consonant of a simple word governs all the rest,"[78] involves strategically placing elements from his purified alphabet at the beginnings of words to which they do not, conventionally speaking, belong. Such a change in the initial consonant, according to him, transforms the word meaning and establishes a poetic link to the universal, peace-bringing language of the future ("[*Zaum*] alone will be able to unite all people. Rational languages have separated them"[79]). By replacing the *p* of *pravo* (right), *pravda* (truth), and *pravitel'stvo* (government), for instance, with the *n* of *nravit'sia* (to like, to love), he arrives at the yet-to-be-realized ideal of *nravitel'stvo* (loverment). By the same process, a tiller of the soil reemerges as a tiller of time, a warrior as a "singgior," and courtiers as "creatiers" of life[80] ("If contemporary man can restock the waters of exhausted rivers with fish, then language husbandry gives us the right to restock the impoverished streams of language with new life, with extinct or *non-existent* words"[81]). In order, however, for this link between present and future, sense and trans-sense, *um* and *za-um*, to appear *as link*, the transformation itself must remain readable, which means, in general, that it must take place at the level of the verse itself.[82] Neologism and original form therefore tend to occur in close proximity to one another, and the effect is a peculiar kind of *non*alliteration, or end rhyme, which directs without alliterating an alliterative intensity toward the meaning-bearing initial. Traditional, positive alliterations occurring in this context—as the result of multiple substitutions making use of the same alphabetic element—depend for their significance on the language-renewing power of the nonalliterative, but nevertheless faithfully Wagnerian, principle of poetic construction.

Khlebnikov compares his letter-based manipulations to the practice of calculating with imaginary numbers. The mathematical symbol *i* refers by definition to the unreal square root of -1, and it therefore relates to its natural integer counterparts, he insists, much like the *zaum* of the poetic alphabet to the *um* of conventional Russian. From the empirically unattested yet poetically derivable roots of ordinary Slavic speech will spring forth a future metalinguistic forest to replace and fulfill the promise of nineteenth-century German language trees ("'Bald language' will cover its fields with new shoots").[83] And the transformation will occur in accordance with new, Futurist sound laws that mirror to the point of parody the time-structuring rhythms of the German *Stabreim*.

Oleg: [. . .] Duality, the division of the ancient world into G and R (Greece and Rome) is matched in the new era by Russians and Germans. In this case the G-R opposition is older than the countries involved. And this is no mere trick of coincidence. "Fate" has the double meaning of destiny and that-which-is-spoken. The initial sound in a word, unlike the subsequent ones, is a wire, a conductor for the current of destiny.

Kazimir: And it has a tubiform structure, and makes use of sound in order to over hear the future in the confusion of everyday speech.[84]

The letters *G* and *R* that here battle their way through conventional historical time bear witness to a future truth, and to the Futurist truth of language-as-fate, because the periodicity of their punctual dissonance gestures toward a *telos* of ultimate ethnolinguistic accord. By colliding repeatedly across regular intervals of an unfolding world history, they acquire the "tubiform structure" of the structuring *Stab*, and thus of the "sound writing [. . .] out of which the tree of universal language may someday grow."[85] Only the scientist of *zaum*, who is acquainted with the mathematical law of sympathetic resonance ("I have discovered that in general a time-period Z separates similar events: $Z = (365 + 48y)x$, where y can have a positive and negative value")[86] can hear the echoes of a harmonious future in the discordant rhythms of these "alliterating" initials. And only the *zaum* musician-poet, whose Russian balalaika chords are intended to replace and fulfill the Wagnerian promise of an *Ur*-Germanic "music of the future,"

can turn this aural-oracular experience of cosmic modulation into the
basis for a new, future-oriented poetry of the *Stab*.[87]

Jakobson's own writings shed little direct light on the question of
how, exactly, such a poetic radicalization of the Wagnerian-Wundtian
program could be said to "anticipate" the achievements of a differential phonological analysis. His most expansive treatment of the
language-poetry relationship—in the late 1960 lecture "Linguistics
and Poetics"—concerns itself exclusively with the superficially more
fundamental question of what the practice of poetry, in general, can
be presumed to signify for the project of language science. From this
perspective, the sound repetitions inherent to Khlebnikov's line-by-line
play with minimal pairs represent merely a special case of the sound
repetitions that constitute poetic rhythm per se: "Measure of sequences
is a device, which, outside of poetic function, finds no application in
language. Only in poetry with its regular reiteration of equivalent units
is the time of the speech flow experienced, as it is—to cite another
semiotic pattern—with musical time. Gerard Manley Hopkins, an out-
standing searcher in the science of poetic language, defined verse as
'speech wholly or partially repeating the same figure of sound.'"[88]

The technique of rhythmic sound repetition represents, in turn, a
special case (though perhaps also the only *true* case) of the poetic *differentia specifica*, which Jakobson characterizes as follows:

> What is the empirical linguistic criterion of the poetic function? In
> particular, what is the indispensable feature inherent in any piece of
> poetry? To answer this question we must recall the two basic modes
> of arrangement used in verbal behavior, *selection* and *combination*. If
> "child" is the topic of the message, the speaker selects one among the
> extant, more or less similar, nouns like child, kid, youngster, tot, all
> of them equivalent in a certain respect, and then, to comment on this
> topic, he may select one of the semantically cognate verbs—sleeps,
> dozes, nods, naps. Both chosen words combine in the speech chain.
> The selection is produced on the base of equivalence, similarity and
> dissimilarity, synonymity and antonymity, while the combination, the
> build up of the sequence, is based on contiguity. *The poetic function
> projects the principle of equivalence from the axis of selection into
> the axis of combination.*[89]

Poetry, in other words, is the mode of language use that makes diachronic utterances out of synchronic analogies. Explicitly conceived
as a version of Saussure's "inner treasury," Jakobson's synchronic
axis of (dis)similarity is less linear than it is planar, an intricate web
of disparate associations that encompasses all thinkable linguistic

categories. Synonymity and antonymity are joined here, under the rubric of equivalence, by relationships of homophony (month/moth) and conceptual proximity (month/year), of morphology (month/months) and (folk) etymology (month/moon). When Jakobson defines poetry as the projection of this plane onto the *vector* of contiguity, he therefore has in mind the temporal development of the metric line, which is punctuated, both internally and in relation to other lines, by the repetition of sounds somehow marked as equivalent (end rhyme, with its repeated vowels, being here only one relatively banal example of a more fundamental structuring technique). Jakobson's poets structure the punctual development of their works around precisely those "point[s] of intersection of several series <of> analogues" (*Notebooks* I 68a / *CLG*, 289:2), those *joints* or *articuli*, that can be presumed to delimit—for Saussure as for the comparative grammarians—the boundaries of the linguistic unit. They thereby manage to realize in rhythmic time the underlying systemic conditions of speech and, in this sense, to perform the ideal of writing *langue*.[90]

Even a careful reader of "Linguistics and Poetics" could be forgiven for failing to find in this 1960 Jakobsonian account of the poetry-language relationship anything other than a particularly cogent reformulation of the familiar nineteenth-century organicist model, according to which the joints, or organs, of both cosmic and linguistic systems find temporal expression in the metric periods, or stages, of a rhythmically unfolding development. Jakobson himself appears to endorse just such an interpretation when he cites Goethe as an illustration of his central point: "Similarity superimposed on contiguity imparts to poetry its thoroughgoing symbolic, multiplex, polysemantic essence, which is beautifully suggested by Goethe's 'Alles Vergängliche ist nur ein Gleichnis' (Anything transient is but a likeness). Said more technically, anything sequent is a simile."[91] Poetry articulates sequentially, and in doing so reflects microcosmically, the equivalences or analogies—the web of relations among relations—that turn the Many into a One, the body parts into an organism, the vibrations into a chord, the sounds into a language, the list into a paradigm set, the substrate into a system. In the context of such an apparently traditional, nineteenth-century understanding, it comes as no surprise to find Russian poetry playing the role of particularly "harmonious" example—in a diagram of superimposed metric waves that recalls the tone writing of Eduard Sievers (see figure 18)—rather than of radical avant-garde deformation.[92]

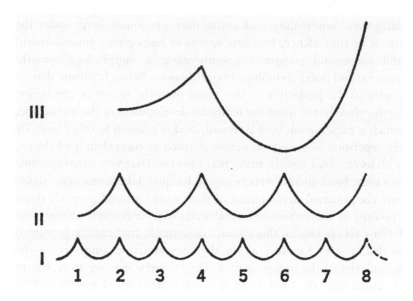

Figure 18. Roman Jakobson, "Linguistics and Poetics," in Jakobson, *Selected Writings*, 3:33.

To stop here, however, with the notion of Jakobson as a nature-philosophical philologist in disguise, is to forget that Jakobsonian phonology seeks ultimately to redefine what it would even mean to *be* nature-philosophically equivalent in the first place, and thus also to be One or The Same. To stop here, in other words, is to forget about the glides, which ground the analogies at all other levels of language in a correlation that takes the ground out of relation. Just as the *telos* of an emptied teleology takes priority, therefore, over every less marginal mode of linguistic interaction, so, too, must the rhythm of zero-degree rhyme be presumed to undergird, albeit quietly, Jakobson's analyses of Gerald Manley Hopkin's sonnets or Alexander Pushkin's iambic pentameter.

Such rhymes will then turn out to anticipate and illumine future linguistic analyses, not in the sense that they themselves practice phonology "avant la lettre," but in the sense that they poeticize according to an eminently phonological principle of structure-defining *emptying*. Jakobson's earliest and most extensive attempt to make this principle available to a theory of verse stems from his very first, prephonological monograph, *Newest Russian Poetry*, which treats the Futurist poetry of Khlebnikov as paradigmatic exemplum. The central idea

of the book is precisely *not*—quite—the famous Formalist concept of
"deformation," which consists, according to the definition of Jakobson's fellow Formalist and occasional collaborator, Yuri Tynianov,
in the "pushing forward of one group of factors at the expense of
another."[93] The early Jakobson clearly believes, with Tynianov, that
"the sensation of form is always the sensation of the flow (and, consequently of the alteration) of correlation between the subordinating, constructive factor and the subordinated factors."[94] The history
of poetry unfolds, for him, in a series of such energizing subordina-
tions, or deformations, performed upon older, more traditional conventions that have lost their ability to surprise and distort through
the inevitable process of assimilation: "A form only exists for us as
long as it is difficult to perceive, as long as we sense the resistance of
the material."[95] Jakobson, however, takes his analysis of this familiar, Formalist dynamic one crucial step further. The notion of a creative, even animatory agon at the heart of historical progression—as
the origin of what Tynianov calls "pure movement"[96] and Jakobson,
citing Khlebnikov, the future-bringing "wind of the gods of poetry"
(*NRP*, 58/299–300)—undergoes, here, a transformation that effectively succeeds in "deforming" the very concept of deformation itself.
It is this discovery of a technique that propels by subtracting propul-
sion ("pushing forward") for which Jakobson, I would argue, considers himself indebted to Khlebnikov.

The deformation of deformation occurs in accordance with the
following (implicit) line of reasoning. The amorphous agon of formative principle and formed matter, which drives the growth and
demise of verse systems, is to be given concrete content through
a description of changes in poetic technique. Poetic technique, in
turn, is to be understood in terms of "an elementary procedure"
called "the juxtaposition of two units" (*NRP*, 77/336), under which
falls the production of similarities both semantic (Jakobson names
metaphor, parallelism, comparison, and metamorphosis as examples) and phonic (assonance, alliteration, and rhyme). Through such
juxtapositions, the poet establishes certain elements as "rhymable,"
in the most general sense of "repeatable," and the result is a decision about poetic meter, conceived here as a punctual recurrence of
the linguistic same. To innovate poetically, in this context, means
always and only to challenge existing conventions about what counts,
rhythmically speaking, as *analogous*. Poetic innovation therefore
doubles, given the significance of the "sameness" conundrum for

the problem of linguistic unithood more generally, as a temporally extended reflection on the deep-structural conditions of comprehensible communication per se. "The form of a word," writes Jakobson just a few lines after declaring the juxtaposition of two units to be the elementary procedure of poetry, "can be apprehended only on the basis of its regular recurrence within a given linguistic system" (*NRP*, 77/336). The poet that plays with the possibility of perceptible repetition at the level of the unfolding line establishes "a kind of linguistic system *in statu nascendi*" (*NRP*, 77/337), in which previously meaningless linguistic "chunks"—the ends of two rhyming lines, for instance, or the beginnings of two alliterating words— acquire over the course of the poem the status of potentially meaningful linguistic "units."

Poetic juxtapositions become perceptible *as* juxtapositions insofar as they brush against the grain of our extrapoetic expectations, which is to say, insofar as they differ from, or "deform," the juxtapositions that undergird ordinary communicative language. Poetic juxtapositions become perceptible as *poetic* insofar as they deform conventional language from within rather than without, which is to say, insofar as they *neglect* to justify themselves by reference to some extrapoetic principle of explanation. Jakobson has in mind, here, as the model against which he positions his endorsement of immanence, an "expressivist" poetics of the kind prevalent amongst Wagnerians, psychophysiologists, and many members of the European avant-garde.[97] To interpret poetic innovation primarily as a means for expressing ever more extreme emotional states, he argues, is to efface the specificity of poetry.

> In emotional language and poetic language, the verbal representa-
> tions (phonetic as well as semantic) concentrate greater attention on
> themselves; the connection between the aspect of sound and that
> of meaning is tighter, more intimate, and language is accordingly
> more revolutionary, insofar as habitual associations by contiguity
> retreat into the background. [. . .] But beyond this there is no affin-
> ity between emotional and poetic language. In emotional expres-
> sion, the affective character of the verbal mass dictates its laws; the
> "turbulent stream of emotion bursts the pipe of the sentence." But
> poetry—which is nothing other than *an utterance oriented toward
> the [linguistic] expression*—is governed, so to speak, by its own
> immanent laws; the communicative function, essential to both practi-
> cal and emotional language, gets reduced here to a minimum. Poetry
> is indifferent to the subject of the utterance. (*NRP*, 62/304–5).

Only an approach, in other words, that seeks to rigorously disentangle the peculiarities of poetic juxtaposition from the peculiarities of a diseased, ecstatic, impassioned, or comedic psyche can bring the true character of the poetry-language agon to light. A polemical footnote follows, in which Jakobson exploits his new definition of poetry—as an utterance oriented toward the structure of the linguistic expression itself, rather than the transmission of expressive content—to turn the terminology of "expressionism" inside out: "If we accept such a definition of poetry, then we can term the method of study that results from it 'expressionist'" (ibid.). The vehemently antiexpressivist approach of *Newest Russian Poetry*, so the implied assertion, can (more) legitimately lay claim to the title of "expressionist" precisely because it eliminates from the interpretive equation the abstraction of a nonappearing yet expressing self, and in doing so gives access to a purer, more literal sense of "expression."

The gesture recalls the structure of the tautological *figurae etymologicae*, by means of which Saussure sought to claim originarity for his own antiexpressivist conception of language ("to signify *wants to say* a sign"; "articulations, *that is to say* articuli").[98] The focus here, however, is no longer on the origin of the system but on the end. The poetic emptying of expression does not precede the formation of expressive language any more than, at the phonological level, the glides precede the formation of the elementary consonants and vowels. Rather, the essence of poetry resides in the tendency to continually *subtract* a preexisting tendency toward goal-oriented communication ("the communicative function gets reduced here to a minimum") according to a teleological dynamic that can ultimately be expected—and herein lies the point where Jakobson leaves his less radical Formalist colleagues behind[99]—to retroactively deform the structure of all teleological dynamism per se. Such an antiteleological *telos* finds its most advanced historical manifestation, Jakobson thinks, in the "sense-transcending speech," or *zaum*, of the "newest Russian poets," whose principle innovation consists, for him, in the "laying-bare of the procedure via the stripping away of every logical motivation" (*NRP*, 65/309). The weight of this dense and somewhat awkward formulation falls on "of *every* logical motivation": if the history of poetry can be understood as a continual process of deforming language in order to reveal the formative principle of linguistic communication, and if the specificity of this process must be sought in the continual elimination of (other) motivating factors that could give (other) meaning to these deformations, then the "future" of poetry, in the sense not of its "tomorrow" but of its terminus, will necessarily deform and demotivate the poet's *own* language use. The communicative content of poetry, which is here being pronounced identical to the logic of linguistic communication "laid bare," must itself, as a motivating principle that "pushes" the poem in the direction of its language-internal meaning or purpose, be actively "stripped away."

The result will look like babble and perform like the outer boundary of sense, as Jakobson makes clear in a footnote that provocatively literalizes his figure of a literalizing divestment: "But insofar as [a practical language] exists, and a phonetic tradition is present, sense-transcending language can no more be compared with prelinguistic onomatopoeias, than an undressed [*entblößter*] European of today with a naked [*nackten*] troglodyte" (*NRP*, 354). The simile provokes, in part, by assimilating a dizzying array of impossible comparisons into a "system" of analogical equation. The "before" and "beyond" of communicative language are first declared incommensurable before being set equal, in their very incommensurability, with the incommensurable (non)relation of primitive and civilized modes of nakedness. The counterintuitive pairings that emerge from this second-order comparison of incomparables—this analogy between two *logoi alogoi*—have much in common with the (non)rhymes and (non)similes of the Russian Futurist poets, in the sense that their elements—linguistic meaning and clothing, on the one hand, Russian *zaum* and an undressed European, on the other—have little to nothing in common with one another. "Like semantic correspondences," writes Jakobson, "rhyme, as a euphonic correspondence, is very approximate in modern poetry" (*NRP*, 80/343). And a few lines later he writes, "[T]he evolution of poetic euphony runs parallel to the evolution of modern music—from tone to noise" (*NRP*, 81/343).[100] Repetition, in the hands of the Russians, has relinquished its exclusive relationship to the same, or, more accurately, the same has relinquished its claim to the rationality of a common denominator. The trajectory of poetic technique, and thus also of poetic juxtaposition, outruns at the moment of its transrational *generalization* ("this enrichment of rhyme"; *NRP*, 349[101]) the restrictive claims of a unifying language spirit, with the result that poetic harmony becomes indistinguishable from dissonance.

This new dissonance does not thereby become consonant with the older dissonance of a preharmonic, or precommunicative noise.

Rather, to add a second, equally Khlebnikovian juxtaposition to the pairing of "European" and "troglodyte": the (non)relationship between trans-sense and pre-sense parallels the (non)relationship of modern irrational numbers, with which one can do mathematics, to the ancient irrationality of the *apeiron*, with which one could not.[102] On Jakobson's interpretation, the properly Futurist poet no longer strives, as Wagner and Wundt had presumed, to reach the "end" of a universal metalanguage by intensifying the formative principle of the beginning. The more capacious harmony of a transrational future, whether poetic, mathematical, linguistic, or musical, does not simply expand on a genesis with which it would otherwise remain continuous according to the teleologic of an actualized potential. Khlebnikov insists, for this reason, on the poetic significance of square roots such as $\sqrt{2}$ and $\sqrt{-1}$, which can be obtained only by *eliminating* all rational content from the activity of "taking a root"; Jakobson insists, for this reason, on the linguistic significance of what he calls "poetic etymologies,"[103] which can come into being only once words have been *emptied* of their original meanings. "It is necessary," writes Khlebnikov, "to universalize one's birth, one's ancestry; necessary to keep white the gleaming root."[104] "Said more technically," writes Jakobson, "*anything* sequent is a simile."[105]

The central message in both cases—on Jakobson's interpretation—is a new concept of systemic time. Khlebnikov's statement responds, with its notion of a genealogy retroactively rendered imaginary, to the tyranny of nineteenth-century German derivational laws, while Jakobson's explicitly translates and travesties the key *geschichtsphilosophical* lines from the end of Goethe's *Faust*:

Alles Vergängliche
Ist nur ein Gleichnis;
Das Unzulängliche,
Hier wird's Ereignis;
Das Unbeschreibliche,
Hier ist's getan
Das Ewig-Weibliche
Zieht uns hinan.

[All that is transitory
is only a likeness;
what seems unachievable
here is seen done;
what's indescribable
here becomes fact;

Woman, eternally,
shows us the way.][106]

The question of whether this latter transformation, performed in
the 1960s by Jakobson the linguist, actually corresponds to the former, performed in the 1920s by Khlebnikov the poet—and so also of
whether Jakobson is "right" to read Khlebnikov's poetry as an emptying rather than a fulfillment of the nineteenth-century *geschichts-*
philosophical model—must here be left open. The point is not to
arrive at any definitive conclusions regarding Khlebnikov's true views
on language essence but rather to illuminate the system-theoretical
kernel hidden within Jakobson's powerful interpretation thereof:[107]
like the oldest Germanic poetry of the *Stabreim* for the nineteenth-
century scientists of *Sprachgeist*, Khlebnikov's Futurist oeuvre represents, for Jakobson, the privileged site of a system articulating its
own essence *at the level of the temporally unfolding speech stream.*
Whereas the *Stabreim*, however, had fulfilled this all-important explicatory function by rhyming the etymological roots of an organically
evolving *Sprachgeist*, Futurist poetry, in Jakobson's terms, actively
strips these roots of their historical priority. It retroactively universal-
izes the dynamic of its birth and "makes white" the propulsive principle of its own forward trajectory, such that *all* sound-sense moments
now become equal candidates for the title of time-structuring "joint."

The resulting (lack of) rhythm simultaneously embodies and evis-
cerates Wagner's eschatological fantasy of total rhyme. What disappears from Goethe's "All that is transitory / is only a likeness" [*Alles
Vergängliche / Ist nur ein Gleichnis*] in the shift to "anything sequent
is a simile" is precisely the "only" that points forward and upward
toward the transcendent, posthistorical telos of every possible similarity: toward the atemporal end-of-times perspective from which the
significance of all temporal events could be definitively judged, and the
systemic unity of world history comprehensively ascertained. Jakobson's "more technical" paraphrase requires no universal equivalent,
no ultimate Logos, no notation-exploding "unachievable" (*Unzulängliche*) and "indescribable" (*Unbeschreibliche*)—whether imagined as
God or as *Bildungsprinzip*—to ground its thoroughly analogical conception of time. It presupposes, instead, an account of poetic essence
that turns essence itself, and with it, the production of poetic unity
or similarity, into a function of the emptying of essence. The capacity
of poetry-in-general to include, at the outer limit of its developmental
trajectory, the parodic deflations of Russian Futurist "verse," ensures

the legibility of this particular history as a meaningful, systemic whole, but it does so only on the condition that the deflation remains readable precisely *as a deflation*. Trans-sense poetry quite literally means nothing without the backdrop of the tradition it transcends, just as the zero phoneme quite literally cannot be written outside the context of the phonological structure it grounds. Systems exist, in time—according to the theory here implicitly expounded—insofar as they succeed in bringing forth their own zero points, which of course also means that system *theories* exist, in time, insofar as they *model* this dynamic.

Jakobson's travesty of a translation of Goethe, in other words, self-reflexively performs his new theory of system as a travestying translation of past theories. In the absence of such a performance, which polemically juxtaposes new and old, it would be possible and perhaps even unavoidable to interpret a statement like "anything sequent is a simile" in one of the following inaccurate ways, each of which corresponds to a familiar, nineteenth-century account of the relationship between the future and the past. On the one hand, it could be assumed that anything sequent now counts as a simile because *all* sameness, and with it all sense, has become recognizable as the result of an arbitrary imposition from outside. According to this perspective, everything can rhyme because nothing "really" does, and every proposed principle of development—poetic or otherwise—necessarily distorts the fundamentally unstructured nature of reality. History appears as a formless heap of happenings (Kant's *coacervatio*), and Futurism itself as one event among infinitely many others, none of which bears any essential, system-internal connection to its predecessors. On the other hand, it could be assumed that anything sequent now counts as a simile because Jakobson believes we have actually arrived *at the end of poetic history*—as envisioned by nineteenth-century theorists of organic unfolding and by the Chorus Mysticus of Goethe's *Faust*—where every transient phenomenon reveals itself to be a microcosmic reflection of an all-encompassing time-structuring principle.[108]

Between these two alternatives, one of which fails to see "anything sequent is a simile" as a translation of *geschichts*-philosophical models, and the other of which fails to see it as a travesty, lies, I believe, Jakobson's long-unrecognized attempt to articulate a viable third way: according to the interpretation I have articulated here, his theories of phonological and poetic formation are themselves a product, and thus also an expression, of the very deflationary processes they seek to articulate. As such, they possess a necessary rather than a contingent connection to the past of which they are nevertheless definitively *not* the teleological fulfillment. They neither actualize nor abandon the spirit-centered potential of their nineteenth-century predecessors, with the result that their proximity to prior systems theories proves neither totally arbitrary nor totally constraining. This third way of relating to history, which gets articulated for the very first time in the context of the structuralist linguistic paradigm, has been almost universally ignored or misunderstood since its inception. The question of what it can contribute to contemporary debates about the epistemic status of history, and about the nature of systemic time, is one that must now be explicitly posed.

Afterword

What does it mean to say that the linguistic sign is arbitrary? The question is seldom still asked, perhaps because we are presumed to have long known the answer: Saussure's famous dictum in the *Course on General Linguistics* rediscovers and restates, according to one firmly established line of thought, the clear-sighted nominalism of John Locke's *Essay Concerning Human Understanding*—"[Words] signify only Men's peculiar Ideas, and that by a perfectly arbitrary Imposition"[1]—which in turn unfolds a logic of lexical relativism already explored by Plato's *Cratylus*. Words do not reflect the essence of things "out there" in the world; natural laws do not determine the human choice of phonic structure or semantic substance; the conceptual categories available within a given language do not correspond precisely to the categories of scientific knowledge, or to those of other tongues. Such claims are as familiar as they are uncontroversial. Historically speaking, most attempts to counter them have arisen less from a desire to comprehend the day-to-day workings of existing languages than from a theologically motivated desire to render plausible the hypothesis of an originary, God-given lexicon. Among contemporary language scientists, for whom the hypothesis of a divine *Ur*-language no longer merits serious consideration, the principle of the arbitrariness of the linguistic sign—understood in the basic sense outlined above—qualifies as a truism.

Clarity, however, does not reign with respect to the conundrum of what, precisely, this truism entails for the relationship between language and world, subject and object, concept and thing. The "basic sense" of arbitrariness, to which we so readily commit ourselves, is both too capacious and too empty to offer much in the way of direction, since it really only affirms that the material of language remains free of some unspecified form of constraint. Yet no material that participates in a self-perpetuating system such as language can ever be free of *every* kind of constraint, because order just *is* a mode of restricting, or (de)limiting, the possibilities inherent to its domain: every system that constrains does so by deciding what will "count" as material in the first place, which means that what we mean by "arbitrary" will depend entirely on how we conceive of this decision. The arbiter, after all, is the "one who decides," and the arbiter who arbitrates arbitrarily, traditionally, is one who decides in accordance with his or her own free will, unfettered by the established laws of either nature or society. If this mysterious, inherently unpredictable decision remains an isolated event of boundary-drawing—as when a judge pronounces on the basis of a purely idiosyncratic whim—then its logic, or rather its lack thereof, can perhaps be ignored without serious consequences for knowledge. The inscrutability of such an act with respect to existing orders might in this case not imply the need for a new explanatory model. If, however, the decision in question manages to call forth an entirely new category of object, which establishes itself at the foundation of an entirely new kind of order—as is unequivocally the case for the "decision" tying linguistic signs to their meanings—then the problem of how the new order relates to the old ones becomes pressing. Every theory of language confronts this problem in its own way, with the result that there are as many conceptually distinct principles of arbitrariness as there are accounts of the language-world relation.[2]

The Writing of Spirit traces the genesis of the twentieth century's best-known and most influential such principle, arguing in the process that it has been widely and deeply misinterpreted, in ways that fundamentally distort its contribution to the debate about how systems, in general, get formed. This debate, closely tied as it is to the question of whether systems are exclusively mental phenomena, takes place from its inception between the proponents of two fundamentally opposed theories of knowledge. Structuralist linguistics, I have argued over the course of this study, manages to bridge—or rather to empty out—the apparently inescapable divide between rationalists and empiricists, mentalists and physicalists, teleologists and mechanists, according to a historically unprecedented interpretation of the mind-world interface of language. Within the book itself, I trace the emergence of this structuralist solution against the backdrop of the particular epistemological positions that would have been familiar to the nineteenth- and twentieth-century protagonists themselves, namely, from the early nineteenth century, the disagreement between preformationists and epigenesists in the life and language sciences, and, from the early twentieth century, the transdisciplinary conflict over the merits of radical positivism. Here in the afterword, I want to supplement this historical account with a brief look at two of the most significant contemporary solutions to the language-as-interface puzzle, in the hope that doing so will provide a clearer picture of how the structuralist proposal—on my understanding—still challenges us to think differently today. Within the constraints of such an overview, I obviously cannot aspire to do justice to the complexity of the solutions in question. The goal is rather to illuminate the present-tense import of the unfamiliar conceptual terrain toward which the structuralist project, alone among its peers, seeks to direct our attention.

The concept of arbitrariness traditionally goes hand in hand with the concept of structure, not as its complement but as its other; the adjective "arbitrary," when applied to linguistic phenomena in particular, emphasizes the distance between the order of our conceptual categories and the order of what we have categorized, which means also between our systems of knowledge and the truth. This distance exists, albeit with different meanings, for both the empiricist and the rationalist positions as conventionally understood. Empiricists such as Locke and Hume, for instance, cite the lack of a natural word-thing bond in support of the claim that most human ideas can only ever impose upon a reality to which they ultimately do not correspond. The "truest" representations of the world, on this view, will be the ones that picture their object with the lightest, least conceptually laden hand: the inherent arbitrariness of the conceptual boundary-drawing process can be tempered, though not eliminated, by confining it to processes of probabilistic generalization (Hume's "habitual association") and pragmatic convention (Locke's "tacit agreement"), which serve the sole purpose of facilitating environmental interaction.

Twentieth-century inheritors of this empiricist position, particularly as it pertains to the theory of language emergence, have tended to focus their attention on providing a more substantive account of how such processes actually operate. The philosopher David Lewis, in his extremely influential work from the final four decades of the

twentieth century, argues, first, that convention can best be understood on the basis of the principles of rational decision making, as set
out in the rules of game theory, and, second, that rationality can best
be understood as the tendency to expect, in any given case, the objectively most probable outcome (slightly more formally, the basic idea—
conceived as a version of Hume's "uniformity principle"—is that a
rational subject will tend, at any given moment, to bring the probability function that governs her beliefs about the future into line with
the probability function that has governed, to the best of her knowledge, similar events in the past). Lewis derives from these two pillars
a rich account of how meaning, in a particular language, can arise
collectively yet nonuniversally from encounters among belief-holding
individuals, without any need for some mysterious prior agreement—
either implicit or explicit—about what sounds should refer to which
ideas.[3] Language, for Lewis, is defined as a mapping that links particular statements of belief to their truth conditions: a sentence is true
in language L if and only if the actual world belongs among the set
of possibilities to which the sentence corresponds. There are infinitely
many possible mappings of this kind, and no compelling reasons,
game-theoretical or otherwise, for a given individual to choose one
such mapping over another. There are, however, compelling game-
theoretical reasons, according to Lewis, for a given individual to
choose *some* mapping over no mapping at all, and, in doing so, to
make her choice conform to that of other individuals with whom she
repeatedly interacts. The result is a convention of verbal communication, a reservoir of communally comprehensible utterances, which
emerges "naturally" without actually needing to be "natural." Language does not belong from its inception to the fabric of the physical
world; neither, however, does it enter this world from some mysterious, transempirical outside. Its existence therefore does not require
the assumption of any additional, transempirical source of knowledge, so long as the individual mental states on which it depends—
Lewis's probability functions of belief and desire, Hume's tendencies
of habitual association—can themselves be reduced to the results of
empirical processes.

The philosophical appeal of such an approach is clear. On the one
hand, it appears to solve, at least for the particular phenomenon of
language, the problem of bridging the *physis/nomos* divide, by showing that the nomological can, in fact, develop on the basis of the physical. On the other hand, it does so without ever calling into question

the common-sense interpretation of language as a *vehicle*—an intermediary term, a mapping function—that links the previously existing spheres of nature and thought. Among contemporary philosophers, with their widespread, eminently defensible preference for empiricist epistemologies, minimalist ontologies, and reductivist methods, Lewis's story about arbitrariness and the origins of language has been enthusiastically (though by no means universally) embraced; his related work on the technical exposition of semantic and pragmatic linguistic phenomena has found fertile ground among natural language semanticists as well.

Linguistically speaking, however, there is a problem, and this problem—as the self-proclaimed neorationalist Noam Chomsky has repeatedly pointed out—goes right to the heart of the traditional empiricist/rationalist debate. An approach like Lewis's, which explains language as a contingent agglomeration of isolated sentence-and-truth-condition pairings (and thus, following the Fregean tradition of "extensional" definition, as a collection of input-output couples for which no "intensional" rule of construction need be given), cannot possibly be called upon to account for the particular structure of its various propositional constituents—to explain why, in other words, certain sentences can be arranged *only* so and not otherwise. Chomsky offers, by way of example, two sentences containing the English phrase "each other," one of which is clearly well formed ("The candidates wanted each other to win") and one of which is not ("The candidates wanted me to vote for each other").[4] No amount of probabilistic reasoning or inductive generalization, Chomsky claims, can explain the ability of every English speaker to make a definitive distinction between the two, since without knowledge of the underlying rule itself, which permits the former and excludes the latter, judgments about the "legitimacy" of particular assertoric strings will necessarily be based solely on the frequency of their distribution within the set of previously experienced utterances. A well-formed but extremely unusual sentence that has never before been heard possesses on Lewis's account precisely the same status as a malformed one that would never *be* heard, which means, in turn, that the conventionalist account of language, unlike language speakers themselves, lacks the resources to meaningfully distinguish between linguistic creativity and error.[5] This inability to account for the emergence of linguistic creativity, one of the manifest possibilities of empirically attested language use, belongs among the most serious disadvantages of the truth-and-trust-based conventionalist model. An extremely idiosyncratic utterance cannot register among listeners, on Lewis's interpretation, as an attempt to tell a truth, which obviates the trust required for the game-theoretical understanding of convention to operate. This implies, in turn, that such utterances cannot acquire their meanings in the manner proposed by Lewis's theory. Since such meanings do, nonetheless, demonstrably exist for the speakers of a given language, some *other* source of knowledge over and above the one represented by the experience of use—so Chomsky's argument—must be available.

Chomsky's own well-known version of this "other source" is the innate language faculty, or universal grammar, within whose framework all empirical usage unfolds. Chomsky affirms as incontrovertible fact the "poverty of the stimulus" principle, which states that competent speakers of a given language have access to infinitely more meaningful utterances than could ever be acquired from the finite data of experience, and he responds to the conundrum thereby raised with an internalist, essentialist model that defines knowledge ("competence") independently of use ("performance"). The Chomskean account of what defines such knowledge has changed drastically over the years: earlier versions presume the presence of a highly articulated and specific set of structural rules and levels, as formalized, from 1970 onward, in the tree-shaped hierarchies of X-bar theory.[6] The more recent minimalist program, on the other hand, envisions a single, recursive operation called "Merge" that is responsible for producing all possible syntactic combinations whatsoever, as (partially) formalized in the newer notation of bare phrase structure.[7] The former perspective is explicitly aligned, by Chomsky himself, with the tradition of Cartesian rationalism, while the latter appears in many ways closer to the "minimalist" transcendentalism of Kant. In both cases, however, the fundamental break with the empiricist perspective remains decisive: some *a priori* form-bestowing principle, or set of principles, which is notably absent from the nonhuman world, must be presupposed as given to all language users, universally and exclusively, if the existence of language is to be explained. On this view, linguistic structure does not emerge, piecemeal, out of the unstructured material of empirical experience. An infinity that admits of definite boundaries (since a language system allows for the actualization of infinitely many, but not all, possible utterances) can be defined only by means of a rule rather than a list. And such a rule, according to

Chomsky, cannot be extrapolated from available empirical evidence, because extrapolation, by definition—based as it is on Hume's uniformity principle, which posits a future identical to the past—must necessarily fail to yield rules that range freely over the domain of the unexpected.

The problem with Chomsky's quasirationalist approach, from an explanatory perspective, is that it reopens the very breach between world and mind that empiricist models have historically worked so hard to close. Chomsky himself would presumably disagree, and it is true that he is not, with respect to the dualism versus monism debate, a traditional rationalist: he believes that the human mind finds its remainderless instantiation in the physical brain, and that we will someday be able to map the neural networks through which this instantiation occurs. His proposed answer, however, to the question of how the brain *acquires* these characteristics—and thus also to the question of how the brain actually *becomes* a mind—simply transfers the template of an unbridgeable abyss to the level of the bodily material itself: a single genetic mutation in a single individual, he suggests, may have been sufficient to call the (neural networks undergirding the) operation of Merge into being, and thus also, in one fell swoop, the capacity for self-reflexive, recursive, higher-order thought, as reflected in the open systematicity of natural language.[8] No individual who does not directly inherit the relevant genetic mutation, on this hypothesis, could ever hope to gradually develop the ability to preside over rule-governed infinities. No continuum of degree, therefore, connects the brain as it existed before this miraculous evolutionary leap with the brain as it exists thereafter, which means also that no continuum of degree connects the human brain as we know it today with the brains of even our closest mammalian relatives. The human brain is, effectively, its very own kind of ontological entity, with a mode of organization that has no counterpart "out there" in the nonhuman world.

It is into this familiar breach, as hypostatized by Chomsky's neorationalism and denied by Lewis's neoempiricism, that linguistic structuralism, on my interpretation, somewhat perversely attempts to leap. Saussure and Jakobson seek to colonize, on behalf of language science, the profoundly inhospitable "space" of the mind-world interface, and they do so by thoroughly rethinking the relationship of the arbitrary to the essential. For Lewis and the British empiricists, the inherent arbitrariness of language implies that language systems, no matter how highly developed, necessarily lack an interior, intensionally specifiable essence. And Lewis's recourse to the game-theoretical model effectively demonstrates, against traditional rationalist arguments to the contrary, that such systems are not, in fact, necessarily a contradiction in terms. Relatively stable, self-perpetuating structures, like the extensionally defined set of utterance-conventions he identifies with language, can indeed arise out of multiple, accidental "collisions" of actor-atoms, independently of any tacit or explicit intention on the part of the agents themselves—which is to say also without any reference to their (or our) ability to specify a particular rule, or set of rules, underlying the collection. Chomsky responds, on behalf of the rationalists, that such empiricist derivations are ultimately irrelevant because the kinds of systems whose emergence they explain bear little relationship to the kinds of systems we presuppose when we speak. Natural language, he insists, does not just self-perpetuate, it also self-*generates* (hence "generative linguistics"), and this productive, creative aspect of structure cannot be captured, he thinks, without a strong, intensionalist understanding of "self." Chomsky certainly does not deny that natural languages have an arbitrary component, but he rejects the idea that this component plays any significant role whatsoever in the formation of the language *system*. The contingent conventions of the speaker's environment determine what lexical "material" gets plugged into the rational, natural, syntactical framework of universal grammar, while the framework itself, which belongs to the very fabric of the human mind, determines the eminently nonarbitrary shape of the whole.

Against these empiricist and rationalist positions, Saussure and Jakobson take seriously the existence of an intermediate, uniquely linguistic realm. For them, as I have tried throughout this book to show, the mind-world interface is neither illusory, in the sense that the two sides could be remainderlessly reduced to one, nor trivial, in the sense that they retain their functional independence from one another. Rather, mind and world come together, and in doing so, "establish an order" (*donnent un ordre*; WGL, 32/51) whose properties definitively transcend those of *both* its component realms. The emphasis on a real, transformative encounter between opposites has its most significant historical precedent in the nature-philosophical orientation of much nineteenth-century German science. Whereas the nature-philosophical position, however, effectively commits its proponents to the inflationary ontology of a quasimonistic trialism, according to which the All is an entity composed of (1) mind, (2) matter, and (3) the event of their union—as represented by the proliferation of mother-father-child metaphors across so many widely differing disciplinary domains—the structuralist solution places parsimonious emphasis on the *single* given of the mind-matter interface itself. Saussure and Jakobson are no more inclined to go in search of an unactualizable, inaudible, purely mental grammatical structure than they are to make inferences about a mentally unmediated, prelinguistic, empirical reality. The sole available phenomenon, on their view, is a phenomenon of meeting, and this meeting therefore takes priority in a more than merely pragmatic sense. The question of what kinds of isolated entities there might be, or, better, might have been, in the absence of such a boundary, arises primarily as a matter of rhetorical awkwardness (Saussure speaks, in this context, of "the two chaoses," *les deux chaos*; ibid.), since the "things that meet" become accessible to the linguist only retrospectively, from the perspective of the language system they ostensibly precede. The result is an approach that turns the problem of linguistic arbitrariness into an eminently language-internal affair.

But if the problem of arbitrariness is language-internal, then so is the problem of language origins, because conclusions about the former are always also conclusions about the relationship between language and its "substrates," out of which (or in spite of which) linguistic structure is presumed to have emerged. And if both of these problems are language-internal, then they are both, in turn, a matter of language essence. From this counterintuitive series of premises—which binds accident to substance, flux to form, chaos to order—everything follows, including, most significantly, a radically unfamiliar account of systemic time. Where empiricists see an illusion of structural constancy supervening upon the ever-shifting contingency of the real, and rationalists an immutable scaffold persisting independently of variations from without, the structuralists follow the nature-philosophers in making change do truly system-*building* work. This work, however, no longer takes the form of a substantive interaction between the opposing poles of mind and world, which means also that it requires no additional premise about the existence of a substantive link: in place of the nature-philosophers' world soul, and the nature-philosophical linguists' language spirit, both of which tie matter to mental life across the mediating trajectory of temporal development, the structuralists advocate for a third term devoid of all content—a spirit that programmatically *refuses to inspire*—with the result that arbitrariness *itself* becomes essence. The "two chaoses" come together to the extent that each fails to impose its own organization on the other, and the domain of their convergence acquires independent validity to the extent that it continues, thereafter, to make a principle of this originary "failure."

The structuralist insistence on the "fundamental fact" of arbitrariness therefore does not entail, as it does for Locke and Lewis, that language arises without reference to any internal rule but rather that it does so spontaneously, which is to say freely, with respect to the language-external logics of both mind and world. System is here reconceived as an entity that tends, teleologically, toward the abolition of all prior mental and material tendencies. Such a singular telos does indeed provide a criterion, or rule, for distinguishing definitively between inside and out: speakers can decide with confidence on the question of "new" versus "wrong" because language bears a goal-oriented rather than probabilistic relationship to the set of its possible futures. This goal, however, remains utterly unlike every traditional candidate for the task of bestowing form on flux, in the sense that it derives its rule-governed, future-constraining power from an emptying of the very "stuff" it organizes. Speakers, on the structuralist view, do not need to be born preprogrammed with a timeless framework that precedes all practical application, into which they can then plug the data they go on to experience over the course of their respective histories. Instead, the framework of language emerges at those liminal points where the data of experience gets "zeroed"— repeatedly and thus temporally—into the form of a system-defining limit. The delimitation of the undelimited, which in turn makes possible the writing of the unwritable, takes place within the mind-world interface rather than beneath it, as a function of use rather than competence, via a process that finds paradigmatic expression, for both Saussure and Jakobson, in the genuinely language-internal *activity* of poetry.

Language-as-system, in other words, acquires here the character of an ongoing event rather than a stable form, but its scientific definition does not thereby become a matter of relative probabilities or provisional negotiations, i.e., of history as conventionally conceived. The truth of language, from the structuralist perspective, is simultaneously temporal and time-transcending, in the sense that it tackles the task of eliminating contingency without leaving the continuum of flux or Becoming behind. The result is a new way of thinking about the role of linguistic creativity, as a boundary-drawing force capable of mediating, structurally, between the traditional poles of nature and culture, natural science and the humanities. The result is also, and relatedly, a set of provocative yet still largely unexamined principles concerning the relationship between systems and time: principles that our contemporary intellectual moment, with its pronounced aversion to thinking "real," scientific knowledge as a function of historical emergence, might do well to take more seriously.

ACKNOWLEDGMENTS

A book of this kind, no matter how idiosyncratic, is always at some level a collective endeavor. I am grateful, first and foremost, to the extraordinary mentoring of Michael Jennings and Barbara Hahn, who have influenced my intellectual development in ways that go far beyond the content of these pages, and whose probing questions left their mark at every stage of the book's development. To the inimitable Stanley Corngold, who introduced me to the riches of the German tradition in the first place, I owe more years of inspired and inspiring guidance than either of us probably cares to contemplate. I am thankful, in addition, for the insightful feedback of my Princeton colleagues, especially Brigid Doherty, Daniel Heller-Roazen, and Nikolaus Wegmann, all of whom read and commented extensively on various iterations of the project; for my students at Stanford and Princeton, whose passion for grappling with big ideas reminds me again and again why I got into this business to begin with; and for productive discussions with the participants in the Works-in-Progress colloquium at Princeton, the Society of the Humanities colloquium at Stanford, and the colloquium of Winfried Menninghaus at the Free University Berlin. Financial support for my research in these varying contexts was provided by the Porter Ogden Jacobus Fellowship, the Mellon Foundation, and the Humboldt Foundation. Financial support for the publication process itself comes from the Barr Ferree Publication Fund and Princeton's University Committee on Research in the Humanities and Social Sciences. The transition from book manuscript to book proper was made immeasurably easier by my indefatigable research assistants, Spencer Hadley and Juan-Jacques Aupiais, who helped me track down and prepare all the English translations, as well as by my marvelous editor, Thomas Lay, with whom it has been a delight and a privilege to work. Special thanks are due, finally, to Catharine Diehl for her stellar work on the index (in addition to our many fruitful conversations regarding the shape of the argument as a whole), and to Kerrie Maynes for her exceptionally rigorous copy editing.

I owe a very different kind of debt—one that can neither be discharged nor adequately articulated here—to my parents, who taught me what it means to teach, and to my husband and two sons, who taught me what it means to go all in. This book is for them.

INTRODUCTION

1. See, in particular, Iain Hamilton Grant, *Philosophies of Nature after Schelling* (London: Continuum, 2006); Robert J. Richards, *The Romantic Conception of Life: Science and Philosophy in the Age of Goethe* (Chicago: University of Chicago Press, 2002); Helmut Müller-Sievers, *Self-Generation: Biology, Philosophy, and Literature around 1800* (Stanford, CA: Stanford University Press, 1997). All three of these scholars seek to modify, in various ways and with varying degrees of explicitness, Michel Foucault's groundbreaking 1966 account of a nineteenth-century "episteme" common to the sciences of "life, labor, and language." Michel Foucault, *The Order of Things: An Archeology of the Human Sciences*, trans. Alan Sheridan (New York: Vintage, 1973). Original French *Les mots et les choses: Une archéologie des sciences humaines* (Paris: Gallimard, 1966). All three also rely on the pioneering work of Timothy Lenoir, *The Strategy of Life: Teleology and Mechanics in Nineteenth-Century German Biology* (Chicago: University of Chicago Press, 1982). More recent and specialized contributions to the conversation include Bruce Matthews, *Schelling's Organic Form of Philosophy: Life as the Schema of Freedom* (Albany, NY: SUNY Press, 2012); and Jennifer Mensch, *Kant's Organicism: Epigenesis and the Development of Critical Philosophy* (Chicago: University of Chicago Press, 2013). Leif Weatherby's work on the metaphysical implications of what he calls "organology" in the work of philosophers and poets from Leibniz to Marx adds a new and exciting dimension to this extremely fruitful line of inquiry. Leif Weatherby, *Transplanting the Metaphysical Organ: German Romanticism between Leibniz and Marx* (New York: Fordham University Press, 2016).

2. Foucault himself, of course, famously relegates the analysis of his own twentieth-century context to a brief final chapter, where he treats it simultaneously as an extension of the nineteenth-century "modern episteme" and as the dawn of a potentially revolutionary return (via the Saussurean theory of the sign) to the classical model. Since Foucault, there has been virtually no attempt to read the history of language science against this broader backdrop of a uniquely nineteenth-century transformation in the concept of science as such. For an enormously powerful interpretation of modern linguistics as the

belated actualization of a quite different, pre-nineteenth-century concept of science—the "Galilean" model as analyzed by Alexandre Koyré—see Jean-Claude Milner, *Introduction à une science du langage* (Paris: Éditions du Seuil, 1973) and *L'amour de la langue* (Paris: Éditions du Seuil, 1978).

3. For a comprehensive exploration of the history of French structuralism and its poststructuralist successors, from the 1950s through the 1990s, see François Dosse, *Histoire du structuralisme*, vol. 1, *Le champ du signe, 1945–1966*, and vol. 2, *Le chant du cygne, 1967 à nos jours* (Paris: La Découverte, 1991 and 1992). For a survey of the influence exerted in other fields by ideas borrowed from structuralist linguistics, see John E. Joseph, "The Exportation of Structuralist Ideas from Linguistics to Other Fields: An Overview," in *History of the Language Sciences: An International Handbook on the Evolution of the Study of Language from the Beginnings to the Present*, ed. Sylvain Auroux, E. F. K. Koerner, Hans-Josef Niederehe, and Kees Versteegh, vol. 2 (Berlin: De Gruyter, 2001), 1880–1908. On Roman Jakobson's interest in the emerging sciences of information theory and cybernetics, which early on suggested to him the possibility of broader ramifications for central structuralist concepts such as binarism and markedness, see Bernard Dionysius Geoghegan, "From Information Theory to French Theory: Jakobson, Lévi-Strauss, and the Cybernetic Apparatus," *Critical Inquiry* 38, no. 1 (2011): 96–126; and Jürgen van de Walle, "Roman Jakobson, Cybernetics, and Information Theory," *Folia Linguistica Historica* 29, no. 1 (2008): 87–123. It was primarily this information theoretical aspect of Jakobson's work, in turn, that made his presentation of phonology appear relevant to sociologists and communications theorists, from Talcott Parsons (with whom he once considered coauthoring a book) to Jürgen Habermas, Niklas Luhmann, and Anthony Giddens.

4. Jacques Derrida, *Of Grammatology*, trans. Gayatri Chakravorty Spivak (1974; repr., Baltimore: Johns Hopkins University Press, 1997). Original French *De la grammatologie* (Paris: Minuit, 1967).

5. Milner, *L'amour de la langue* and *Le périple structural: Figures et paradigmes* (Paris: Éditions du Seuil, 2002). As will become clear in what follows, my understanding of linguistic structuralism diverges radically from the interpretations offered by both Derrida and Milner, who together represent perhaps the most influential and insightful readers of the early structural linguistic corpus. I disagree with Derrida's central claim that a hidden prioritization of voice and presence over writing and absence shapes the development of Saussurean linguistic theory; and I disagree with Milner that Saussure does little more than explicate the epistemological conditions of the nineteenth-century German, historical linguistic tradition he inherits (a tradition that, according to Milner, itself does little more than begin the long-delayed process of formalizing the study of language according to preexisting Galilean principles). However, as will also become clear in what follows, my account owes to both of these thinkers—so different from one another—a debt more profound than any divergence. The work I do in this book is not recognizably Derridean or Milnerian, but it would not have been possible, in its current form, without the writings of Derrida and Milner.

6. Noam Chomsky, *Cartesian Linguistics: A Chapter in the History of Rational Thought* (New York: Harper & Row, 1966).

7. Among the most significant works in this category—or, indeed, any category of scholarship devoted to the history of Saussurean structuralism—are Boris Gasparov, *Beyond Pure Reason: Ferdinand de Saussure's Philosophy of Language and Its Early Romantic Antecedents* (New York: Columbia University Press, 2013); Ludwig Jäger, *Ferdinand de Saussure: Zur Einführung* (Hamburg: Junius, 2010); Patrice Maniglier, *La vie énigmatique des signes: Saussure et la naissance du structuralisme* (Paris: Éditions Léo Scheer, 2006); and Simon Bouquet, *Introduction à la lecture de Saussure* (Paris: Payot & Rivages, 1997). See also Johannes Fehr, "Ein einleitender Kommentar," in Ferdinand de Saussure, *Linguistik und Semiologie: Notizen aus dem Nachlaß: Texte, Briefe, und Dokumente*, collected and trans. Johannes Fehr (Frankfurt am Main: Suhrkamp, 1997), 17–226. On the surface, at least, the approach closest to my own is probably Gasparov's, since he explicitly seeks to understand Saussure's project against the backdrop of the German Romantic tradition of *Sprachgeist*. Gasparov, however, focuses primarily on early Romantic language philosophy, which precedes the early nineteenth-century discovery, by Jacob Grimm and others, of a rigorously systemic approach to language time. Since it is this early nineteenth-century breakthrough with which I argue that Saussure was principally concerned, Gasparov's conclusions turn out to have relatively little in common with mine. For a much older, precedent-setting attempt to interpret Saussure's work as continuous with the work of his historical precursors, see E. F. K. Koerner, *Ferdinand de Saussure: Origin and Development of His Linguistic Thought in Western Studies of Language* (Braunschweig: Vieweg, 1973). For a recent genealogy of both linguistic *and* anthropological structuralism, which provides a fascinating overview of the various ways in which the boundary between the articulable and the inarticulable has been conceived over the past two centuries, see Markus Wilczek, *Das Artikulierte und das Inartikulierte: Eine Archäologie strukturalistischen Denkens* (Berlin: De Gruyter, 2012).

8. This extraordinary find, comprising pages and pages of previously unknown Saussurean writings, has since been catalogued among the manuscripts held in the Bibliothèque de Genève, and published, though not in full, as part of the *Écrits de linguistique générale*. General consensus dates the notes to the early 1890s, the period during which Saussure first began to concern himself more intensely with the foundations of his discipline. See the "Préface des éditeurs," in Ferdinand de Saussure, *Écrits de linguistique générale*, ed. Simon Bouquet and Rudolf Engler (Paris: Gallimard, 2002), 7–14, esp. 13–14.

9. For the definitive modern articulation of this position by one of Jakobson's students, see Paul Kiparsky, "The Phonological Basis of Sound Change," in *The Handbook of Phonological Theory*, ed. J. A. Goldsmith (Oxford: Blackwell, 1995), 640–70. Kiparksy proposes to account for the relationship between "mechanical" variation and structural logic via a "two-stage theory of sound change according to which the phonetic variation inherent in

speech, which is blind in the Neogrammarian sense, is selectively integrated into the linguistic system and passed on to successive generations of speakers through language acquisition" (315). The conclusions in this paper strike me as both intuitively persuasive and profoundly faithful to the essence of the Jakobsonian project; in what follows, I attribute an incipient, still largely inchoate version of this position to Saussure as well. In doing so, however, I intend only to lay the groundwork for my own interpretive conclusions regarding the true foundations of Saussurean and Jakobsonian theories of system. I make no attempt to intervene in a contemporary debate about the *actual* causes and consequences of historical sound change, a debate that must clearly remain the province of practicing historical linguists.

10. For the exception that proves—and draws its strength from—the rule, see, of course, Gérard Genette, *Mimologics*, trans. Thäis E. Morgan (Lincoln: University of Nebraska Press, 1995); original French: *Mimologiques: Voyage en Cratylie* (Paris: Éditions du Seuil, 1976). Scholarly consensus has long held that even Plato did not affirm the (modified) Cratylan position—according to which language is fundamentally natural with an unavoidable conventional component—that he attributes to Socrates in the dialogue of the same name. For two recent and convincing attempts to argue otherwise, see David Sedley, *Plato's Cratylus* (Cambridge: Cambridge University Press, 2003); and Rachel Barney, *Names and Nature in Plato's Cratylus* (New York: Routledge, 2001).

11. My understanding of "literalization," which insists on the continuing significance of spirit for the modern conception of language as system, is thus importantly different from, albeit related to, the concept developed by Jean-Claude Milner in his work on the epistemology of linguistics, and picked up by Simon Bouquet in his interpretation of Saussure. Milner speaks of "literalization" where Alexander Koyré had used "formalization," to emphasize the fact that the Galilean mathematization of science can be productively reconceived as a matter of scientific writing. This writing, however, becomes "formal" and "literal," for Milner, as for Koyré's Galileo, to precisely the extent that it manages to completely despiritualize the phenomena it describes. Milner, *Introduction*; Bouquet, *Introduction*.

12. I am thinking here, again, primarily of Jean-Claude Milner's approach in *Introduction à une science du langage*, which understands itself, among other things, as a Koyré-inspired corrective to Noam Chomsky's exclusive prioritization of the seventeenth-century French rationalist quest for a timeless, universal grammar. Milner, *Introduction;* Chomsky, *Cartesian Linguistics*. Milner's assimilation of comparative grammar to the scientific project of formalization he finds at work from Galileo onward fails to adequately explicate, in my view, the particularly nineteenth-century challenge of writing down *Sprachgeist*. The technique of formal notation takes on a whole new valence when its object begins to self-articulate. On the relationship between Galilean science and the nineteenth-century German phenomenon of *Wissenschaft*, see also Jean-Claude Milner, Ann Banfield, and Daniel Heller-Roazen, "Interview with Jean-Claude Milner," trans. Chris Gemerchak, *S: Journal of the Jan Van Eyck Circle for Lacanian Ideology Critique* 3 (2010): 4–21.

13. See, of course, Foucault, *Order of Things*, but also, following Foucault, Ian Hacking, *Historical Ontology* (Cambridge, MA: Harvard University Press, 2002), esp. chapters 8 and 9: "How, When, and Where Did Language Go Public?" and "Night Thoughts on Philology," 121–51.

14. Both Ian Hacking and Alan Schrift suggest that Foucault's implicit point of departure can be found in Kant's postcritical anthropological writings, which explicitly broach the question, "What is man?" on the basis of a philosophy that poses the question, "What is knowledge?." Ian Hacking, "Night Thoughts on Philology," in *Michel Foucault: Critical Assessments*, ed. Barry Smart (London: Routledge, 1994), 266–77, here 272; Alan Schrift, "Foucault and Derrida on Nietzsche and the End(s) of Man," in *Michel Foucault: Critical Assessments*, 277–92, here 281.

15. Foucault, *Order of Things*, 373–86 / *Les mots et les chose*, 385–98.

16. The specifically nature-philosophical heritage of the nineteenth-century linguistic theories I explore in this book distinguishes them in decisive ways, I would argue, from the theories of their most important German predecessors, Wilhelm von Humboldt and Johann Gottfried Herder (with whom Jacob Grimm, in particular, is so often compared), both from the perspective of the quest for a formal science of language and, more generally, from the perspective of a history of system theories. Humboldt and Herder remain firmly in the Kantian camp with regard to the exclusively human provenance of reason— despite their emphatically un-Kantian penchant for construing reason in linguistic terms—which means that, for them, the study of language and its structure has little in common with the science of nature and its laws. The history of language does not run parallel to the history of the universe, and the historical development of language systems can tell us nothing useful about the emergence of worldly order. Under such conditions, the larger question of the relationship between time, in general, and system, in general, which is the question that primarily concerns me here, need not and does not arise.

17. It is Jakobson who is generally credited with having introduced the term "structuralism," in the sense in which I use it here, to refer to the approach he and his colleague Nikolai Trubetzkoy—cofounder, with Jakob-son, of the linguistic discipline of phonology—considered themselves to share with Saussure.

18. Jakobson first begins to elaborate his notion of the linguistic zero toward the end of the 1930s, concurrently with his first general formulations of a phonological system theory. See, for instance, Roman Jakobson, "Signe Zéro," in *Selected Writings*, vol. 2, *Word and Language* (The Hague: Mouton, 1971), 211–19; and "Das Nullzeichen," ibid., 220–22. The concept of a zero-degree sign has been the object of many sophisticated reflections, in particular by French structuralists and poststructuralists. See, in particular, Claude Lévi-Strauss, *Introduction to the Work of Marcell Mauss*, trans. Felicity Baker (London: Routledge, 1987); original French: "Introduction à l'œuvre de Marcel Mauss" (1950), in Marcel Mauss, *Sociologie et anthropologie*, 4th ed. (Paris: PUF, 1968), vii–lii; Jacques Derrida, "Structure, Sign, and Play in the Discourse of the Human Sciences," in *Writing and Difference*, trans. Alan Bass (London: Routledge, 1978), 278–94; original French: "La structure, le signe, et le jeu dans le discours des sciences humaines," in

L'écriture et la différence (Paris: Seuil, 1967), 409–28; and Gilles Deleuze, "How Do We Recognize Structuralism?," in *Desert Islands and Other Texts, 1953–1974,* trans. Michael Taormina, ed. David Lapoujade (Los Angeles: Semiotext[e], 2004), 170–92; original French: "À quoi reconnaît-on le structuralisme?" (1972), in *Le XXe siècle, Histoire de la philosophie: Idées, doctrines,* vol. 8, ed. Françoise Châtelet (Paris: Hachette, 2000), 299–335. My argument seeks to demonstrate that these past treatments are correct to raise the zero to the level of keystone for a structuralist concept of language-as-system, but that they are wrong—or rather, crucially ahistorical—in their interpretations of its actual, foundational role.

19. "Panchronic" is Jakobson's own word, which he introduces in the context of his work on child language development. See Roman Jakobson, *Child Language, Aphasia, and Phonological Universals,* trans. Allan R. Keiler (The Hague: Mouton, 1968); original German: *Kindersprache, Aphasie, und Allgemeine Lautgesetze* (1941), in *Selected Writings,* vol. 1, *Phonological Studies* (The Hague: Mouton, 1962), 328–401. I discuss this usage in chapter 6, 218–19.

PART I. "THE ETERNAL ETYMOLOGY": FROM *SPRACHGEIST* TO FERDINAND DE SAUSSURE

1. Ferdinand de Saussure, *Writings in General Linguistics,* trans. Carol Sanders and Matthew Pires (Oxford: Oxford University Press, 2006), 11; original French: *Écrits de linguistique générale,* ed. Simon Bouquet and Rudolf Engler (Paris: Gallimard, 2002), 26. Hereafter cited as *WGL*, with the English pagination followed by the French. I have modified the translation where necessary to preserve aspects of the original French formulations that are relevant to my argument.

2. See "Introduction," note 8.

CHAPTER 1. LANGUAGE ENSOULED

1. Historians of linguistics, on the whole, have tended to accept this claim of a revolution in method, although they have also tended to qualify its scope. See, for two classic examples, Winfred Lehmann, *Historical Linguistics: An Introduction* (London: Routledge, 1992), 27ff., and E. F. K. Koerner, "The Concept of 'Revolution' in Linguistics: Historical, Philosophical, and Methodological Issues," in *Linguistic Historiography: Projects and Prospects* (Amsterdam: Benjamins, 1999), 85–96. See also, more recently, Jean Rousseau, "La genèse de la grammaire comparée," in *History of the Language Sciences: An International Handbook on the Evolution of the Study of Language from the Beginnings to the Present,* ed. Sylvain Auroux, E. F. K. Koerner, Hans-Josef Niederehe, and Kees Versteegh, vol. 2 (Berlin: De Gruyter, 2001), 1197–210. The notion of a linguistic paradigm shift has been most extensively and influentially defended by Michel Foucault, in his account of the early nineteenth-century transition from a classical-representational to a historical-empirical

"episteme." See Michel Foucault, *The Order of Things: An Archaeology of the Human Sciences*, trans. Alan Sheridan (New York: Vintage, 1973), 217–343 / *Les mots et les choses: Une archéologie des sciences humaines* (Paris: Gallimard, 1966), 229–313. For an opposing view, which denies to Rasmus Rask, Bopp, and Grimm the status of founders of a discipline, see Roy Harris's openly polemical take in his introduction to *Foundations of Indo-European Comparative Philology, 1800–1850*, ed. Roy Harris (1816; repr., London: Routledge, 1999), 1:2–18.

2. See, for instance, the following definition of linguistic perfection by the eighteenth-century Swiss philosopher Johann Georg Sulzer: "For the perfection of any language, one requires primarily three different things. 1) An adequate supply of good words and idioms, through which every concept is clearly and definitely [*deutlich und bestimmt*] expressed. 2) A sufficient number of clear and well-differentiated modifiers of nouns and verbs, through which the concepts receive their particular circumstantial definition [*Bestimmung*] [. . .]. 3) A flexibility in the composition of many words into a sentence, such that a whole thought can be presented definitely [*bestimmt*], correctly, and in accordance with the nature of the matter at hand, either delicately or emphatically"; Johann Georg Sulzer, *Kurzer Begriff aller Wissenschaften und anderer Theile der Gelehrsamkeit, worin jeder nach seinem Inhalt, Nuzen, und Vollkommenheit kürzlich beschrieben wird* (Leipzig: Johann Christian Langenheim, 1759), 11–12. See also John Locke, "On the Abuse of Words," in *An Essay Concerning Human Understanding* (1690), ed. Peter Nidditch (Oxford: Oxford University Press, 1975), 490–508.

3. A particularly high-profile example of lay interest can be found in Johann Wolfgang Goethe, whose *West-östlicher Divan* was influenced by his reading of Jones. Goethe himself cites Jones as a source in his "Noten und Abhandlungen zu besserem Verständnis des West-östlichen Divans," in *Werke: Weimarer Ausgabe*, ed. Gustav von Loeper, Erich Schmidt, and Paul Raabe (Weimar: Hermann Böhlau, 1888), I:7, 218–20. See also Walter Veit, "Goethe's Fantasies about the Orient," *Eighteenth-Century Life* 26, no. 3 (2002): 164–80.

4. Sir William Jones, "The Third Anniversary Discourse, on the Hindus, Delivered to the Asiatic Society, 2 February 1786," in *Sir William Jones: Selected Poetical and Prose Works*, ed. Michael Franklin (Cardiff: University of Wales Press, 1995), 355–70, here 361. On the power of Jones's formulation, and its effect on the imaginations of his fellow scholars, see Garland Cannon, "Jones's 'Sprung from Some Common Source': 1786–1986," in *Sprung from Some Common Source: Investigations into the Prehistory of Languages*, ed. Sydney M. Lamb and E. Douglas Mitchell (Stanford, CA: Stanford University Press, 1991), 23–47.

5. Friedrich Schlegel, *On the Language and Wisdom of the Indians*, in *A Reader in Nineteenth-Century Historical Indo-European Linguistics*, trans. and ed. Winfred P. Lehmann (Bloomington: Indiana University Press, 1967); original German *Über die Sprache und Weisheit der Indier: Ein Beitrag zur Begründung der Alterthumskunde* (1808; repr., Amsterdam: Benjamins, 1977). Schlegel's brother, August Wilhelm Schlegel, eventually followed suit,

publishing the three-volume journal *Indische Bibliothek* (1823–30) and editing both the *Bhagavad Gita* (1823) and the *Ramayana* (1829).

6. F. Schlegel, *Language and Wisdom*, 25 / *Sprache und Weisheit*, 28. On Schlegel's own understanding of language science, and its relationship to the comparative anatomical perspective, see Michael Eggers, "Von Pflanzen und Engeln: Friedrich Schlegels Sprachdenken im Kontext der frühen Biologie," in *Die Lesbarkeit der Romantik: Material, Medium, Diskurs*, ed. Erich Kleinschmidt (Berlin: De Gruyter, 2009), 159–83.

7. Franz Bopp, *Über das Conjugationssystem der Sanskritsprache in Vergleichung mit jenem der griechischen, lateinischen, persischen, und germanischen Sprache*, in Harris, *Foundations of Indo-European Comparative Philology*, vol. 1. Bopp studied under the polymathic Karl Joseph Windischmann—court doctor to the Elector of Mainz, friend of Friedrich Schelling, passionate student of ancient Indian texts, translator of Plato's *Timaeus*—during his time as a gymnasium student in Aschaffenburg.

8. The phrase stems from Bopp's book-length 1827 review of Jacob Grimm's *Deutsche Grammatik*, and is intended to designate the approach that he and Grimm have in common. Franz Bopp, *Vocalismus, oder sprachvergleichende Kritiken über J. Grimm's deutsche Grammatik und Graff's althochdeutschen Sprachschatz, mit Begründung einer neuen Theorie des Ablauts* (Berlin: Nicolaische Buchhandlung, 1836), 83.

9. Precursors from earlier in the eighteenth century include Gottfried Wilhelm Leibniz, the Scottish linguist Lord Monboddo, and the students of Dutch philologist Tiberius Hemsterhuis (founder of an approach known as the "Schola Hemsterhusiana"), all of whom practiced some form of linguistic genealogy prior to Jones's "discovery" of Sanskrit. On Monboddo and the Dutch philologists, see Jan Noordegraaf, "Dutch Philologists and General Linguistic Theory: Anglo-Dutch Relations in the Eighteenth Century," in *Linguists and Their Diversions: A Festschrift for R. H. Robins on His 75th Birthday*, ed. Vivian A. Law and Werner Hüllen (Münster: Nodus Publikationen, 1996), 211–43. On Leibniz, see Tullio De Mauro and Lia Formigari, eds., *Leibniz, Humboldt, and the Origins of Comparativism* (Amsterdam: Benjamins, 1990).

10. Bopp, *Vocalismus*, 1. This passage has been previously translated by Otto Jespersen, *Language: Its Nature, Development, and Origin* (New York: Henry Holt and Company, 1922), 65; and E. F. K. Koerner, *Professing Linguistic Historiography* (Philadelphia: Benjamins North America, 1995), 50.

11. As already mentioned above, the relationship between contemporaneous developments within the new science of "comparative anatomy" and the broader domain of late eighteenth- and early nineteenth-century philosophical thought has received a great deal of attention in recent years. Michel Foucault's classic treatment of this nexus focuses on the comparative anatomical theories of the French figure of Georges Cuvier. Most of the more recent approaches, however, focus instead, as I will do here, on the earlier German figures of Johann Friedrich Blumenbach and Carl Friedrich Kielmeyer. See Foucault, *The Order of Things*, esp. 263–79 / *Les mots et les choses*, 275–92; Timothy Lenoir, *The Strategy of Life: Teleology and Mechanics in Nineteenth-Century German Biology* (Chicago: University of Chicago

Press, 1982); Helmut Müller-Sievers, *Self-Generation: Biology, Philosophy, and Literature around 1800* (Stanford, CA: Stanford University Press, 1997); Robert J. Richards, *The Romantic Conception of Life: Science and Philosophy in the Age of Goethe* (Chicago: University of Chicago Press, 2002); Iain Hamilton Grant, *Philosophies of Nature after Schelling* (London: Continuum, 2006); and Leif Weatherby, *Transplanting the Metaphysical Organ: German Romanticism between Leibniz and Marx* (New York: Fordham University Press, 2016). For an attempt to understand the role played by the concept of an articulated system during this period, *without* reference to the notion of life, see Markus Wilczek, *Das Artikulierte und das Inartikulierte: Eine Archäologie strukturalistischen Denkens* (Berlin: De Gruyter, 2012), 92–113.

12. On the origins of the debate in the first half of the eighteenth century, see Shirley T. Roe, *Matter, Life, and Generation: 18th Century Embryology and the Haller-Wolff Debate* (Cambridge: Cambridge University Press, 1981).

13. "Develop," *Oxford English Dictionary Online*, Oxford University Press, accessed September 19, 2016, http://dictionary.oed.com/. "Evolve," *Oxford English Dictionary Online*, Oxford University Press, September 19, 2016, http://dictionary.oed.com/.

14. The notion of multiple kinds of causality, along with the link between mechanism and meteorology, goes back to Aristotle, whose discussion of causation in the *Physics* declares both explanatory modes to be necessary: the "for the sake of which" of final ends, as the ultimate goal of all knowledge, should take priority for the observer of nature, Aristotle asserts, *except* in cases, such as a rainstorm or an eclipse, where final ends or purposes do not exist. With regard to such phenomena, a mechanical "source of change or rest" may be accepted as the primary cause. See Christopher Byrne, "Aristotle on Physical Necessity and the Limits of Teleological Explanation," *Apeiron* 35 (2002): 20–46; Monte Ransom Johnson, *Aristotle on Teleology* (Oxford: Oxford University Press, 2005); and Mariska Leunissen, *Explanation and Teleology in Aristotle's Science of Nature* (Cambridge: Cambridge University Press, 2010).

15. Leibniz, famously, was simultaneously a teleologist and a mechanist, and his notion of matter as animated by a teleological *vis viva* provided an important precedent for the comparative anatomical theories that followed. For an account of the Leibnizian position, see Jonathan Bennett, "Leibniz's Two Realms," in *Leibniz: Nature and Freedom*, ed. Donald Rutherford and J. A. Cover (Oxford: Oxford University Press, 2005), 135–55. On the Rationalist relationship to science more generally, see Rocco Gennaro and Charles Huenemann, eds., *New Essays on the Rationalists* (Oxford: Oxford University Press, 1999). On Cartesian science, see Daniel Garber, *Descartes' Metaphysical Physics* (Chicago: University of Chicago Press, 1992).

16. With regard to the origins and outlines of the eighteenth-century mechanist account, I follow here the account provided by Roe in *Matter, Life, and Generation*. The seventeenth- and eighteenth-century scholars themselves tended to speak not in terms of mechanism and teleology but of preformation (or evolution) and epigenesis. For an account of the significance

and history of these terms, see Roe *Matter, Life, and Generation*, 1–20; and Müller-Sievers, *Self-Generation*, 1–47.

17. The phrase is from the Leibnizian philosopher Caspar Friedrich Wolff, who insists in his *Theoria generationis* of 1759 that embryonic development can be explained only with reference to a teleological "principle of generation, or essential force" (*vis essentialis*). Cited in Richards, *Romantic Conception of Life*, 215ff. Wolff's *vis essentialis* is a descendant of Leibniz's *vis viva* and a clear progenitor of Blumenbach's later *Bildungstrieb*. For an extensive discussion of the Leibnizian position, see Justin E. H. Smith, *Divine Machines: Leibniz and the Sciences of Life* (Princeton, NJ: Princeton University Press, 2011); on the relationship between the Leibnizian position and the Kantian one, see Weatherby, *Transplanting the Metaphysical Organ*, chapters 1 and 2.

18. Johann Friedrich Blumenbach, *Über den Bildungstrieb und das Zeugungsgeschäft* (Göttingen: Johann Christian Dieterich, 1781), 31–32. An English translation of the entire essay, which is so loose that I was unable to use it as the basis for my own, was published in 1792; Johann Friedrich Blumenbach, *An Essay on Generation*, trans. A. Crichton (London: printed for T. Cadell, Strand, 1792).

19. Blumenbach's discussion of the epistemological status of his formative drive includes a footnote on Newton. Johann Friedrich Blumenbach, *Über den Bildungstrieb*, 2nd ed. (Göttingen: Johann Christian Dieterich, 1789), 25–26.

20. On the role played by Blumenbach for Kant, see Robert J. Richards, "Kant and Blumenbach on the *Bildungstrieb*: A Historical Misunderstanding," *Studies in the History and Philosophy of Biological and Biomedical Science* 31, no. 1 (2000): 11–32; and Timothy Lenoir, "Kant, Blumenbach, and Vital Materialism in German Biology," *Isis* 71 (1980): 77–108. On the role played by the notion of a teleological organicism in the development of Kant's philosophy, see Müller-Sievers, *Self-Generation*, 48–64; Philippe Huneman, "Reflexive Judgment and Wolffian Embryology: Kant's Shift between the First and the Third Critiques," in *Understanding Purpose: Kant and the Philosophy of Biology*, ed. Philippe Huneman (Rochester, NY: University of Rochester Press, 2007), 75–100; and Weatherby, *Transplanting the Metaphysical Organ*, chapter 2. On the role played by Kant's notion of a teleological organicism in post-Kantian philosophy, see Eckart Förster, "Die Bedeutung von §§76, 77 der *Kritik der Urteilskraft* für die Entwicklung der nachkantischen Philosophie [Teil 1]," *Zeitschrift für philosophische Forschung* 56 (2002): 185–86. For a lucid reconstruction of the relationship between Kant's mechanist understanding of epigenesis and his theory of rationality in the context of the *Critique of Pure Reason*, see Jennifer Mensch, *Kant's Organicism: Epigenesis and the Development of Critical Philosophy* (Chicago: University of Chicago Press, 2013).

21. Immanuel Kant, *Critique of the Power of Judgment*, trans. Paul Guyer and Eric Matthews, ed. Paul Guyer (Cambridge: Cambridge University Press, 2007), 246 (translation very slightly modified); original German: *Kritik der Urteilskraft*, in *Gesammelte Werke*, ed. Königlich Preußische Akademie

der Wissenschaften (Berlin: De Gruyter, 1908), 5:374. The organism-clock antithesis is Kant's version of a traditional trope. For another famous example, see the following remarks by Bernard de Fontenelle: "Do you say that beasts are machines just as watches are? Put a male dog-machine and a female dog-machine side by side, and eventually a third little machine will be the result, whereas two watches will lie side by side all their lives without ever producing a third watch"; Bernard Le Bovier de Fontenelle, *Lettres diverses de M. le chevalier d'Her . . .* , reprinted as *Lettres galantes* in *Oeuvres* (Paris: Libraires Associés, 1766), 1:312. Translated in Roe, *Matter, Life and Generation*, 1.

22. Kant, *Critique of the Power of Judgment*, 204 / *Kritik der Urteilskraft*, 5:376.

23. "Joint" is the original meaning of the German *Glied*, which shares a root with *Gelenk*. Jacob Grimm and Wilhelm Grimm, "Glied," in *Deutsches Wörterbuch* (Leipzig: S. Hirzel, 1854–1960), vol. 8, columns 5–6.

24. "In such a product of nature each part is conceived as if it exists only through all the others, thus as if existing for the sake of the others and on account of the whole, *i.e.*, *as an instrument (organ)*, which is, however, not sufficient (for it could also be an instrument of art, and thus represented as possible at all only as an end); rather it must be thought of as an organ that produces the other parts (consequently each produces the others reciprocally), which cannot be the case in any instrument of art, but only of nature, which provides all the matter for instruments (even those of art)"; *Critique of the Power of Judgment*, 245 / *Kritik der Urteilskraft*, 5:373–74; emphasis mine.

25. See, for example, Johann Friedrich Blumenbach, *Handbuch der Naturgeschichte*, 12th ed. (Göttingen: Dieterichsche Buchhandlung, 1830), 17. Cited in Richards, "Kant and Blumenbach," 11. Whether Blumenbach actually subscribed to the Kantian interpretation of his central idea, independently of his willingness to propagate it, is a matter of debate. Traditionally, it has been assumed that he did. For the argument that he did not, or at least, did not do so wholeheartedly, see Richards, "Kant and Blumenbach."

26. Carl Friedrich Kielmeyer, *Gesammelte Schriften*, ed. F. H. Holler (Berlin: Keiper, 1938), 56; emphasis in the original. Translations of several passages from Kielmeyer's writings, including this one, are provided by Grant, *Philosophies of Nature after Schelling*, here 120; others can be found in Robert J. Richards, *Romantic Conception of Life*. Translations that follow are my own, produced on the basis, wherever possible, of the ones included in these two works. In my account of Kielmeyer's contribution, I will follow the broad outlines provided by Richards and Grant, both of whom argue, against the earlier narrative in Lenoir, *Strategy of Life*, that Kielmeyer's approach must be understood as genuinely teleological, and thus nature-philosophical, rather than merely "teleomechanist," and thus Kantian, in its thrust. For statements of the respective positions, see Lenoir, *Strategy of Life*, 37–53; Richards, *Romantic Conception of Life*, 216–51; and Grant, *Philosophies of Nature after Schelling*, 119–57.

27. Kielmeyer, *Gesammelte Schriften*, 92ff.

28. "The previous material [*Stoff*] of possible observations related to developments, which would be at the same time the material [*Stoff*] of a description of those developments, acquires new growth and order [*Zuwachs und Anordnung*] as a result of the time in which it is given to us [. . .]. Only through this supplement [of time] does dead description become living history, and only in this way does the object itself, with which the description is concerned, acquire life"; Kielmeyer, *Gesammelte Schriften*, 115. The passage stems from Kielmeyer's introductory notes for a never completed comprehensive treatment of the principles of organic development ("Ideen zu einer allgemeineren Geschichte und Theorie der Entwicklungserscheinungen der Organisationen" [Ideas for a more universal history and theory of the developmental phenomena of organizations]). The notes end with a comprehensive classification of *kinds* of change, which is intended to underpin the traditional classification of animal types. Kielmeyer, *Gesammelte Schriften*, 110–24.

29. Reprinted in Kielmeyer, *Gesammelte Schriften*, 59–101.

30. Kielmeyer, *Gesammelte Schriften*, 46.

31. Kielmeyer, "Über die Verhältnisse," in *Gesammelte Schriften*, 93–94.

32. Kielmeyer, *Gesammelte Schriften*, 206. Peculiarly, Kielmeyer claims in the sentence directly following this passage to have expressed himself even more clearly on this subject in the *Speech* itself, despite the fact that the *Speech* nowhere explicitly draws the conclusions in question. His reasoning is presumably that the *Speech* lays out, more systematically than in the unpublished manuscripts, the steps that led him to this position.

33. Kielmeyer, *Gesammelte Schriften*, 57–58.

34. The phrase stems from an unpublished text titled "Entwurf einer vergleichenden Zoologie" [Sketch of a comparative zoology], ibid., 27.

35. Ibid., 213.

36. Richards speculates that Kielmeyer may have published so little during his lifetime in part out of concern for the conservative sensibilities of Duke Karl Eugen, at whose whim he held his Karlsschule teaching position. Richards, *Romantic Conception of Life*, 238.

37. For Goethe's own account of the meeting, see the diary entry from September 9, 1797, in Johann Wolfang von Goethe, *Tagebücher* in *Werke* (Weimar: Hermann Böhlau, 1888), II:3, 129–30. On Goethe's relationship to Kielmeyer, see also Gabrielle Bersier, "Visualizing Carl Friedrich Kielmeyer's Organic Forces: Goethe's Morphology on the Threshhold of Evolution," *Monatshefte* 97, no. 1 (2005): 18–32.

38. Cited in Richards, *Romantic Conception of Life*, 238

39. See Friedrich von Hardenberg, vol. 3 of *Novalis' Schriften: Historisch-kritische Ausgabe* (hereafter *HKA*), ed. Richard Samuel, Hans-Joachim Mähl, and Gerhard Schulz (Stuttgart: W. Kohlhammer, 1960–2006), 3:432.

40. On the subject of Kielmeyer's significance for fellow zoologists and physiologists, including Georges Cuvier, see Lenoir, *Strategy of Life*, 54–71; Volker Hess, "Das Ende der 'Historia naturalis'?," in *Philosophie des Organischen in der Goethezeit: Studien zu Werk und Wirkung des Naturforschers Carl Friedrich Kielmeyer (1765–1844)*, ed. Kai Torsten Kanz

(Stuttgart: Franz Steiner, 1994), 153–73; and Grant, *Philosophies of Nature after Schelling*, 12–14 and 124–28. It is certainly true that Cuvier was the better known and more influential of the two scientists; nevertheless, the tendency to see in Cuvier's *Leçons* from 1800–1805 the origin of an epistemic shift in the direction of function and time, over against a timeless taxonomy of visible bodies—a shift that is clearly already well under way by the time of Kielmeyer's significantly earlier and demonstrably influential *Rede*—undoubtedly derives from the treatment Cuvier receives at the hands of Foucault, who neglects to mention either Blumenbach or Kielmeyer. Foucault's treatment may also be indirectly responsible for the quite improbable but nonetheless widespread assumption that Schlegel's use of the phrase "vergleichende Anatomie" derives directly from his reading of Cuvier. As should by now be clear, the notion of comparative anatomy as a powerful new disciplinary paradigm had become widespread in German intellectual circles well before the end of the eighteenth century, and could have reached Schlegel from any number of other sources, including his one-time friend Schelling's *Naturphilosophie*. In the contrast between Kielmeyer's understanding of system as a temporally unfolding typology, and Cuvier's conception of fixed Aristotelian classes, Grant sees an earlier version of the opposition that would later characterize the debate between Cuvier and Geoffroy Saint-Hilaire, on which Gilles Deleuze reflects in *Différence et répetition* (Paris: Presses Universitaires de France, 1968), as well as in *Mille Plateaux* (Paris: Éditions de Minuit, 1980). See Grant, *Philosophies of Nature after Schelling*, 124ff.

41. Friedrich Schelling, *Von der Weltseele: Eine Hypothese der höhern Physik zur Erklärung des allgemeinen Organismus*, in *Werke: Historisch-kritische Ausgabe* (hereafter *HKA*), ed. Thomas Buchheim, Jochem Hennigfeld, Wilhelm G. Jacobs, Jörg Jantzen und Siegbert Peetz (Stuttgart: Friedrich Frommann Verlag, 1976f.), I:6 (2000), 253. A short but important section of the work has been translated by Iain Hamilton Grant, "F. W. J. Schelling, 'On the World Soul,' Translation and Introduction," in *Collapse*, vol. 6, *Geo/Philosophy*, ed. R. Mackay (Bristol: University of West of England, 2010), 58–95. On the significance of Kielmeyer for Schelling, see Thomas Bach, *Biologie und Philosophie bei C. F. Kielmeyer und F. W. J. Schelling* (Stuttgart-Bad Cannstatt: Frommann-Holzboog, 2001); and Weatherby, *Transplanting the Metaphysical Organ*, chapter 5.

42. On the relationship between Goethe and Schelling, see Erwin Jäckle, "Goethes Morphologie und Schellings Weltseele," *Deutsche Vierteljahrsschrift für Literaturwissenschaft und Geistesgeschichte* 15 (1937): 295–440; and, more recently, Jeremy Adler, "The Aesthetics of Magnetism: Science, Philosophy and Poetry in the Dialogue between Goethe and Schelling," in *The Third Culture: Literature and Science*, ed. Elinor S. Shaffer (New York: De Gruyter, 1997), 66–102; Richards, *Romantic Conception of Life*, 463–71; Olaf Breitbach, *Goethes Metamorphosenlehre* (Munich: Fink, 2006), 214–25; Gabriel Trop, "Poetry and Morphology: Goethe's 'Parabase' and the Intensification of the Morphological Gaze," *Monatshefte* 105, no. 3 (2013): 389–406; and Dalia Nassar, *The Romantic Absolute: Being and Knowing in Early German Romantic Philosophy* (Chicago: University of Chicago

Press, 2014), 157–211 (Nassar argues, against the prevailing tendency to presume the influence of Schelling on Goethe, for a more nuanced picture of mutual determination). Friedrich von Hardenberg, according to his study notes, devoted the late summer of 1798 to intensive study of Schelling's *Von der Weltseele*. See Hardenberg, *HKA*, 3:102–3. The effects of this encounter become abundantly clear in the fragments of the "Das allgemeine Brouillon," which he began writing in that same year, and which are permeated by nature-philosophical reflections. See Friedrich von Hardenberg, "Das allegemeine Brouillon," *HKA*, 3:242–478. For a classic account of Schelling's relationship to the early Romantic project that stresses their affinities, see Frederick Beiser, *The Romantic Imperative: The Concept of Early German Romanticism* (Cambridge, MA: Harvard University Press, 2003), esp. 131–70. For an account from the other side of the spectrum, see Manfred Frank, *Unendliche Annäherung: Die Anfänge der philosophischen Frühromantik* (Frankfurt am Main: Suhrkamp, 1997). Frank argues that both Hardenberg and Friedrich Schlegel remained dedicated proponents of the Kantian critical position, and that, despite Hardenberg's early interest in *Naturphilosophie*, he will ultimately reject *all* Idealist attempts, including Schelling's, to conceptualize the Absolute.

43. Some of the most significant figures among the working scientists include, along with Goethe himself, the philosopher and physiologist Karl Joseph Windischmann, the geologist Henrik Steffens, the physicist Johann Wilhelm Ritter, and the naturalist polymath Alexander von Humboldt, all of whom avowedly discovered in *Naturphilosophie* a perspective more congenial to their own natural scientific praxis than the Kantian one. Kielmeyer himself, in a letter to the skeptically minded Cuvier, credits Schelling with a fundamentally correct understanding of nature, but disapproves of the latter's transempirical methods. Kielmeyer, "Über Kant und die deutsche Naturphilosophie (Ein Schreiben an Cuvier)," *Gesammelte Schriften*, 249–52. For an extensive treatment of the natural scientific backdrop that informs Schelling's own nature-philosophical writings, see Manfred Durner, Francesco Moiso, and Jörg Jantzen, *Ergänzungsband zu Werke Band 5 bis 9: Wissenschaftshistorischer Bericht zu Schellings Naturphilosophische Schriften, 1797–1800*, *HKA*.

44. Friedrich Schelling, *First Outline of a System of the Philosophy of Nature*, trans. Keith R. Peterson (Albany, NY: SUNY Press, 2004), 141; original German: *Erster Entwurf eines Systems der Naturphilosophie*, *HKA* I:7, 210.

45. Schelling, "Über das Verhältnis des Realen und Idealen in der Natur, oder Entwicklung der ersten Grundsätze der Naturphilosophie an den Prinzipien der Schwere und des Lichts" [On the relationship of the real and the ideal in nature, or a development of the first principles of nature-philosophy out of the principles of weight and light], in *Von der Weltseele*, 2nd ed. (Hamburg: Perthes, 1806), xvii–liv, here xxi. The treatise, which was originally published as an introduction to the revised second edition of *Von der Weltseele*, will be included in the as-yet unpublished volume 16 of *HKA*.

46. See Schelling, *First Outline*, 141 / *Erster Entwurf, HKA*, I:7, 210.
47. Schelling, *Von der Weltseele, HKA*, I:6, 70–71.
48. Friedrich Schelling, *Timaeus* (1794), ed. Hartmut Buchner (Stutt-gart: Frommann-Holzboog, 1994). Schelling's early Plato commentary remained unpublished until the appearance of this 1994 edition, which has given rise to a great deal of exciting new scholarship. On the role of Plato's physics for the development of nature-philosophy, see Hermann Krings, "Genesis und Materie—Zur Bedeutung der 'Timaeus'-Handschrift für Schellings Naturphilosophie," published together with the commentary itself; Michael Franz, *Schellings Tübinger Platon-Studien* (Göttingen: Vanderhoeck & Ruprecht, 1996); John Sallis, "Secluded Nature: The Point of Schelling's Reinscription of the Timaeus," *Pli* 8 (1999): 71–85; Tanja Gloyna, *Kosmos und System: Schellings Weg in die Philosophie* (Stutt-gart: Frommann-Holzboog, 2002); Christoph Asmuth, *Transformation: Das Platonbild bei Fichte, Schelling, Hegel, Schleiermacher, und Scho-penhauer und das Legitimationsproblem der Philosophiegeschichte* (Göt-tingen: Vandenhoeck & Ruprecht, 2006), 47–124; Grant, *Philosophies of Nature after Schelling*, esp. 26–58 and 119–57; and Matthews, *Schelling's Organic Form of Philosophy*, 103–36. On the role of Platonic philoso-phy for Romanticism and German Idealism more generally, see, in addi-tion to Christoph Asmuth's *Transformation*; Beiser, *Romantic Imperative*, 56–72; and Burkhard Mojsisch and Orrin Summerell, eds., *Platonismus im Deutschen Idealismus: Die platonische Tradition in der klassischen deutschen Philosophie* (Munich: K. G. Saur, 2003). For important treat-ments of Schelling's relationship to Plato from the decade prior to the pub-lication of the *Timaeus* commentary, see Dieter Henrich, "Der Weg des spekulativen Idealismus: Ein Resumé und eine Aufgabe," in *Jakob Zwill-ings Nachlaß: Eine Rekonstruktion*, ed. Dieter Henrich and Christoph Jamme (*Hegel-Studien Beiheft* 28, 1986), 77–96; and Birgit Sandkaulen-Bock, *Ausgang vom Unbedingten: Über den Anfang in der Philosophie Schellings* (Göttingen: Vandenhoeck & Ruprecht, 1990).
49. For hints that Kielmeyer does indeed understand conceptual develop-ment in terms of organic development, and rational organization in terms of life, even if he does not himself seek to elaborate on the parallel, see Kiel-meyer, *Gesammelte Schriften*, 52–53 and 95ff.
50. See, for instance, Plato, *Philebus*, 29d–30a, trans. Dorothea Frede, in *Complete Works*, ed. John M. Cooper and D. S. Hutchison (Indianapo-lis: Hackett, 1997), 417–18: "SOCRATES: To the combination of all these elements taken as a unit [gathered up in a "one"] we give the name 'body,' don't we? PROTARCHUS: Certainly. SOCRATES: Now, realize that the same holds in the case of what we call the ordered universe. It will turn out to be a body in the same sense, since it is composed of the same elements. [. . .] SOCRATES: Of the body that belongs to us, will we not say that it has a soul [*psyche*]? PROTARCHUS: Quite obviously that is what we will say. SOCRATES: But where does it come from, unless the body of the universe which has the same properties as ours, but more beautiful in all respects, happens to possess a soul?"

51. At stake here is Plato's famous account of the "four kinds"—*apeiron*; *peras*; the mixture of *apeiron* and *peras*; and the cause of the mixture—by means of which the Socrates of the *Philebus* represents the genesis of harmonious worldly structure. In what follows, I will provide an overview of this Platonic theory of harmony, and its Pythagorean-mathematical backdrop, which seeks primarily to make available the basic ideas I believe Schelling to have inherited therefrom. My overview should thus *not* be interpreted as an attempt to summarize the "true meaning" of Plato's extraordinarily enigmatic cosmology. On the role of Pythagorean concepts in the *Philebus*, see Carl A. Huffman, "The Philolaic Method: The Pythagoreanism behind the *Philebus*," in *Essays in Ancient Greek Philosophy*, vol. 6, *Before Plato*, ed. A. Preus (Albany, NY: SUNY Press, 2001), 67–85. For a sustained investigation into the extent and significance of Plato's Pythagoreanism more generally, see Phillip Sydney Horky, *Plato and Pythagoreanism* (Oxford: Oxford University Press, 2013).

52. Plato, *Philebus*, 16c–d, in *Complete Works*, 404.

53. Plato, *Phaedrus*, 265e, trans. Alexander Nehamas and Paul Woodruff, in *Complete Works*, 542.

54. *Philebus*, 18b–c, in *Complete Works*, 406.

55. *Philebus*, 18c–d, in *Complete Works*, 406 (translation slightly modified).

56. *Philebus*, 17c–d, in *Complete Works*, 405.

57. *Philebus*, 23d, in *Complete Works*, 404.

58. Plato, *Timaeus*, 31b–c, trans. Donald J. Zyle, in *Complete Works*, 1237.

59. Plato, *Timaeus*, 43a, in *Complete Works*, 1246. The question of Plato's own relationship to the cosmological principles expounded by the character of Timaeus in this dialogue is famously fraught, particularly since Socrates's voice remains all but absent throughout. Schelling's commentary, however, ignores the potentially ironic implications of this silence, treating the substance of the dialogue as a more or less straightforward expression of Platonic principles.

60. Schelling's understanding of induction and deduction is complex and—complexly—indebted to the Kantian concepts of transcendental deduction and reflective judgment. The details of the theory can be left aside here.

61. Schelling, *Von der Weltseele*, *HKA*, I:6, 257.

62. Schelling, *Timaeus*, 28. For the original Platonic version of this equation, which is phrased and affirmed as a question, see Laws 896a: "Do you mean the entity which we all *call* soul is precisely that which is *defined* by the expression 'self-generating motion'?" Plato, *Laws*, trans. Trevor J. Saunders, in *Complete Works*, 1552.

63. Plato, *Timaeus* 37d, in *Complete Works*, 1241, and Plato, *Timaeus*, 27d–28c, in *Complete Works*, 1234–35.

64. This is not to suggest, of course, that Schelling simply inherits his notion of a teleological temporality from Kielmeyer. On the complicated emergence of teleology as a category of German Idealist thought (in connection with the categories of "striving" and "potency"), see Frederick Beiser, *German Idealism: The Struggle against Subjectivism, 1781–1801*

(Cambridge, MA: Harvard University Press, 2002). On Schelling's under-standing thereof as an interpretation of the Fichtean notion of imaginative power, or *Einbildungskraft*, see Christoph Asmuth, "'Das Schweben ist der Quell aller Realität': Platner, Fichte, Schlegel, Novalis und die produktive Einbildungskraft," in *System and Context: Early Romantic and Early Ide-alistic Constellations/System und Kontext: Frühromantische und Frühide-alistische Konstellationen*, ed. Rolf Ahlers (New York: New Athenaeum / Toronto: Neues Athenaeum, 2004), 349–74. For a thorough treatment of the development of Schelling's early, nature-philosophical thought and its (other) sources, see Grant, *Philosophies of Nature after Schelling*; and Weatherby, *Transplanting the Metaphysical Organ*, chapter 5.

65. Schelling, *Von der Weltseele, HKA*, I:6, 255.

66. "Such an oscillation [*Schweben*] of Nature, therefore, between pro-ductivity and product, will necessarily appear as a universal duplicity of prin-ciples, whereby Nature is maintained in continual activity, and prevented from exhausting itself in its product"; Schelling, *First Outline*, 197 / "Einlei-tung zu seinem Entwurf eines Systems der Naturphilosophie," in *Sämtliche Werke*, ed. K. F. A. Schelling, section I, vol. 3 (Stuttgart: Cotta'scher Verlag, 1858), 277 (translation very slightly modified). Schelling adapts his notion of *Schweben* ("to hover," "to oscillate") from Fichte. On the role played by this concept in Fichte and Schelling, and its relationship to the formative drive of *Einbildungskraft*, see Asmuth, "'Das Schweben ist der Quell aller Realität'."

67. Schelling, *First Outline*, 18 / *Erster Entwurf, HKA*, I:7, 276 (transla-tion slightly modified).

68. For the traditional etymology, see Julius Pokorny, "*sreu-*," *Indoger-manisches Etymologisches Wörterbuch*, 11th ed. (Berlin: A. Francke AG, 1957), 1003, where the Greek *ritmos* is cited as a derivative and provided with the definition "regular movement, rhythm (aligned with the wave-action of the sea)." For an argument against this traditional derivation, which returns to Plato in order to insist on the nonidentity—conceptual as well as mor-phological—of flux and (rhythmic) form, see Émile Benveniste, "La notion de 'rythme' dans son expression linguistique," in *Problèmes de linguistique générale* (Paris: Gallimard, 1966), 1:327–35.

69. English Standard Version Bible (London: Crossway, 2012), Gen. 1:2 and 1:4. Martin Luther's German version gives these verses as follows: "Und der Geist Gottes schwebet auff dem Wasser. [. . .] Da scheidet Gott das Liecht vom Finsternis." Martin Luther, *Die gantze Heilige Schrifft: Der komplette Originaltext von 1545 in modernem Schriftbild*, ed. Hans Volz and Heinz Blanke (Munich: Rogner & Bernhard, 1972).

70. Schelling, *First Outline*, 53ff. / *Erster Entwurf, HKA*, I:7, 117ff.

71. Friedrich Schelling, *Ideas for a Philosophy of Nature*, trans. Errol E. Harris and Peter Heath (Cambridge: Cambridge University Press, 1988), 30 (translation slightly modified); original German: *Ideen zu einer Philosophie der Natur*, in *Sämtliche Werke*, ed. K. F. A. Schelling, section I, vol. 5 (Stutt-gart: Cotta'scher Verlag, 1858), 93.

72. For Schelling's version of this story, see, for instance, Schelling, *Philos-ophie der Kunst (aus dem handschriftlichen Nachlaß)* (1802–3), in *Sämtliche*

Werke, ed. K. F. A. Schelling, partition I, vol. 5 (Stuttgart: Cotta'scher Verlag, 1859), 484.

73. Jacob Grimm, *On the Origin of Language*, trans. Raymond A. Wiley (Leiden, Netherlands: Brill Publishers, 1984), 7; original German: "Über den Ursprung der Sprache," in *Kleinere Schriften*, ed. Otfrid Ehrismann, in Jacob Grimm und Wilhelm Grimm, *Werke*, ed. Ludwig Erich Schmitt (Hildesheim: Olms-Weidmann, 1991), 1:256–99, here 265. I have modified the translation where necessary to preserve aspects of the original German that are relevant to my argument. Grimm claims in the opening lines to have written the essay in response to a (subsequently retracted) suggestion of Schelling's (Grimm, *On the Origin of Language*, 1 / "Über den Ursprung der Sprache," 256). The question of the ultimate origins of language is one that both Grimm himself and his fellow language-scientific pioneers, Bopp and Rasmus Rask, tend to leave explicitly untreated in the context of their actual scientific research. The tone of the essay is therefore speculative, but the notion of a teleological development from physical wind to language spirit—and the accompanying assumption of a singular formative principle governing the entirety of natural history—will turn out to undergird the most significant scientific innovations of the period. For a different understanding of the essay, according to which Grimm merely exploits the discoveries of the new nineteenth-century language science in order to retell the story of Johann Gottfried Herder's 1772 "Abhandlung über den Ursprung der Sprache" ["Treatise on the Origin of Language"], see Ulrich Wyss, *Die wilde Philologie: Jacob Grimm und der Historismus* (Munich: C. H. Beck Verlag, 1979), 174–75. For a concise treatment of the Herderian perspective Wyss clearly has in mind, see Wilczek, *Das Artikulierte*, 92–98.

74. Grimm, *On the Origin of Language*, 7 / "Über den Ursprung der Sprache," 265–56.

75. Ibid.

76. Jacob Grimm, *Deutsche Grammatik, Erster Theil*, 1st ed. (Göttingen: Dieterichsche Buchhandlung, 1819), x. Grimm's reference is to Johann Christoph Adelung, author of the *Grammatisch-kristisches Wörterbuch der Hochdeutschen Mundart, mit beständiger Vergleichung der übrigen Mundarten, besonders aber der oberdeutschen* (Leipzig: Johann Gottlob Immanuel Breitkopf, 1774–86), along with multiple grammatical writings both pedagogical and scholarly. Adelung's dictionary was by far the most comprehensive such project prior to Grimm's own, and as such it exercised a normative, unifying influence on the scattered German dialects of the eighteenth century. On Adelung as a predecessor to Grimm, see Wyss, *Die Wilde Philologie*, 96–100.

77. Jacob Grimm, *Deutsche Grammatik, Erster Theil*, 1st ed. (1819), xv.

78. Grimm and Grimm, "Vorrede," in *Deutsches Wörterbuch*, vol. 1, xii–xiii. The Grimms go on to envision their dictionary assuming a position that had previously been reserved for the Bible, at the heart of the family's collective evening reading: "Why shouldn't the father select a few words and go through them in the evenings with his sons, testing their language skills

while at the same time sprucing up his own? The mother would listen with pleasure"; ibid.

79. Franz Bopp, *A Comparative Grammar of the Sanskrit, Zend, Greek, Latin, Lithuanian, Gothic, German, and Slavonic Languages*, trans. Edward B. Eastwick (London: Williams and Norgate, 1885), 1:124; original German *Vergleichende grammatik des sanskrit, send, armenischen, griechischen, lateinischen, litauischen, altslavischen, gothischen, und deutschen* (Berlin: F. Dümmler, 1857–61), 1:242. I have modified the translation where necessary to preserve aspects of the original German that are relevant to my argument.

80. Bopp, *Comparative Grammar*, v / *Vergleichende Grammatik*, iii.

81. Johann Christoph Adelung, *Umständliches Lehrgebäude der deutschen Sprache zur Erläuterung der Deutschen Sprachlehre für Schulen* (Leipzig: Johann Gottlob Immanuel Breitkopf, 1782), 179.

82. Bopp, *Comparative Grammar*, v / *Vergleichende Grammatik*, iii.

83. Jacob Grimm, *Deutsche Grammatik, Erster Theil*, 3rd ed. (Göttingen: Dieterichsche Buchhandlung, 1840), 30. An earlier version of this passage can be found in the second edition as well, toward the end of the discussion of the letters rather than the beginning. Grimm, *Deutsche Grammatik, Erster Theil*, 2nd ed. (Göttingen: Dieterichsche Buchhandlung, 1822), 580.

84. See, for instance, August Ferdinand Bernhardi, *Anfangsgründe der Sprachwissenschaft* (Berlin: Heinrich Frölich, 1805), 72–73; and Baruch Spinoza, *Abrégé de grammaire hébraique*, ed. J. Askénazi and J. Askénazi-Gerson (Paris: Librairie Philosophique J. Vrin, 1968), 35–36. Spinoza attributes the account of language origins as a union of vocalic flow and consonantal contour to the Hebrew commentary tradition, where it does indeed play a significant role. But the idea also has ancient Greek roots. According to Plato (*Sophist*, 253a), the vowels operate as the underlying "bond" (*desmos*) of spoken language, without which the more rigid, less fluid forms of the consonants would never be able to combine. It is this Platonic version to which Schelling then more or less explicitly alludes in his own reflections on the vowel-consonant relation as a version of the *apeiron-peras* dichotomy. Friedrich Schelling, *Philosophie der Kunst*, in *Sämtliche Werke*, ed. K. F. A. Schelling, partition I, vol. 5 (Stuttgart: Cotta'scher Verlag, 1859), 484–85. The notion that such an understanding of vowels and consonants first begins to shape linguistic thinking around 1800, as frequently presumed, therefore cannot be substantiated, though it may indeed be the case that the ancient binary of feminine matter and masculine form acquires additional rhetorical force in the context of a new emphasis on organic emergence. What *does* appear to change around 1800 is the *valuation* of the categories in question, with the Rationalists almost unilaterally privileging the consonants and the nature-philosophers frequently attempting to rehabilitate the vowels.

85. Jacob Grimm, "Über Etymologie und Sprachvergleichung," in *Kleinere Schriften*, in *Werke* 1:299–326, here 1:308.

86. Ibid.

87. Rasmus Rask, "On Etymology in General" (1818), in *Investigation of the Origin of the Old Norse or Icelandic Language*, trans. Niels Ege (Copenhagen: Linguistic Circle of Copenhagen, 1993), 11–53, here 16. Ege has here "analysis" where Rask has *forklarung*. *Forklarung*, however, as Ege himself points out in his foreword (*Investigation*, iii), has a more or less exact English counterpart in the word "explanation." I have therefore adapted the translation to reflect Rask's original vocabulary. The treatise was originally published as *Undersøgelse om det gamle Nordiske eller Islandske Sprogs Oprindelse* (Copenhagen, 1818), just two years after Bopp's *Conjugationssystem* and one year before the first volume of the first edition of Grimm's *Deutsche Grammatik*. For a parallel to Grimm's rejection of earlier forms of etymology as insufficiently systematic, see Rask, "On Etymology in General," 43–44.

88. Franz Bopp, *Analytical Comparison of the Sanskrit, Greek, Latin, and Teutonic Languages, Showing the Original Identity of their Grammatical Structure* (1820; repr., Amsterdam: Benjamins, 1974), 21. This work is a revised and expanded English version of the *Conjugationen*, in which the hypothesis of grammatical origins is made far more explicit than in the earlier German work of 1816. On the relationship between the original and the translation, and for a brief reception history of the latter work, see E. F. K. Koerner, "Preface to the 1974 Edition," in Bopp, *Analytical Comparison*, vii–x.

89. Bopp, *Comparative Grammar*, 102 / *Vergleichende Grammatik*, 113.

90. Ibid. For the argument that Bopp's perspective must in fact be considered a "mechanist" rather than an "organicist" one, since it locates the origins of grammatical structure in the "mechanical" accumulation of external parts rather than in the "organic" unfolding of an interior principle, see Pieter A. Verburg, "The Background to the Linguistic Conceptions of Bopp," *Lingua* 2 (1950): 438–68. While it is true that Bopp rails against an "unscientific" notion of organicism, which he associates above all with Friedrich Schlegel's attempt to distinguish *ahistorically* between organic and inorganic "types" of languages—to distinguish, that is, between languages that "live" from the very beginning and languages that will never do so, no matter how radically they may transform over time—I do not see any reason to interpret this polemic as a rejection of teleological, organicist principles. Rather, Bopp's insistence on the fundamental *unity* of the entire domain of language forms, whether "organic" or "inorganic," and his commitment to the discovery of the "mechanical" laws by which more primitive, "inorganic" structures *develop into* more complex "organic" ones, strikes me as teleological in a deeper, more encompassing sense than Schlegel's restricted notion of language life. The rhetoric of mechanistic explanation functions for Bopp, in accordance with the nature-philosophical perspective advocated by Schelling (for whom mechanistic causality is always only the insufficiently comprehended flip side of a cosmos-governing teleologism), to lay the groundwork for a truly rigorous teleological science. For an account of Bopp's organicism that aligns broadly with my own, see Anna Morpurgo Davies, "'Organic' and 'Organism' in Franz Bopp," in *Biological Metaphor*

and Cladistic Classification: An Interdisciplinary Perspective, ed. Henry M. Hoenigswald and Linda F. Weiner (Philadelphia: University of Pennsylvania Press, 1987), 81–107.

91. Bopp's formulation of his revolutionary new thesis in fact unfolds by way of an explicit disagreement with Schlegel: "Mr. Frederic Schlegel, in his excellent work on the language and philosophy of the Hindus, very judiciously observes, that language is constructed by the operation of two methods; by inflection, or the internal modification of words, in order to indicate a variation of sense, and secondly, by the addition of suffixes, having themselves a proper meaning. But I cannot agree with his opinion, when he divides languages, according as he supposes them to use exclusively the first or second method, into two classes [. . .]"; Bopp, *Analytical Comparison*, 10. On the opposition to Bopp's thesis among the Schlegels and their associates in the "Bonner school," see Verburg, "Linguistic Conceptions of Bopp," 445–47. For Friedrich Schegel's own version of his thesis, see F. Schlegel, *Sprache und Weisheit*, 41.

92. Grimm, *On the Origin of Language*, 16 / "Über den Ursprung der Sprache," 283.

93. The new linguistic model thus also contains an implicit Schellingian critique of the Kantian perspective, according to which *articulatio*, as the mode of being proper to organic life and rational systems, must remain forever distinct from *coacervatio*, as the mode of being proper to all else. See the following famous definition of rational systems from Kant's *Critique of Pure Reason*: "Under the government of reason our cognitions cannot at all constitute a rhapsody but must constitute a system, in which alone they can support and advance its essential ends. I understand by a system, however, the unity of manifold cognitions under one idea. This is the rational concept of the form of a whole, insofar as through this the domain of the manifold as well as the position of the parts with respect to each other is determined a priori. The scientific rational concept thus contains the purpose and the form of the whole that is congruent with it. The unity of the purpose, to which all parts are related and in the idea of which they are also related to each other, allows the absence of any part to be noticed in our knowledge of the rest, and there can be no contingent addition or undetermined magnitude of perfection that does not have its boundaries determined a priori. The whole is therefore articulated (articulatio) and not heaped together (coacervatio); it can, to be sure, grow internally (per intus susceptionem) but not externally (per appositionem), like an animal body, whose growth does not add a limb but rather makes each limb stronger and fitter for its end without any alteration of proportion"; Immanuel Kant, *Critique of Pure Reason*, trans. and ed. Paul Guyer and Allen W. Wood (Cambridge: Cambridge University Press, 2007), 691; original German: *Kritik der reinen Vernunft*, 2nd ed., in *Gesammelte Werke*, ed. Königlich Preußische Akademie der Wissenschaften (Berlin: De Gruyter, 1904), 3:538–39.

94. Grimm, *On the Origin of Language*, 16 / "Über den Ursprung der Sprache," 283.

95. Bopp, *Vocalismus*, 1. See also Grimm, *On the Origin of Language*, 18 / "Über den Ursprung der Sprache," 286–87.

96. Grimm, *On the Origin of Language*, 3 / "Über den Ursprung der Sprache," 259.

97. Grimm, *On the Origin of Language*, 22 / "Über den Ursprung der Sprache," 294. Grimm earlier makes clear that this future fusion will take a specifically "Indo-Germanic" form; Grimm, *On the Origin of Language*, 3 / "Über den Ursprung der Sprache," 259.

98. Critics of the comparative grammarians' self-proclaimed revolutionary status have tended to point to this fact as evidence that the comparative grammatical method did not, after all, succeed in overcoming the normative tendencies of its predecessors, which sought to formalize empirical language by comparing it against a nonempirical ideal. The first and most vitriolic expressions of this critique can be found in the works of the late nineteenth-century linguists known as the "Neogrammarians" (*Junggrammatiker*). See, for instance, Karl Brugman, *Zum heutigen Stand der Sprachwissenschaft* (Strassbourg: K. J. Trübner, 1885), 23–30.

99. Grimm, *Deutsche Grammatik, Erster Theil* (1819), xii.

100. Rask's presentation appears in his *Investigation*, 161–64. Grimm's encounter with Rask's treatise, shortly after the publication of his ground-breaking *Deutsche Grammatik* in 1819, inspired him to begin his second edition with a six-hundred-page discussion of Germanic phonemes, titled *On Letters* (*Von den Buchstaben*). This book is then followed by a revised version of the original text, under the new title, *On Word Formations* (*Von den Wortbildungen*). Grimm, *Deutsche Grammatik*, 2nd ed. (1822). On the relationship between the morphological and phonetic "camps" within early historical linguistics, and Grimm's position as an intermediary between them, see N. E. Collinge, "The Introduction of the Historical Principle into the Study of Languages: Grimm," in *History of the Language Sciences: An International Handbook on the Evolution of the Study of Language from the Beginnings to the Present*, ed. Sylvain Auroux, E. F. K. Koerner, Hans-Josef Niederehe, and Kees Versteegh, vol. 2 (Berlin: De Gruyter, 2001), 1210–22, here 1219.

101. Grimm, *Deutsche Grammatik, Erster Theil*, 2nd ed. (1822), 588; emphasis mine.

102. Ibid., 584. Greek stands here as the representative of an originary Indo-European state, Gothic as the representative of what linguists now call proto-Germanic.

103. Ibid., 580. Grimm goes on to suggest that the consonantal sound laws may in fact lay the groundwork for this discovery: "Should it not be possible, supported by [the relationship of the consonants] to simultaneously trace the points of contact of the vowels?"; ibid., 592.

104. Up until this point in the history of language study, the traditional notion of language as a grammatical system supplemented by a lexicon (or a lexicon supplemented by a grammar) had tolerated few competitors; it remained prevalent well into the nineteenth century. My reading of the table and its implications is not intended to suggest that Grimm himself transcends this model in the direction that his table first makes visible. This he in fact does only sporadically.

105. Grimm, *Deutsche Grammatik, Vierter Theil,* 2nd ed. (Göttingen: Dieterichsche Buchhandlung, 1837), vi.

106. Though it is therefore technically true, as has been frequently pointed out, that Rask deserves credit for the *discovery* of the sound laws, I would argue that Grimm's presentation made the more profound contribution to the future of language science. On the subject of the Grimm-Rask relationship, see Uwe Petersen's introduction to his translation of Rask's essay on etymology, in *Von der Etymologie überhaupt: Eine Einleitung in die Sprachvergleichung,* ed. and trans. Uwe Petersen (Tübingen: Gunter Narr Verlag, 1992), 28–29. See also E. H. Antonsen, "Rasmus Rask and Jacob Grimm: Their Relationship in the Investigation of Germanic Vocalism," *Scandinavian Studies* 34 (1962): 183–94.

107. Grimm, *Deutsche Grammatik, Erster Theil,* 2nd ed. (1822), 588.

108. Ibid.

109. August Friedrich Pott, *Etymologische Forschungen auf dem Gebiete der Indo-Germanischen Sprachen mit besonderem Bezug auf die Lautumwandlung im Sanskrit, Griechischen, Lateinischen, Littauischen, und Gothischen,* in Roy Harris, *Foundations of Indo-European Comparative Philology,* ed. Roy Harris (1833; repr., London: Routledge, 1999), 12:xii. This passage is quoted and translated in Berthold Delbrück, *Introduction to the Study of Language: A Critical Survey of the History and Methods of Comparative Philology of the Indo-Europeans Languages,* trans. Eva Channing (1882) and ed. E. F. K. Koerner (Philadelphia Benjamins, 1974), 34 (translation modified).

110. August Wilhelm Schlegel, "Rezension von Altdeutsche Wälder" (1815), in A. W. Schlegel, *Sämtliche Werke,* ed. Eduard Böcking, vol. 12 (1847; repr., Hildesheim: Olms, 1971), 383–426, here 400. Though versions of this *bon mot,* which is always attributed to Voltaire, have appeared in countless works of or about etymology over the course of the last two centuries, no reliable information as to its original source has ever been uncovered.

111. "We permit ourselves in the process absolutely no rules for the changing or replacing of letters, but rather demand complete equivalence of the word as proof of descent"; F. Schlegel, *Language and Wisdom,* 24 / *Sprache und Weisheit,* 6.

112. Pott, *Etymologische Forschungen,* xii.

113. For Schleicher's own perception of this shift, which accuses the nature-philosophical linguists of the very body-spirit dualism they criticized in the philosophies of their predecessors, see August Schleicher, "On the Significance of Language for the Natural History of Man," in *Linguistics and Evolutionary Theory: Three Essays,* trans. J. Peter Maher and ed. E. F. K. Koerner (Philadelphia: Benjamins, 1983), 75–82, here 76–78; original German *Über die Bedeutung der Sprache für die Naturgeschichte des Menschen* (Weimar: Hermann Böhlau, 1865), 8–10. I have modified the translation where necessary to preserve aspects of the original German that are relevant for my argument.

114. Schleicher, "On the Significance of Language," 76 / *Über die Bedeutung der Sprache,* 8–9.

115. "It is possible, maybe even probable, that such an investigation would lead to no satisfactory results. Such negative results could nevertheless hardly shake our conviction about the presence of material-bodily conditions of speech. For who would want to deny the existence of material relations, which for the time being still elude our immediate perception, and which we will perhaps also never be able to make into objects of direct observation"; ibid.

116. Schleicher, "On the Significance of Language," 77 / *Über die Bedeutung der Sprache*, 10.

117. See also August Schleicher, "The Darwinian Theory and the Science of Language," in Schleicher, *Linguistics and Evolutionary Theory: Three Essays*, trans. Alexander V. W. Bikkers (1869), ed. E. F. K. Koerner (Philadelphia: Benjamins, 1983), 13–69, here 25–26; original German *Die Darwinische Theorie und die Sprachwissenschaft: Offenes Sendschreiben an Herrn Dr. Ernst Häckel* (Weimar: Hermann Böhlau, 1873), 10: "A necessary consequence of the monistic perspective, which seeks nothing behind the things, but rather considers the thing to be identical with its appearance, is the significance which observation has today acquired for science, and particularly for natural science. Observation is the foundation of contemporary knowledge. [. . .] Every *a priori* construction, every form of pure speculation [*ins Blaue hinein Gedachte*], can be considered at best an ingenious game; for science it is worthless rubbish." Here and in the citations that follow, I have modified the translation where necessary to preserve aspects of the original German that are relevant for my argument.

118. The scientist responsible for investigating it is the "chemist." Schleicher, "On the Significance of Language," 77 / *Über die Bedeutung der Sprache*, 10.

119. Schleicher, "Darwinian Theory," 20–21 / *Die Darwinische Theorie*, 7.

120. "Now observation teaches us that all living organisms, which fall within the range of sufficient observability, undergo change according to definite laws. These changes, their life, is their true essence; and we know these organisms only when we know the totality of these changes, when we know their whole essence"; Schleicher, "Darwinian Theory," 26 / *Die Darwinische Theorie*, 10.

121. Schleicher, *Compendium der vergleichenden Grammatik der indogermanischen Sprache* (Weimar: Böhlau, 1861). For a partial translation, see Schleicher, *A Compendium of the Comparative Grammar of the Indo-European, Sanskrit, Greek, and Latin Languages*, trans. Herbert Bendall (1874; repr., Cambridge: Cambridge University Press, 2014). The table of vowels, which traces vocalic development from the "Indo-German *Ur*-language" through to Gothic, thereby fulfilling Grimm's demand for a formalization of vowel shifts to correspond to the "stepwise" progression of the consonants, can be found on page 134. The table of consonants is on page 268. And the tables of the individual languages appear at the beginnings of each chapter under the title "Übersicht der Laute" [Overview of sounds].

122. See, for instance, Bopp, *Vocalismus*, 15; Grimm, *Deutsche Grammatik, Erster Theil*, 1st ed. (1819), xiii–xv; and Grimm, *On the Origin of*

Language, 22 / "Über den Ursprung der Sprache," in *Kleinere Schriften*, 294.

123. "After we have separated out that part which is still doubtful, there remains a rich store of knowledge, embracing the different sides which language offers to scientific treatment: this knowledge will, in my opinion, stand unshaken for all time," Schleicher, *Compendium*, vii / *Ein Compendium*, i.

124. "*Avis akvāsas ka.* Avis, jasmin varnā na ā ast, dadarka akvams, tam, vāgham garum vaghantam, tam, bhāram magham, tam, manum āku bharantam. Vais akvabhjams ā vavakat: kard aghnutai mai vidanti manum akvams agantam. Akvāsas ā vavakant: krudhi avai, kard aghnutai vividvantsvas: manus patis varnām avisāms karnauti svabhjam gharmam vastram avibhjams ka varnā na asti. Tat kukruvants avis agram ā bhugat"; August Schleicher, "Fabel in indogermanischer ursprache" [Fable in the Indo-German Ur-language], *Beiträge zur vergleichenden Sprachforschung auf dem Gebiete der arischen, celtischen und slawischen Sprachen* (1868), 5:206–8. The (approximate) English translation is as follows: "*The Sheep and the Horses*: A sheep that had no wool saw horses, one of them pulling a heavy wagon, one carrying a big load, and one carrying a man quickly. The sheep said to the horses, 'My heart pains me, seeing a man driving horses.' The horses said, 'Listen, sheep, our hearts pain us when we see this: a man, the master, makes the wool of the sheep into a warm garment for himself. And the sheep has no wool.' Having heard this, the sheep fled into the plain." On the function of the asterisk and the peculiar status of the starred forms, particularly as realized in the *Ur*-fable, see Daniel Heller-Roazen, *Echolalias: On the Forgetting of Language* (New York: Zone Books, 2005), 99–112.

125. Schleicher, "On the Significance of Language," 79 / *Über die Bedeutung der Sprache*, 17–18.

126. Schleicher, "On the Significance of Language," 79 / *Über die Bedeutung der Sprache*, 19–20.

127. "Formation [*Ausbildung*] of language, however, is for me synonymous with development [*Entwicklung*] of the brain and the language organs"; Schleicher, "On the Significance of Language," 80 / *Über die Bedeutung der Sprache*, 21. See also Schleicher, "Darwinian Theory," 26 / *Darwinische Theorie*, 10.

128. Schleicher, "On the Significance of Language," 81/*Über die Bedeutung*, 26.

129. Schleicher, "Darwinian Theory," 64 / *Darwinische Theorie*, 31.

130. Schleicher, "On the Significance of Language," 82 / *Über die Bedeutung der Sprache*, 28.

131. Schleicher, "Darwinian Theory," 42–43 / *Darwinische Theorie*, 18–19.

132. Schleicher, "Darwinian Theory," 45 / *Darwinische Theorie*, 21. On the role of writing, see 43ff/19ff.

133. Hermann Osthoff and Karl Brugman, "Foreword to *Morphological Investigations in the Sphere of the Indo-European Languages*," in *A Reader in Nineteenth-Century Historical Indo-European Linguistics*, trans. and ed. Winfred P. Lehmann (Bloomington: Indiana University Press, 1967),

198–209, here 205; original German "Vorwort," in *Morphologische Unter-*
suchungen auf dem Gebiete der indogermanischen Sprachen (1878) 1:iii–xx,
here 1:xiv–xv. I have modified the translation as necessary to preserve aspects
of the original German that are relevant to my argument.

134. The name Junggrammatiker (literally "Young Grammarians")
stems from the Leipzig-based Germanist Friedrich Zarncke, a critic of the
Leipzig school that had formed around the linguist August Leskien. Zarncke
intended it as an insult; Brugman, however, quickly formally adopted it, and
other members of the circle eventually followed suit. Linguists who identified
as *Junggrammatiker* include Hermann Paul, Berthold Delbrück, and Eduard
Sievers.

135. August Schleicher is in this respect, as in others, the Neogrammarians' most obvious forebear. Berthold Delbrück, who examines Schleicher's
stance on the subject, concludes that Schleicher's position was similar to that
of his own circle, but too ambiguously formulated to have had much effect.
The Neogrammarian critique was thus, from his perspective, a matter of (re)
drawing the former's key conclusions with greater rhetorical force. Berthold
Delbrück, *Introduction to the Study of Language,* 40–55 / *Einleitung in das*
Sprachstudium: Ein Beitrag zur Geschichte und Methodik der vergleichen-
den Sprachforschung (Leipzig: Breitkopf & Härtel, 1893), 40–54.

136. This is of course not to suggest that the Neogrammarian contribution to the existing body of linguistic knowledge was insignificant: on the
contrary, many of the most important developments of the second half of the
nineteenth century can be credited to members of this circle, or to linguists
associated with it. These developments, however, do not in general testify to
the presence of a revolutionary new method but rather to a particularly rigorous application of existing techniques.

137. Osthoff and Brugman, "Foreword," 204 / "Vorwort," xiii; first
emphasis mine.

138. Osthoff and Brugman, "Foreword," 205 / "Vorwort," xv.

139. The Neogrammarian emphasis on the criterion of exceptionlessness echoes the definition of natural law provided by one of the nineteenth
century's most famous and pugnacious mechanists, the physicist Hermann
von Helmholtz, who will play an important role in chapter 5: "To know a
law of nature completely means to demand from it *exceptionless validity,*
and to make this exceptionlessness into the criterion of its correctness";
Hermann von Helmholtz, "The Aim and Progress of Physical Science," in
Selected Writings of Hermann von Helmholtz, trans. and ed. Russell Kahl
(Middletown, CT: Wesleyan University Press, 1971), 223–45, here 228;
original German: "Über das Ziel und die Fortschritte der Naturwissen-
schaft," *Vorträge und Reden,* 5th ed. (Braunschweig: Vieweg, 1903), 1:3
(translation modified). Helmholtz's remark is cited in connection with the
Neogrammarian perspective by Jean Leroux, "An Epistemological Assessment of the Neogrammarian Movement," in *History of Linguistics 2005:*
Selected Papers from the Tenth International Conference on the History
of the Language Sciences, ed. Douglas Kibbee (Amsterdam: Benjamins,
2007), 262–73.

140. Osthoff and Brugman, "Foreword," 200 / "Vorwort," vi. "Is not, after all, the credibility, the scientific probability, of the foundational Indo-german forms, which are of course all purely hypothetical constructions, primarily dependent on whether they accord with the proper conception of the development of linguistic forms in general, and on whether they are constructed according to correct methodological principles?"; ibid.

141. Osthoff and Brugman, "Foreword," 202 / "Vorwort," ix–x.

142. Osthoff and Brugman, "Foreword," 203 / "Vorwort," x.

143. Osthoff and Brugman, "Foreword," 201–2 / "Vorwort," ix–x.

144. "When the linguist can hear how things occur in the life of a language, why does he prefer to form his ideas about the consistency and inconsistency of the phonological system solely on the basis of the inexact and unreliable written transmission of older languages?" Osthoff and Brugman, "Foreword," 202 / "Vorwort," ix.

145. Osthoff and Brugman, "Foreword," 202 / "Vorwort," ix.

146. Eduard Sievers, *Grundzüge der Phonetik*, 3rd ed. (Leipzig: Breitkopf & Härtel, 1885), 1. The title of the first edition from 1876 was *Grundzüge der Lautphysiologie* (Foundations of sound physiology). A few brief selections from this work, including this passage, have been translated in Eduard Sievers, "Foundations of Phonetics," in *A Reader in Nineteenth-Century Historical Indo-European Linguistics*, trans. and ed. Winfred P. Lehmann (Bloomington: Indiana University Press, 1967), 258 (translation modified).

147. Hermann Paul, *Principles of the History of Language*, trans. Herbert Augustus Strong (London: Longmans, Green, and Co., 1891), 39; original German: *Prinzipien der Sprachgeschichte*, 3rd ed. (Halle: Max Niemeyer, 1898), 48 (translation modified). I have included in brackets a fair amount of the German phrasing to make clear how precisely the impossibility of language analysis dovetails, for both Sievers and Paul, with the impossibility of analyzing—or arithmetically enumerating—the continuum, a problem that had become urgent among mathematicians during the latter half of the nineteenth century, and that reached a climax in the years directly preceding the publication of their major works. Here Paul credits Sievers with having opened the eyes of a generation to the futility of linguistic dissection: "Sievers was the first to emphasize the significance of the transitional sounds [*Übergangslaute*]"; ibid. And Sievers himself inveighs in this connection against attempts to establish a universal phonetic alphabet (*ein allgemeines Lautsystem*), citing the necessity of abstracting anew for every particular investigative context. See Sievers, *Grundzüge der Phonetik*, 32–48, esp. 47.

CHAPTER 2. SAUSSURE'S DREAM

1. For a nuanced overview of Saussure's relationship to his Neogrammarian contemporaries and teachers, see Anna Morpurgo Davies, "Saussure and Indo-European Linguistics," in *The Cambridge Companion to Saussure*, ed. Carol Sanders (Cambridge: Cambridge University Press, 2004), 9–29. Saussure would later claim to have attended only one lecture of Brugman's and

two of Osthoff's, though he was certainly a regular student of equally significant, Neogrammarian-associated figures such as Georg Curtius and August
Leskien. See Tullio De Mauro, "Notes biographiques et critiques," in *Cours
de linguistique générale*, édition critique, ed. Tullio De Mauro (Paris: Payot,
1972), 380–94. Johannes Fehr cites the following comments by Brugman
(from a letter to Wilhelm Streitberg, a fellow linguist) as evidence for the
strain in the relationship between Saussure and his most significant linguistic contemporaries: "On the evening of the day of his oral doctoral exam,
Johannes Braunack and I were invited by him to an extremely fine dinner—
just we three—in the Hotel Hauffé [. . .]. Clearly he wished at all costs to
avoid the appearance of having been "made in Germany" [in English in the
original], and I now understand why he so often emphasized to you that
he came to Leipzig completely formed [*fertig*]." Johannes Fehr, "Ein einleitender Kommentar," in Ferdinand de Saussure, *Linguistik und Semiologie:
Notizen aus dem Nachlaß: Texte, Briefe, und Dokumente*, collected and
trans. Johannes Fehr (Frankfurt am Main: Suhrkamp, 1997), 99n98. On
the subject of Saussure's linguistic predecessors more generally, see E. F. K.
Koerner, *Ferdinand de Saussure: Origin and Development of His Linguistic Thought in Western Studies of Language* (Braunschweig: Vieweg, 1973),
72–209.

2. For two of the most expansive and rigorous attempts to reread the
entirety of the Saussurean project from the perspective of precisely this
question—with a focus on new insights gleaned from the 1996 manuscripts—
see Simon Bouquet, *Introduction à la lecture de Saussure* (Paris: Payot &
Rivages, 1997); and Patrice Maniglier, *La vie énigmatique des signes: Saussure et la naissance du structuralisme* (Paris: Éditions Léo Scheer, 2006).
Both works take their point of departure from Jean-Claude Milner's understanding of Saussurean structuralism as an exercise in Galilean formalization. Whereas for Milner, however, the Saussurean approach transforms the
acoustic paradigm of the Neogrammarians into a mental model that will
ultimately be superseded by Chomsky's transformational grammar, Bouquet
and Maniglier both claim for the Saussurean intervention—albeit only in
the fully reconstructed form they each take their own works, respectively,
to represent—the status of a still-viable theoretical position. As will become
clear in what follows, my own reading follows a different trajectory. While
I believe that Saussure's meditations still have much to teach us about the
development of a system theory of language, particularly when read against
the backdrop of the nineteenth-century theories he rejects, and while I agree
with Bouquet and Maniglier that the problems Saussure articulates do not
disappear with Chomsky—who effectively displaces the question of system
formation to a domain beyond the purview of linguistics—I also think Milner is right that Saussure himself has no actual solution to offer. The coherence of the Saussurean vision stands and falls, I will argue here, with the
coherence of its realization at the level of a post-Saussurean linguistic writing.

3. In particular Durkheim's notion of the social fact as belonging to a collective realm neither individual nor universal finds strong parallels in Saussure's understanding of the linguistic space. See Jonathan Culler, *Saussure*

(Ithaca, NY: Cornell University Press, 1976), 85–94; and Christine Bierbach, *Sprache als "fait social": Die linguistische Theorie F. de Saussures
und ihr Verhältnis zu den positivistischen Sozialwissenschaften* (Tübingen:
Niemeyer, 1978). The "transcendental" thrust of Saussure's project has historically been used to accuse him of precipitating a turn away from historical particularity; this turn, however, if indeed it can be said to be one, is so
characteristic of an entire epoch that it cannot with any real legitimacy by
ascribed to Saussure. See Fredric Jameson, *The Prison-House of Language:
A Critical Account of Structuralist and Russian Formalism* (Princeton, NJ:
Princeton University Press, 1972); and Frank Lentricchia, *After the New
Criticism* (Chicago: Chicago University Press, 1980).

   4. See, however, his nagging doubts about this stability, and his consign-
ment of the problem of physical contours to a metaphysics of which he wants
no part: "In other domains, unless I am mistaken, one can speak of the
objects envisaged as things existing in themselves, or at least as things that
involve positive things or entities of some kind to be formulated differently
*(unless perhaps facts are taken to the very limits of metaphysics, or of the
question of knowledge, which we cannot go into)*"; WGL, 42/65; emphasis mine. The rhetorical breakdown that precedes the parenthetical rejection
of "metaphysical" questions, pertaining to the limits of knowledge as such,
bears witness to the extraordinary difficulty of drawing this particular line
at this particular historical moment, and bestows new meaning on the word
foi in the phrase *profession de foi*, from the passage cited in the text above.

   5. Ferdinand de Saussure, *Cours de linguistique générale*, édition critique,
ed. Rudolf Engler, vol. 1 (Wiesbaden: Otto Harrassowitz, 1967 and 1974),
436, column 2. Hereafter cited as *CLG*, which will refer exclusively to the
three "tomes" or fascicles of volume 1, followed by page and column numbers. Engler's edition juxtaposes the text of the published *Cours* with relevant sections from both the available students' notebooks and Saussure's
own manuscripts (excluding those first discovered in 1996). Each "page"
actually comprises two pages, with the *Cours* text itself printed as a run-
ning column on the far left-hand side and the various manuscript versions
arranged across the remaining five columns. I will cite almost exclusively
from the actual student manuscripts rather than the student-compiled text
of the *Cours*, and the cases in which I do not do so will be parenthetically
designated as *Course* in addition to *CLG*, in reference to the following English translation of the published text: Ferdinand de Saussure, *Course in General Linguistics*, trans. Roy Harris, ed. Charles Bally and Albert Sechehaye
(Chicago: Open Court, 1983). In these cases, the English pagination will be
followed by the pagination from the Engler edition. The notebooks of Saussure's students, Albert Riedlinger, Charles Patois, and Emile Constantin,
on which the published text of the *Course* is based, have been translated and
published in three bilingual volumes: Ferdinand de Saussure, *Saussure's First
Course of Lectures on General Linguistics (1907): From the Notebooks of
Albert Riedlinger*, ed. and trans. Eisuke Komatsu and George Wolf (Oxford:
Pergamon Press, 1996); *Saussure's Second Course of Lectures on General
Linguistics (1908–9): From the Notebooks of Albert Riedlinger and Charles*

Patois, ed. and trans. Eisuke Komatsu and George Wolf (Oxford: Pergamon Press, 1997); and *Saussure's Third Course of Lectures on General Linguistics (1910–11): From the Notebooks of Emile Constantin*, ed. and trans. by Eisuke Komatsu and Roy Harris (Oxford: Pergamon Press, 1993). Wherever relevant, I will make use of these translations, which will be cited as *Notebooks* I, II, and III, and the corresponding pagination will appear prior to that of the Engler edition. Saussure's own manuscripts will continue to be cited as *WGL*.

6. On the contemporary context of Saussure's reflections regarding the relationship between language and geographical distance, see Rudolf Engler, "La géographie linguistique," in *Histoire des idées linguistiques*, vol. 3, *L'hégémonie du comparatisme*, ed. Sylvain Auroux (Brussels: Mardaga, 2000), 239–52.

7. Saussure notes with interest that several peoples have words for "foreigner" that spring from the experience of a foreign language as unarticulated speech. "Every people gives superiority to its own language [*idiome*]. A person who does not speak its language is easily considered to be incapable of speaking [*barbaros—balb[utire]—bégayer*]"; *CLG*, 437:2. See also *WGL*, 215/307–8.

8. The editors of the English translation use angled brackets to designate marginal or interlinear additions in the original manuscripts.

9. For a catalogue of Saussurean observations regarding the character of this "neither . . . nor," see Rudolf Engler, "Ni par nature ni par intention," in *Recherches de linguistique: Hommages à Maurice Leroy*, ed. Jean Bingen, André Coupez, and Francine Mawet (Brussels: Éditions de l'Université de Bruxelles, 1980), 74–81.

10. See, for instance, *WGL*, 51–52 /77–78.

11. See also *WGL*, 106/159: "[T]he notion of consciousness [*conscience*] is eminently relative, so that we in fact have two degrees of consciousness, the higher of which is still <pure> unconsciousness [*inconscience*] compared to the degree of thought which accompanies most of our acts." For a reading of the shared space of *conscience* as the space of an "inconscience," and for a contrastive comparison with Chomsky's rationalist notion of speaker competence, see Maniglier, *La vie énigmatique*, 161–85; and Patrice Maniglier, "Les choses du langage: de Saussure au structuralisme," *Figures de la psychanalyse* 12 (2005): 27–44. On the "unconscious of language" as a space where ontology and epistemology interpenetrate, see Sémir Badir, "Ontologie et phénoménologie dans la pensée de Saussure," in *Cahier de l'Herne Ferdinand de Saussure*, ed. Simon Bouquet (Paris: L'Herne, 2003), 108–20. On the question of Saussure's relationship to psychological notions of the unconscious, and in particular on the role of association as a kind of unconscious foundation of speech, see Fehr, "Ein einleitender Kommentar," in Saussure, *Linguistik und Semiologie*, 163–200; and André Green, "Linguistique de la parole et psychisme non conscient," in *Cahier de l'Herne Ferdinand de Saussure*, 272–84.

12. See chapter 1, 57–58.

13. For a very different, Deleuze-inspired understanding of the relationship between systematicity and historical transformation—one that

attempts, ambitiously but, I think, ultimately unsuccessfully, to interpret Saussure's notion of language system as a *corollary* to his notion of language flux—see Maniglier, *La vie énigmatique* and "Les choses du langage." Maniglier acknowledges as crucial the distinction Saussure draws between the analytical activity of the historical linguist and the analytical activity of the speaker (*La vie énigmatique*, 168ff), but he interprets this distinction as an expression of the gap that separates, on his reading, the fundamental instability and openness of the speaker's involuntary analyses from the (artificially imposed) rigidity of the grammarian's conscious categories and laws. As will become clear in what follows, I do not think that Saussure has any interest in the construction of such "open" systems, or even that the kinds of protean, serial networks Maniglier describes—as interesting as they may be in themselves—would have counted for Saussure as legitimate scientific objects. For a Deleuzian account of the kinds of objects Maniglier presumably has in mind, see Gilles Deleuze, *The Logic of Sense*, trans. Mark Lester and Charles Stivale (New York: Columbia University Press, 1990); original French *Logique du sens* (Paris: Minuit, 1969).

14. Aristotle uses the term in "Categories" 1a6–1a12 to refer to two different things that share a name by virtue of participating in a single, more general concept of being: "When things have the name in common and the definition of being which corresponds to the name is the same, they are called *synonymous*. Thus, for example, both a man and an ox are animals." Things that share a name without participating in the same definition of being are called homonymous. "Categories," trans. J. L. Ackrill, in *Complete Works of Aristotle: The Revised Oxford Translation*, ed. J. Barnes, vol. 1 (Princeton, NJ: Princeton University Press, 1983), 3.

15. "[. . .] *nothing is ellipsis*, simply because signs of language are always adequate for what they express"; *WGL*, 67/102.

16. See *CLG*, 149:2: "Thus, the acoustic image is not the material sound, but the psychic imprint of this sound." The notes on phonology contain evidence that Saussure came close to extending this emphasis on sense-based sound to phonemes, despite the fact that he had no means of accounting for the relationship between phonemes and meaning. He explicitly declares that all purely phonetic phenomena must remain linguistically irrelevant so long as they do not contribute to the delimitation of units in the spoken chain. Language is for him a matter of acoustic impressions (over and against the physiological act of articulation, whose significance for linguistics he rejects), but he also clearly recognizes that the listener only "hears" differentially relevant information. See *Notebooks* I 24a / *CLG*, 133:2: "But we only take account in the phonatory act of what is salient <for the ear>, differential, and capable of serving to delimit units in the spoken chain." On Saussure's phonology, see Rudolf Engler, "À propos de la réflexion phonologique de F. de Saussure," *Historiographia Linguistica* 30 (2003): 99–128.

17. The text of the published *Cours* transforms the phrase "to create an intermediary realm" (*de créer un milieu intermédiaire*) into the phrase "to act as an intermediary between thought and sound" (*de server d'intermédiaire entre la pensée et le son*) (*Course* 132 / *CLG*, 253:1), which has radically

different implications for the question of how, and whether, language mediates. It is difficult to see how the *Cours* version of the Saussurean alternative differs from the conception of a phonic vehicle, which Saussure here so
explicitly rejects.

18. English Standard Version Bible (London: Crossway, 2012), Gen. 1:2.

19. Figure 2 represents the diagram as it appears in the course notes of
Saussure's students: an unstructured upper hemisphere of air/thought intersects, here, with an equally unstructured lower hemisphere of water/sound
to establish an intermediary realm of orderly divisions. The published *Cours*
diagram, on the other hand (figure 3), represents the two "amorphous
masses"—A and B, referring to air and water, respectively—as though they
were both *already* divided in their entirety, even before they converge. The
figure makes its own structural incoherence explicit by picturing the relevant
units *twice*—once as the waves with which they are, according to Saussure's
analogy, identical, and once as the vertical vectors of an animating force.
The following formulation from Wilhelm von Humboldt provides an exemplary statement of the traditional position, for which the *Cours* diagram
could easily serve as illustration, but which—I am arguing—Saussure himself could not possibly have held: "There are hence two spheres united in man
which are capable of being divided into a determinate [*übersehbare*] number
of stable elements [*fester Elemente*], which are in turn capable of connection unto infinity [*bis ins Unendliche*] [. . .]. When man's consciousness has
become powerful enough to allow these two spheres to be permeated with
a power [*Kraft*] which can create the same effect in the hearer, then he is
also in possession of both in their entirety. Their mutual permeation [*wech-
selseitige Durchdringung*] can only occur as a result of one and the same
power, and this power can only have its origins in reason [*Verstande*]. Likewise the articulation of tones, and the great difference between the muteness
[*Stummheit*] of the animal and the speech of the human, cannot be physically
explained. Only the strength of self-consciousness wrests from our physical
nature the sharp separation [*scharfe Theilung*] and stable delimitation [*feste
Begrenzung*] of sounds which we call articulation"; Wilhelm von Humboldt,
"On the Comparative Study of Language and Its Relation to the Different
Periods of Language Development," trans. John Wieczorek and Ian Roe, in
Essays on Language, ed. Theo Harden and Dan Farrelly (New York: Peter
Lang Publishers, 1997), 1–22, here 3–4; original German: "Über das ver-
gleichende Sprachstudium in Beziehung auf die verschiedenen Epochen der
Sprachentwicklung" (1820), in *Gesammelte Schriften,* ed. Königlich Preussischen Akademie der Wissenschaften, vol. 4, ed. Albert Leitzmann (Berlin:
Behr, 1905), 1–34, here 4; translation modified. Partially cited in Jürgen Trabant, "Signe et articulation: La solution humboldtienne d'un mystère saussurien," *Cahiers Ferdinand de Saussure* 54 (2001): 269–88, here 283. Trabant
understands the animating spirit of an articulating consciousness, or *Geist,*
as the "solution" to the riddle of Saussure's "mysterious fact." I have tried
to show here that Saussure himself considers this mystery—of a space from
which the form-giving spirit has been subtracted, and thus of an articulation
that can not be "explained" with reference to an animating force—to be the

whole point of his ruminations. For another, more recent analysis of Saussure's concept of articulation that remains grounded in the description and diagram provided by the *Cours*, see Markus Wilczek, *Das Artikulierte und das Inartikulierte: Eine Archäologie strukturalistischen Denkens* (Berlin: De Gruyter, 2012), 150–53.

20. On Saussure's relationship to the neuroscientific discoveries of his time, and particularly to the aphasia research of the psychiatrist Paul Broca, see Ludwig Jäger, "Neurosemiologie," *Cahiers Ferdinand de Saussure* 54 (2001): 289–337.

21. Emphasis mine. The published *Cours* text reads, "Moreover, linguistic signs are, so to speak, tangible: writing can fix them in conventional images, whereas it would be impossible to photograph acts of speech in all their details"; *Course* 17 / *CLG*, 44:1. Rudolf Engler cites at this juncture a note from Robert Godel, the first editor of the Saussurean manuscripts, to the effect that "in this entire passage, the word 'writing' does not figure a single time in the manuscripts." This is true, but seems to me trivial. Saussure frequently uses the language of "fixed images" (*images fixes*) in contexts where he quite clearly has writing in mind (see, for instance, *CLG*, 72, in the section "Language and Writing"), and the notion of a translation from language signs to fixed images would imply writing—understood in its broadest, most interesting sense—even if this were not the case. Saussure will go on to cite the possibility of dictionaries and grammars as evidence for his position, and as examples of an "an admissible image of language [*langue*]"; *CLG*, 44:2.

22. These are, of course, the premises that Jacques Derrida so persuasively uncovers and undermines in his analysis of the published *Cours*. Jacques Derrida, "Linguistics and Grammatology," in *Of Grammatology*, trans. Gayatri Chakravorty Spivak (1974; repr., Baltimore: Johns Hopkins University Press, 1997), 27–73; original French "Linguistique et grammatologie," in *De la grammatologie* (Paris: Minuit, 1967), 42–108. The truly marginal role which such premises turn out to play, according to my own interpretation, in Saussure's actual understanding of language (as opposed to that of his student editors), suggests that the link drawn by Derrida between the self-sufficiency of a systemic enclosure (like that of Saussure's differential geometry) and the prioritization of speech, spirit, and presence, might need to be rethought.

23. A relationship between the two does, however, exist. Saussure understands the breakdown of empirical writing in certain forms of speech aphasia as evidence for the inherent writability of linguistic units. See *WGL*, 120/78 and 147/212, and *CLG*, 35. The role of such breakdowns thus precisely mirrors the role of incomprehension (effectively a breakdown at the level of speech) in bearing witness to the transempirical reality of *langue*. The (im)possibility of graphic memorialization can testify to the insignificance of mere sound, and thus to the priority of other, more sense-based units, but not to the shapes of the "fixed images" themselves, on which collective comprehension and therefore all speech depends.

24. Both Derek Attridge and, more recently, Patrice Maniglier, have argued that it is reductive to impose the Saussurean opposition of *langue* and

parole onto the Saussurean opposition of associative and syntagmatic, as I am implicitly doing here. For Saussure, after all, the syntagmatic axis of language as realized in time (and thus as opposed to the paradigmatic or associative "reservoir" of potentially realizable signs) belongs quite explicitly, albeit problematically, to the realm of *langue* in addition to that of *parole*: "The principle type of syntagm is the phrase. [. . .] But the phrase belongs to *parole*, not to *langue*. Does not the syntagm therefore belong exclusively to the phrase and are we not confusing the two orders? [. . .] The second order of relations seems to evoke facts of *parole*, and we are occupied with the facts of *langue*. But we respond: up until a certain degree, *langue* itself knows its own relationships"; *CLG*, 283–84:2. As sympathetic as I am, however, to both Attridge's and Maniglier's attempts to read history back into Saussure—and thus also, in Attridge's case, to read the origins of structuralism against the grain of the ahistorical structuralist strawmen constructed by two of its most influential adversaries, Fredric Jameson and Frank Lentricchia—I do not believe that the insistence on a syntagmatic *langue* has any real interpretive power. Saussure himself admits to lacking any substantive account of what distinguishes the linearity of *langue*-in-discourse from the linearity of parole ("It must be admitted <that here, in the realm of syntax> *parole* and *langue*, the social and the individual fact, execution and association are more or less mingled together"; *CLG*, 286:2), and in fact goes on to make the understanding of the syntagmatic axis a derivative function of the associative (*CLG*, 293). This result is no accident. In a system based on the smallest possible unit of sounding *sense*, which is the only real notion of system that Saussure has to offer, all sequential, syntagmatic unfolding will depend for its comprehensibility on the differential geometry that alone determines the sequential acoustic boundaries that are capable of being heard. The fact that Saussure nonetheless recognizes the importance of a transempirical linearity testifies to the sophistication of his linguistic intuition—an intuition that the methods and units of phonology will eventually substantiate—but not to any real function for such linearity within his own understanding of *langue*. For this and other reasons, I would argue that Saussure is also not the proper point of reference for Attridge's perceptive reflections on the possibility of a self-reflexive, ideology-undermining etymology. Derek Attridge, "Language as History / History as Language: Saussure and the Romance of Etymology," in *Post-Structuralism and the Question of History*, ed. Derek Attridge, Geoffrey Bennington, and Robert Young (Cambridge: Cambridge University Press, 1987), 183–211, here 184–86; Maniglier, *La vie énigmatique*, especially the chapter titled "Linguistique sérielle," 161–85; Jameson, *Prison-House of Language*; Lentricchia, *After the New Criticism*.

25. See, for instance, *WGL*, 82/123: "Elements and characteristics are the same thing. It is a feature of language [*langue*], as of every system in general, that there can be no difference in it between what distinguishes a thing and what constitutes it."

26. Saussure is here speaking of traditional disciplinary distinctions among the fields of grammar, morphology, and lexicology—and insisting that they are illusory.

27. "And thus I ask myself: in order to have a concrete unit, I will have one time *mois* [*mwa*] and another time *mwaz* [*mois*]. One sees that there is in consequence a straining of the principle. There is already a combination of units. <That is to say: either we don't have a unit any more at all, or we no longer have a concrete unit.> Here we see the experiences that impress themselves when one takes the words as concrete units"; *CLG*, 239:4.

28. That Saussure himself understands the phrase this way is made explicit a few lines further on: "The morphologist himself *must* split *kalb/ir*, since that is what the analysis of/by language [*l'analyse de la langue*] does, and this analysis is its only guide"; *WGL*, 126/185.

29. "Observation: *Etymology*, which is sometimes said to form a branch of the science of language, does not represent a determinate domain [*un ordre déterminé*] of research and even less a determinate domain [*un ordre déterminé*] of facts"; *WGL*, 123/181. See also *WGL*, 57–58/84.

30. On the problem of a Saussurean "corpus," and the notion of different editorial and interpretive paradigms, see Simon Bouquet, "La linguistique générale de Ferdinand de Saussure : Textes et retour aux textes," *Historiographia Linguistica* 27 (2000): 265–77.

31. Rasmus Rask, "On Etymology in General," in *Investigation of the Origin of the Old Norse or Icelandic Language* (1818), trans. Niels Ege (Copenhagen: Linguistic Circle of Copenhagen, 1993), 25.

32. Examples include the medieval scholastic formula *natura naturans*, Shakespeare's "love is not love / Which alter when it alteration finds" (Sonnet 116), and the late Heideggerian phrase *das Ding dingt* (the thing things).

33. The passage in question can be found at *Iliad* 6:168.

34. On the role of *semata* in Homer, see Raymond Adolph Prier, *Thauma Idesthai: The Phenomenology of Sight and Appearance in Archaic Greek* (Tallahassee: Florida State University Press, 1989); and John Miles Foley, *Homer's Traditional Art* (University Park: Pennsylvania State University Press, 1999).

35. The line stems from the following famous passage in a letter to Saussure's friend and student, Antoine Meillet: "But I am completely disgusted with it all, and with the general difficulty of writing even ten lines of common sense in the matter of linguistic facts. Having occupied myself for so long with the logical classification of these facts, and with the classification of the points of view from which we treat them, I see more and more, on the one hand, the immense labor that would be necessary to show the linguist *what it is he does* when he reduces each operation to its preordained category; and, at the same time, the great vanity of everything one can ultimately do in linguistics. [. . .] This will all culminate, against my own inclinations, in a book where, without passion or enthusiasm, I will explain why there is not a single term employed in linguistics to which I accord any sense whatsoever. And only after that, I'm afraid, will I be able to take up my work again at the point where I left it." Émile Benveniste, "Lettres de Ferdinand de Saussure à Antoine Meillet," *Cahiers Ferdinand de Saussure* 21 (1964): 93–135, here 95. A partial translation of this letter, including passages not cited here, can also be found in Jameson, *Prison-House of Language*, 13.

36. For a very different understanding of the significance of this passage—namely, as evidence that Saussure took seriously the problematic of individual intention with respect to the dynamic of the sign—see Rudolf Engler, "Die Zeichentheorie F. de Saussures und die Semantik im 20. Jahrhundert," in *History of the Language Sciences: An International Handbook on the Evolution of the Study of Language from the Beginnings to the Present,* ed. Sylvain Auroux, E. F. K. Koerner, Hans-Josef Niederehe, and Kees Versteegh, vol. 2 (Berlin: De Gruyter, 2001), 2130–52, here 2138.

37. These two "items" are part of a long series of "miniscule paragraphs" in which Saussure tackles the problem of renaming the word. *Aposème, contra-sôme, inertôme, parasème,* and *parasôme* are some of the other options he considers. See *WGL*, 70–82/104–19. Saussure's reflections on the relationship between *sème* and *sôme* are clearly also informed by the ancient Greek wordplay *soma sema,* meaning "the body is a tomb [i.e., of the soul]," which can be found in Plato's *Gorgias,* among other places.

38. "The *sôme* will be like the cadaver, divisible into *organized* parts, which is wrong"; *WGL,* 77/113. Saussure goes on to explicitly compare the structure of the organism, which remains accessible to the anatomist in the form of the cadaver even once the life force has disappeared, with the structure of the sign, which depends entirely on the differential principle of systemic relation for its existence: "In the word, there is absolutely nothing anatomical. [. . .] [T]here is only a series of acoustic productions [*phonations*] which are entirely similar to one another, in that nothing more properly constitutes the lung of the word than its foot"; ibid.

CHAPTER 3. VERSE ORIGINS

1. Friedrich von Hardenberg, "Teplitzer Fragmente," published as part of the *Vorarbeiten zu verschiedenen Fragmentsammlungen, HKA,* 2:387. A selection of the *Teplitz Fragments,* which does not include this one, has been translated in Friedrich von Hardenberg, *Philosophical Writings,* trans. and ed. Margaret Mahony Stoljar (Albany, NY: SUNY Press, 1997).

2. For a reading of this fragment in the context of a Romantic, pre-Grimmian mode of etymology, see Stefan Willer, "Haki Kraki: Über romantische Etymologie," in *Romantische Wissenspoetik: Die Künste und die Wissenschaften um 1800,* ed. Gabrielle Brandstetter and Gerhard Neumann (Würzburg: Königshausen & Neumann, 2004), 393–412, here 393ff.

3. See chapter 1, 47–51.

4. Hermann Usener, "Philologie und Geschichtswissenschaft" (1882), in *Vorträge und Aufsätze* (Leipzig: B. G. Teubner, 1907), 1–35, here 4 and 21, respectively. Usener is speaking here about Richard Bentley, a British philologist often cited among the Germans (particularly for his research on Greek metrics) as one of the founders of the new philological discipline.

5. Jacob Grimm, *Teutonic Mythology,* trans. James Steven Stallybrass, vol. 1 (London: George Bell and Sons, 1888), 134; original German: *Deutsche Mythologie,* 2nd ed. (1844), ed. Helmut Birkhan, in Jacob Grimm

und Wilhelm Grimm, *Werke*, ed. Ludwig Erich Schmitt, vol. 26 (Hildesheim: Olms-Weidemann, 2003), 122.

6. Jacob Grimm, *Teutonic Mythology*, xxi–xxii / *Deutsche Mythologie*, xix–xx.

7. Jacob Grimm, *Deutsche Grammatik, Erster Theil*, 3rd ed. (Göttingen: Dieterichsche Buchhandlung, 1840), 559.

8. Jacob Grimm, *Teutonic Mythology*, 960 / *Deutsche Mythologie*, 912.

9. Usener, "Philologie und Geschichtswissenschaft," 28.

10. Ferdinand de Saussure, "Légendes et récits d'Europe du Nord: De Sigfrid à Tristan," sel. and ed. Béatrice Turpin, in *Ferdinand de Saussure: Cahiers de l'Herne*, ed. Simon Bouquet (Paris: Éditions de l'Herne, 2003), 351–429, here 381–82. Hereafter cited as *LEG*. On the dating of the *Nibelungen* notes, and for a general overview of their contents, see Turpin's introduction, 351–59. The notes have also been published in more complete (and less edited) form in Ferdinand de Saussure, *Le leggende germaniche*, ed. Anna Marinetti and Meli Marcello (Este: Zielo, 1986).

11. On the institutional and disciplinary factors at play in the *Nibelungenstreit*, see Rainer Kolk, *Berlin oder Leipzig? Eine Studie zur sozialen Organisation der Germanistik im "Nibelungenstreit"* (Tübingen: Niemeyer, 1990).

12. "The critical spirit of a Lachmann is in such matters clearly anticritical, consisting in personally deciding what is ancient or recent, in depriving the reader of an unmediated view of the elements, and in substituting himself for the public as judge"; *LEG*, 427.

13. Or even Swiss! In his "tentative reconstitution of a Burgund protopoem," Saussure locates Sigfrid's historical prototype in "a little prince, Sigéric," who "grows up in Geneva"; *LEG*, 380. On the "francophilic" theses of the *Nibelungen* notes, see Rudolf Engler, "Heureux qui comme Ulysse a fait un beau voyage . . ." *Cahier Ferdinand de Saussure* 45 (1991): 151–65.

14. "The number of divinities is significant enough to construct the entire scaffolding of other mythology around them; where such pillars stand, one can assume the presence of auxiliary works and ornaments in abundance"; Grimm, *Teutonic Mythology*, xxi / *Deutsche Mythologie*, xix. It is of course not a coincidence that this image appears in the paragraph directly preceding the one in which Grimm compares the gods to the law-abiding letters of the sound shift.

15. Hermann Osthoff and Karl Brugman, "Foreword to *Morphological Investigations in the Sphere of the Indo-European Languages*," in *A Reader in Nineteenth-Century Historical Indo-European Linguistics*, trans. and ed. Winfred P. Lehmann (Bloomington: Indiana University Press, 1967), 205. Original German "Vorwort," in *Morphologische Untersuchungen auf dem Gebiete der indogermanischen Sprachen* (1878), xiv.

16. Saussure's formulation here—and particularly, as will become clear, his choice of the Germanic rune to exemplify the alphabetic letter—is more than a little polemical. As a German-trained linguist, he rejects neither the validity of the sound laws for the study of historical phonetics nor the notion of rule-governed "letters" as implied by the existence of such laws. What he

does reject is the idea that it is possible to move from a rule-governed phoneme to an articulatory element of language or text: the diachronic phonemes, with their regular, traceable transformations, have, for him, *absolutely nothing* to do with the meaning of the words they also do not actually constitute.

17. Hermann Paul, *Principles of the History of Language*, trans. Herbert Augustus Strong (London: Longmans, Green, and Co., 1891), 39; original German *Prinzipien der Sprachgeschichte*, 3rd ed. (Halle: Max Niemeyer, 1898), 48.

18. For studies that read the *Nibelungen* notes in a manner more clearly continuous with the project of the general linguistics, and thus as an opportunity to reflect on the question of diachrony in Saussure, see Rudolf Engler, "Sémiologies Saussuriennes," *Cahiers Ferdinand de Saussure* 29 (1974–75): 45–73; Michel Arrivé, "Saussure: le temps et la symbolization," in *Sprachtheorie und Theorie der Sprachwissenschaft: Geschichte und Perspektiven: Festschrift für Rudolf Engler zum 60. Geburtstag*, ed. Ricarda Liver, Iwar Werlen, and Peter Wunderli (Tübingen: Narr, 1990), 37–47; Johannes Fehr, "'La vie sémiologique de la lange': Esquisse d'une lecture de notes manuscrites de Saussure," *Langages* 107 (1992): 73–82; Béatrice Turpin, "Légendes—Mythes—Histoire: La circulation des signes," in *Ferdinand de Saussure: Cahiers de l'Herne*, 307–16; and Patrice Maniglier, *La vie énigmatique des signes: Saussure et la naissance du structuralisme* (Paris: Éditions Léo Scheer, 2006), 361–69.

19. See, for instance, *LEG*, 385.

20. "The most historical is probably W. Müller, but he is very insufficiently historical, and is even in principle opposed to seeking explications in history.—The inventor of 'historical myth,' he wants history to remain mythical"; *LEG*, 383.

21. Saussure makes this affinity particularly clear in the same letter to Meillet in which he declares his disgust with the question of linguistic foundations: "In the last analysis, it is only the picturesque side of a language [*langue*]—that which makes it different from all others, insofar as it belongs to a particular people with particular origins, this almost ethnographic side—that still holds my interest: and yet I of all people no longer have the pleasure of being able to give myself over to this study without an ulterior motive, of enjoying a particular fact tied to a particular milieu." Émile Benveniste, "Lettres de Ferdinand de Saussure à Antoine Meillet," *Cahiers Ferdinand de Saussure* 21 (1964): 95.

22. On the dating of Saussure's first Saturnian texts, see Peter Wunderli, "Ferdinand de Saussure: 1er cahier à lire préliminairement: Ein Basistext seiner Anagrammstudien," *Zeitschrift für französische Sprache und Literatur* 82 (1972): 193–216, here 201–3.

23. For a characteristic presentation of this widely held position, see Edmund Stengel, *Romanische Verslehre*, in *Grundriss der romanischen Philologie*, ed. Gustav Gröber, vol. 2 (Strassbourg: Karl Trübner, 1902), 77–78, and 233–34. For Stengel, accent-based poetry belongs to a *Volkstradition* that grows organically out of the Latin language itself, while the elitist emphasis on quantity forces living Latin to conform to the foreign models of

a dead Greek tradition. See also Otto Keller, *Der saturnische Vers als ryth-misch erwiesen* (Leipzig: G. Freytag, 1881), 26ff.

24. Westphal was the first and most influential scholar to move in this direction, in his article, "Zur vergleichenden metrik der indogermanischen völker," *Zeitschrift für vergleichende Sprachforschung auf dem Gebiete des Deutschen, Griechischen und Lateinischen* 9 (1860): 437–58. An apparently independent attempt was made shortly thereafter by Karl Bartsch, *Der sa-turnische Vers und die altdeutsche Langzeile: Ein Beitrag zur vergleichenden Metrik* (Leipzig: G. Teubner, 1867). Westphal's model provoked a series of similarly inclined studies, of which the most significant were Fredric Allen, "Ueber den ursprung des homerischen versmasses," *Zeitschrift für verglei-chende Sprachforschung auf dem Gebiete des Deutschen, Griechischen und Lateinischen* 24 (1879): 556–92; and Hermann Usener, *Altgriechischer Vers-bau: Ein Versuch vergleichender Metrik* (1887) (Osnabrück: Otto Zeller, 1965). Edmund Stengel's *Romanische Verslehre* was also written under Westphal's influence.

25. Westphal, "Zur vergleichenden metrik," 437–38: "The question now arises: If the Indo-Germans shared the content of their oldest poetry, did they not perhaps also share a form, which developed in their originary home [*urheimath*] and then was modified in their new locations, but in such a way that the common point of departure can still be recognized? This ques-tion, in the event that it can be answered in the affirmative, would lead to a field analogous to that of comparative historical grammar, to a compara-tive metrics of the Indo-German peoples." On the role of metrical studies in establishing the principles of textual critique which are the foundation of a rigorous philology, see Usener, "Philologie und Geschichtswissenschaft," 4.

26. See Allen, "Ursprung des homerischen versmasses," 591.

27. Benveniste, "Ferdinand de Saussure à Antoine Meillet," 110.

28. English translations of several important passages, including this one, from Saussure's writings on ancient poetics have appeared in a translation of Jean Starobinski's influential 1971 work, *Les mots sous les mots*, which intersperses select passages from the manuscripts with extended, ruminative commentaries. See Jean Starobinski, *Words upon Words: The Anagrams of Ferdinand de Saussure*, trans. Olivia Emmet (New Haven, CT: Yale Univer-sity Press, 1979), 9; original French *Les mots sous les mots: Les anagrammes de Ferdinand de Saussure* (Paris: Éditions Gallimard, 1971), 21. Hereafter cited as *WW*, with the pagination of the English followed by that of the origi-nal French; emphasis in the final line here is mine. Starobinski was the first to publish selections from these manuscripts—known collectively as the ana-gram studies—which were discovered in the process of preparing a critical edition of the *Cours*.

29. Saussure discusses his theory of the *Stab* in the letter to Meillet from September 23, 1907. Benveniste, "Ferdinand de Saussure à Antoine Meillet," 114. Two other, longer passages from the manuscripts are transcribed by Jean-Michel Rey in a review of Starobinski's *Les mots sous les mots*: Jean-Michel Rey, "Saussure avec Freud," *Critique* 29, no. 309 (1973): 136–67, here 155–57. The fourth passage, which I cite here, comes from a notebook

titled "1er Cahier à lire préliminairement," which was intended for Antoine Meillet. The passage itself first appeared in Starobinski, *Les mots sous les mots*; the entire contents of the notebook first appeared in Wunderli, "1er cahier." For other passages that draw similar consequences about the derivative nature of German alliteration, see *WW*, 5/16 and 9–10/20–21.

30. The final section (from "Toute le question" to "Tacitus") was originally typeset in the manner of a footnote. I have reproduced it here as it appears in Wunderli, "1er cahier," 215, not in Starobinski.

31. Saussure's work on the origins of Indo-European poetic meter has tended to be read *either* as a "logocentric" regression in the direction of motivated, nonarbitrary names, *or* as a protopostmodernist aberration with respect to the more straightforwardly "structuralist" *Cours*. The reception history of the anagram studies begins with Jean Starobinski's commentaries, and his approach, while evocative, is characteristic of what will become a general problem with the discourse, in the sense that Saussure's theory of anagrammatic activity becomes for him the object upon which he can project his own Mallarmé-colored poststructuralist theory of poetic language. With regard to "the curious speculation on the beech-rods," for instance, he remarks only that such a hypothesis, taken together with that of phonic symmetry, attributes "to the poet an acute attention to the phonetic substance of words"; *WW*, 26/40. At another point, after acknowledging that Saussure himself deals only in questions of phonic repetition, he takes what he describes as the interpretive next step by turning the theory of the anagrams into a theory of poetic emanation—and thus of a meaning that *unfolds* in the manner of a flower from the phonic, as opposed to the conceptual, seed ("ancient flowers," "vital germ"); *WW*, 42–46/61–65. This understanding of the anagrams—conceived as a challenge to what Samuel Kinser calls the "logocentric theory of signification," which locates the essence and origin of language in the referring concept, or signifying sign—characterizes an entire tradition of interpretation, culminating in Julia Kristeva's theory of a paragrammatic poetic language. Aside from Starobinski, the most substantial studies in this tradition are, in order of appearance, Sylvère Lotringer, "Flagrant délire (introduction)," and "Le 'complexe' de Saussure," *Semiotext(e)* 1 (1974): 7–24 and 90–112; Rey, "Saussure avec Freud"; Michel Dupis, "A propos des anagrammes saussuriens," *Cahiers d'analyse textuelle* 19 (1977): 7–24; Julia Kristeva, "Pour une sémiologie des paragrammes," in *Semeiotike: Recherches pour une sémanalyse* (Paris: Seuil, 1969), 113–46; Samuel Kinser, "Saussure's Anagrams: Ideological Work," *Modern Language Notes* 94 (1979): 1105–38, here 1107; and Paul de Man, "Hypogram and Inscription," in *The Resistance to Theory* (Minneapolis: University of Minnesota Press, 1986), 27–53. Occasionally, as is the case, for instance, in both Lotringer and Rey, an attempt is made to demonstrate that such a reading goes against the grain of Saussure's own predilections, and to discern a complementary movement of recuperative logocentrism in the anagrams themselves; it is this perspective that generates the oft-referenced notion of the "two Saussures," and of the anagram notes as the site of a battle for priority between the unconscious Saussure of decentered phonic play and the

conscious Saussure of the chaos-controlling name. In light of my argument in chapter 2, it should be obvious that I do not see in Saussure's linguistics a "logocentrism" of any kind, and it will therefore come as no surprise that the real "ideological work" of the anagram notes has, for me, little to do with *either* a mystical-theological recuperation of the Name, *or* its deconstructive undoing. More recent and less poststructuralist approaches to the anagram notes have continued to focus on the question of a subterranean psychic content. On the anagrams and Lacan, see David Shepheard, "Saussure et la loi poétique," in *Présence de Saussure: Actes du Colloque international de Genéve (21–23 mars 1988)*, ed. René Amacker and Rudolf Engler (Geneva: Droz, 1990), 235–46. On the relationship between the phonological theory of double articulation and an anagrammatic Saussurean theosophy, see Francis Gandon, "Le dernier Saussure: Double articulation, anagrammes, brahmanisme," *Semiotica* 133 (2001): 69–78.

32. Tacitus, *Germania*, trans. (with introduction and commentary) J. B. Rives (Oxford: Clarendon Press, 1999), 81. Translation slightly modified. Rives, in his commentary, argues for the possibility of interpreting Tacitus's "certain marks" as a reference to runes, but notes that "Tacitus' brief remark does not allow for a decision one way or the other" (165–66).

33. His review accuses Jacob Grimm, in particular, of letting unrelated letters and words flow together like an "etymological Heraclitus." August Wilhelm Schlegel, "Rezension von Altdeutsche Wälder" (1815), in A. W. Schlegel, *Sämtliche Werke*, ed. Eduard Böcking, vol. 12 (1847; repr., Hildesheim: Olms, 1971), 403. Ulrich Wyss includes a brief discussion of the review in his perceptive treatment of the *Altdeutsche Wälder*, a short-lived journal founded by the brothers as a depository for the odds and ends of their research into German origins. Ulrich Wyss, *Die wilde Philologie: Jacob Grimm und der Historismus* (Munich: C. H. Beck Verlag, 1979), 227–33.

34. Jacob and Wilhelm Grimm, "Bedeutung der Blumen und Blätter," in *Altdeutsche Wälder*, ed. Otfrid Ehrismann, in Jacob and Wilhelm Grimm, *Werke*, ed. Ludwig Erich Schmitt, vol. 37 (Hildesheim: Olms-Weidmann, 1999), 131–45, here 143.

35. Ibid., 141.

36. "In the end, *Buch* [book] would indeed be one with *Buche* [beech tree] [. . .] the proffered external reason [for the similarity of the words], i.e., that people wrote on tree bark, disproves this [inner identity] just as little as the similarity between shield and linden leaf [since *Lind* means "shield" in Icelandic] is undermined by the explanation that bast [the inner bark of a tree] was used for fastening shields [. . .]"; ibid., 142.

37. Ibid., 143.

38. Wilhelm Grimm, *Über deutsche Runen* (1821) (Berlin: Akademie Verlag, 1987), 23.

39. Ibid., 37.

40. "One can probably assume that [the use of runic writing] began with the immigration of Odin."; ibid., 36. The poem is the "Hávamál" (loosely translated, "sayings of the High One"), from the collection of ninth- through twelfth-century poems known as the *Poetic Edda* (preserved primarily in a

thirteenth-century Icelandic manuscript called the *Codex Regius*), which is the primary source for all knowledge regarding Old Norse mythology.

41. Ibid., 312.

42. Ibid., 67–73.

43. Ibid., 298.

44. Ibid., 314.

45. Rochus v. Liliencron, "Zur Runenlehre," in Liliencron and Karl Müllenhoff, *Zur Runenlehre* (Halle: Schwetschke & Sohn, 1852), 1–25, here 21.

46. Ibid., 18.

47. Evidence for the hypothesis that the Grimms may already have had in mind precisely this nexus of *loosen* (to cast lots) and *lesen* (to read) when they engaged in the cryptic speculations of the *Altdeutsche Wälder* can be found in the second volume of Jacob Grimm's *Deutsche Grammatik*, where he describes the origins of *lesen* as follows: "the mental [*geistige*] combining of letters and words was originarily a sensual collecting, binding, counting of sticks [*sinnliches sammeln, binden, zählen der stäbe*] (λέγειν, to say, to collect)"; *Deutsche Grammatik, Zweiter Theil,* 1st ed. (Berlin: Dümmlers Verlag, 1878 [1826]), 83.

48. Liliencron, "Zur Runenlehre," 19.

49. "If one already had a means for designating the initial sound of words, and now became acquainted with the nature of the Greek-Latin alphabet, one would necessarily immediately perceive that one had, oneself, long possessed the elements of an alphabet"; ibid., 24.

50. Ibid., 19.

51. *The Havamal: With Selections from Other Poems of the Edda, Illustrating the Wisdom of the North in Heathen Times,* ed. and trans. D. E. Martin Clarke (Cambridge: Cambridge University Press, 1923), 81. For a mid-nineteenth-century German translation of the Old Icelandic original, by a contemporary of Lilioncron's whose work was also important for Richard Wagner, see "Hávamál," in *Die Edda, die ältere und jüngere nebst den mythischen Erzählungen der Skalda* (1851), ed. and trans. Karl Simrock (Stuttgart: Cotta'scher Verlag, 1882), 37–58, here 55. I have modified the available English translation to correspond more closely to this German version, which was the one most familiar to, and hence most relevant for, nineteenth-century German scholars. Simrock's "older" and "younger" *Eddas* correspond to the Icelandic compilations known today as the *Poetic* and *Prose Eddas,* respectively. Simrock also cites Liliencron's study in his commentary on the poem. See Simrock, *Die Edda,* 383.

52. Liliencron, "Zur Runenlehre," 19.

53. Ibid., 20.

54. Ibid.

55. Jacob Grimm, "Anmerkung über die prosodie" and "Anmerkung über den accent," in *Deutsche Grammatik, Erster Theil,* 2nd ed. (1822), 12–20 and 20–24; Wilhelm Grimm, *Zur Geschichte des Reims* (Berlin: Akademie der Wissenschaften, 1852).

56. Karl Lachmann, *Über althochdeutsche Prosodie und Verskunst mit Beiträgen von Jacob Grimm,* ed. and intro. Ursula Hennig (Tübingen:

Niemeyer, 1990). On Lachmannn's correspondence with Grimm, the history of the Lachmann manuscript, and its influence on the metric reflections of Karl Müllenhoff and Wilhelm Scherer, see Hennig's introduction, 1–60.

57. Early attempts to confront the metrical form of the *Stabreim* borrowed the prosodic category of syllable "quantity" from classical metrics, resulting in explanations based on vowel length (albeit often in combination with a very *un*classical emphasis on dynamic accent, which plays no role in ancient Greek poetry). Within the German sphere, the most influential break with this approach—one that led to the abandonment of classical quantity and the rise of a "metrics" exclusively based on dynamic accent—did not occur until the relatively late studies of Ferdinand Vetter, Max Rieger, and Eduard Sievers. Only with the shift to the accent could German philologists legitimate their long-standing claims to a uniquely Germanic metric ground. See Ferdinand Vetter, *Zum Muspilli und zur germanischen Alliterationspoesie: Metrisches—Kritisches—Dogmatisches* (Vienna: Carl Gerold's Sohn, 1872); Max Rieger, "Die alt- und angelsächsische Verskunst," *Zeitschrift für deutsche Philologie* 7 (1876): 1–64; and Eduard Sievers, *Altgermanische Metrik* (Halle: Niemeyer, 1893). For a detailed study of the trajectory of *Stabreim* scholarship from Rasmus Rask (whose early work on the accent-based rhythms of old Icelandic verse went virtually unnoticed in Germany) to Andreas Heusler, see Jürgen B. Kühnel, *Untersuchungen zum germanischen Stabreimvers* (Göppingen: Kümmerle Verlag, 1978).

58. The example is taken from Sievers, *Altgermanische Metrik*, 48.

59. Ibid., 36.

60. This is the work known among nineteenth-century philologists as the *Younger Edda*, to distinguish it from the *Older*, or *Poetic Edda*.

61. The opposition of rhythmic freedom to metrical calculation is a trope that precedes the "rediscovery" of the *Stabreim* meter. See, for instance, Wilhelm Grimm, "Der Nibelungen Lied, herausgegeben durch Friedrich Heinrich von der Hagen," in *Kleinere Schriften*, ed. Otfrid Ehrismann, in Jacob Grimm und Wilhelm Grimm, *Werke*, vol. 32, 78–79.

62. Stengel, *Romanische Verslehre*, 25.

63. See Ludwig Tieck, *Minnelieder aus dem schwäbischen Zeitalter* (1803; repr., Hildesheim: Olms, 1966), xiii–xiv: "It is not at all a drive toward artifice, or toward difficulty, that first introduces rhyme into poetry, but rather the love of tone and sound, the feeling that similar-sounding words must stand in a relationship of clear or more mysterious kinship to one another, the striving to transform poetry into music, into something definite-indefinite. For the rhyming poet, the measure of long and short [syllables] disappears entirely; he joins together individual sounds in accordance with his striving, which seeks melodious sound [*Wohllaut*] in the uniform consonance [*gleichförmigen Zusammenklang*] of words, without regard for the prosody of the ancients"; cited in Willer, *Poetik der Etymologie*, 215. On the broader implications of the Romantic celebration of rhyme, see Winfried Menninghaus, *Unendliche Verdopplung: Die frühromantische Grundlegung der Kunsttheorie im Begriff absoluter Selbstreflexion* (Frankfurt am Main: Suhrkamp, 1987), 9–29.

64. From a German perspective, then, the late nineteenth-century French inventors of a "vers libre" would not have been seen as revolutionizing poetry per se but as casting off the superficial unity that had cobbled together previous French verse with the "glue" of accidental sound similarities—thereby giving newly rigorous, modern expression to a very old national propensity for inner formlessness.

65. Rudolf Westphal, *Philosophisch-historische Grammatik der deutschen Sprache* (Jena: Mauke, 1869), 6–7. The ancient Germans were of course not the only people to accent etymologically; according to Westphal, however, they were the only ones to do so stringently and consistently (ibid.).

66. Since German is the only etymologically accentuating language, other linguistic traditions (such as the French) that structure their poetry according to the principle of the accent do not have this same advantage of inherent meaning.

67. Westphal, *Philosophisch-historische Grammatik*, 7.

68. See the discussion of grammatical inflection in chapter 1, 42–51.

69. "[T]he letter—as the tangible linguistic element, which, although it is admittedly not stable, nonetheless moves along a comparatively calm track—is on the whole a more certain thread in the dark labyrinth of etymology than the meaning of words, which often jumps boldly around"; August Pott, *Etymologische Forschungen auf dem Gebiete der Indo-Germanischen Sprachen mit besonderem Bezug auf die Lautumwandlung im Sanskrit, Griechischen, Lateinischen, Littauischen, und Gothischen*, in Roy Harris, *Foundations of Indo-European Comparative Philology*, ed. Roy Harris (1833; repr., London: Routledge, 1999), 12:xii.

70. Ibid., xi.

71. See, for instance, Stengel: "The assumption of influence from the direction of German alliterative poetry suggests itself forcefully"; *Romanische Verslehre*, 61. Westphal, in an opinion that dissents from that of most linguists and for which effectively no evidence exists, grants ancient Latin a brief (Saturnian) period of "etymological" accentuation. That the ancient Italians so quickly left this phase behind again suggests, however, that an etymological orientation toward the origins of language is less firmly entwined with the Roman essence than it is with the German one. See Rudolf Westphal, *Allgemeine Metrik der indogermanischen und semitischen Völker auf Grundlage der vergleichenden Sprachwissenschaft* (Berlin: S. Calvary & Co., 1892), 220–34.

72. Benveniste, "Ferdinand de Saussure à Antoine Meillet," 112.

73. Ibid., 111. Jean-Michel Rey, who devotes significant attention to the role played by Saussure's *Stab* story, interprets this prioritization of remainderless purity as a fetishization of the voice—and therefore also of nature, presence, and life. The logic, which leads Rey to compare Saussure unfavorably to the proto-poststructuralist sophistication of Sigmund Freud, is thus very similar to the one employed by Derrida in *On Grammatology*, and it offers, for all its interpretive subtlety, just as distorted a perspective of the Saussurean project. See Rey, "Saussure avec Freud."

74. "In effect one understands the superstitious idea that could have suggested that in order for a prayer to have an effect, it was necessary for the syllables themselves of the divine name to be indissolubly mixed in it: one rivets, so to speak, God to the text [. . .]"; Benveniste, "Ferdinand de Saussure à Antoine Meillet," 114. Already in this letter to Meillet from 1907, we find Saussure raising the possibility that the phonic symmetry he finds so fascinating actually derives, chronologically speaking, from the "superstitious" practices of the anagrams: "It is probable that the different phonetic games of versification are derived from the anagram"; ibid. At other points, he will insist on the possibility that the anagrams derive, in turn, from an attempt to project meaning onto the phonic symmetries (see, for instance, WW, 94–96/125–26). The question of originary motivations, however, is for him ultimately as meaningless as it is unanswerable: "[W]hat do we know of the reason which interwove anagram into the short lyric pieces we place at the foundation [of epic poetry]? The reason *might have originated* in the religious idea that the invocation, prayer, or hymn would have power only if the syllables of the divine name were worked into the text. [. . .] The reason *might have been* nonreligious, and purely poetic [. . .]. And so on. So that the desire to say—for any period—*why* something exists reaches beyond the fact"; WW, 42/60. It is easy to hear echoes of the notes on general linguistics in this declaration of indifference toward the ostensible priority of the chronologically first.

75. Benveniste, "Ferdinand de Saussure à Antoine Meillet," 110.

76. Peter Wunderli makes this same point in his very early study of Saussure's anagrams, though he later goes on to align the alliterating phonemes with the sounding substance Saussure banishes from *langue* in his linguistic notes. See Peter Wunderli, *Ferdinand de Saussure und die Anagramme* (Tübingen: Niemeyer, 1972), 74–78 and 92–95.

77. "I would not be surprised if Indian grammatical knowledge, from both the phonic and the morphological points of view, were a continuation of Indo-European traditions of poetic procedures necessary to the construction of a *carmen,* based on the *forms* of the divine name"; WW, 24/38.

78. Benveniste, "Ferdinand de Saussure à Antoine Meillet," 114: "Every *vātes* was above all a specialist of phonemes." There follows one of the three versions of the Saussurean *Stab* story.

79. Saussure goes on to point out the phonic asymmetries in the line he analyzes, and to admit that a perfectly symmetrical example is rare. He insists, however, that the numbers remain significant: "But there is already strong pressure to wait until all the words are combined so that one arrives at an even number for 2/3rds of the letters, and it is more than 3/4ths [of the examples] that at any moment realize this 'performance,' to use the language of the turf"; WW, 21/34.

80. The motto inverts a cryptic line from Vergil's eighth Eclogue: "Numero deus impare gaudet" (God loves uneven numbers). Saussure studied Vergil's texts in search of anagrammatic residues that could help support his theories about the phonic structure of the more ancient Saturnian fragments.

PART II. TENDING TOWARD ZERO: FROM RUNES TO
PHONEMES

1. The other is Nikolai Trubetzkoy, a Russian aristocrat who fled the
1915 revolution for Vienna, and with whom, among others, Jakobson
founded the Prague Linguistic Circle in 1926. Trubetzkoy died of a heart
attack in 1938, only hours after the Gestapo broke into his Vienna home,
and without having completed his synthetic masterwork, *Principles of Phonology* (*Grundzüge der Phonologie*), which Jakobson subsequently prepared
for publication in 1939. See Nikolai Trubetzkoy, *Principles of Phonology*,
trans. Christiane A. M. Baltaxe (Berkeley: University of California Press,
1969); original German: *Grundzüge der Phonologie* (= *Travaux du Cercle
Linguistique de Prague* 7), 7th ed. (Göttingen: Vandenhoeck & Ruprecht,
1989). For a succinct overview of Trubetzkoy's extremely significant role in
the development of Prague Circle theories about language, and of his close
relationship to Jakobson, see Josef Vachek, "Prolegomena to the History of
the Prague School of Linguistics," trans. Z. Kirschner, in *Prague Linguistic Circle Papers*, vol. 4 (Philadelphia: Benjamins, 2002), 3–51, here 24–30.

2. Roman Jakobson, "Zur Struktur des Phonems" ("On the Structure of
the Phoneme," 1939), in *Selected Writings*, vol. 1, *Phonological Studies* (The
Hague: Mouton, 1962), 280–310, here 280.

3. The scholarly treatments that come closest, in my opinion, to articulating the true system-theoretical contribution of Jakobson's methodological
perspective are those that focus on its relationship to contemporaneous developments in mathematics and cybernetics. In such cases, however, the question
of the relationship between the phonological method and the nineteenth-
century linguistic traditions out of which it emerges—and on which, in my
view, it counterintuitively turns out to depend—tend, for obvious reasons, to
receive little or no attention. See, for instance, Edna Andrews, *Markedness
Theory: The Union of Asymmetry and Semiosis in Language* (Durham, NC:
Duke University Press, 1990); see also the previously cited articles by Bernard
Dionysius Geoghegan, "From Information Theory to French Theory: Jakobson, Lévi-Strauss, and the Cybernetic Apparatus," *Critical Inquiry* 38, no. 1
(2011): 96–126; and Jürgen Van de Walle, "Roman Jakobson, Cybernetics,
and Information Theory, *Folia Linguistica Historica* 29, no. 1 (2008): 87–123.

4. Aristotle, *History of Animals*, 486b. These and many other examples are
collected and discussed in Mary Hesse's classic "Aristotle's Logic of Analogy,"
Philosophical Quarterly 15, no. 61 (1965): 328–40. For a more recent survey
of the various ways in which the term was used in the classical tradition, see
Carl Huffman, *Archytas of Tarentum: Pythagorean, Philosopher, and Mathematician King* (Cambridge: Cambridge University Press, 2005), 179–81.

5. On the form that the notion of "substantive Oneness" assumes in the
case of Aristotle himself, where it appears primarily as a sameness of cause or
conceptual genus, see Hesse, "Aristotle's Logic of Analogy," 338ff.

6. Jakobson himself comes closest to explicitly reflecting on the epochal
status of his approach when he aligns it with the group theoretical concept
of "invariance," which he begins to do with increasing frequency toward the

end of the 1950s. See, for instance, Roman Jakobson, "Verbal Communication" (1972), in *Selected Writings*, vol. 7, *Contributions to Comparative Mythology* (The Hague: Mouton, 1985), 81–92.

CHAPTER 4. WAGNER'S POETRY OF THE SPHERES

1. Based on the heroic poems of the *Poetic Edda*, the thirteenth-century Icelandic saga tells the story, in prose, of the Volsung clan (from Sigi, son of Oðinn, to Sigurd, the German Siegfried) in relation to the Burgundian Giukings (German Niflungs or Nibelungs). The Fouqué trilogy met with both popular and scholarly success, receiving enthusiastic attention from, among others, Friedrich Schlegel, who treated it in his 1812 essay "Über die nordische Dichtkunst." A detailed reception history can be found in Wolf Gerhard Schmidt, *Friedrich de la Motte Fouqués Nibelungen-Trilogie 'Der Held des Nordens': Studien zu Stoff, Struktur und Rezeption* (St. Ingbert: Röhrig, 2000).

2. The argument has also been made that Wagner's reading of *Der Held des Nordens* influenced his restructuring of the epic content, despite the fact that Wagner himself never mentions Fouqué. See Friedrich Panzer, "Richard Wagner und Fouqué," *Jahrbuch des freien deutschen Hochstifts* (1907): 157–94; and, more recently, Wolf Gerhard Schmidt, "Der ungenannte Quellentext: Zur Wirkung von Friedrich de la Motte Fouqués, 'Held des Nordens' auf Richard Wagners 'Ring'-Tetralogie," *Jahrbuch der Fouqué-Gesellschaft Berlin-Brandenburg* (2000): 7–42. For a more general investigation of Wagner's reading background, both literary and philological, see Elizabeth Magee, *Richard Wagner and the Nibelungs* (Oxford: Clarendon Press, 1990), 25–56.

3. Ludwig Ettmüller, "Einleitende Vorrede," in *Die Lieder der Edda von den Nibelungen, stabreimende Verdeutschung nebst Erläuterungen*, trans. Ludwig Ettmüller (Zürich: Orell, Füssli & Co., 1837), vii–xlii; and Karl Simrock, "Erläuterungen," in *Die Edda, die ältere und jüngere Nebst den mythischen Erzählungen der Skalda* (1851), ed. and trans. Karl Simrock (Stuttgart: Cotta'scher Verlag, 1882), 331–462. Shortly after the first edition of his *Edda* translation in 1851, Simrock published a scholarly treatise titled *Die Nibelungenstrophe und ihr Ursprung: Beitrag zur deutschen Metrik* (Bonn: Eduard Weber, 1858). On Ettmüller's understanding of the *Stabreim* as a rhythm founded exclusively on accent, and thus as an anticipation of the later, more influential theories of Vetter, Rieger, and Sievers (see 299n57), see Jürgen B. Kühnel, *Untersuchungen zum germanischen Stabreimvers* (Göppingen: Kümmerle Verlag, 1978), 245–46. On Ettmüller's influence on Wagner, see Hermann Wiessner, *Der Stabreimvers in Richard Wagners "Ring des Nibelungen"* (Berlin: Matthiesen Verlag, 1924).

4. Wilhelm Jordan, *Die Nibelunge* (Frankfurt am Main: W. Jordans Selbstverlag, 1869 and 1875). See Richard Wagner, "Epilogue to the 'Nibelung's Ring,'" trans. William Ashton Ellis, in *Judaism in Music and Other Essays* (1894; repr., Omaha: University of Nebraska Press, 1994), 255–73, here 261; original German: "Epilogischer Bericht über die Umstände und Schicksale, welche die Ausführung des Bühnenfestspieles 'Der Ring des Nibelungen' bis zur Veröffentlichung der Dichtung desselben begleiteten," in

Sämtliche Schriften und Dichtungen, ed. Richard Sternfeld (Leipzig: Breit-kopf & Härtel / C. F. W. Siegel, 1911–16), 6:257–72, here 6:262. Jordan is for Wagner a "literary charlatan" with whom he has no desire to be compared.

5. See the preface to Sievers, *Altgermanische Metrik* (Halle: Niemeyer, 1893), vii–xii.

6. See Richard Wagner, *Opera and Drama,* trans. William Ashton Ellis, in *Richard Wagner's Prose Works,* 8 vols. (1895–99; repr., Omaha: University of Nebraska, 1995), 2:242; original German: *Oper und Drama,* in *Sämtliche Schriften und Dichtungen,* vol. 4, 106. Hereafter cited as OD, with the English pagination followed by the German. I have heavily modified (and modernized) the language of the nineteenth-century Ellis translation throughout, both in order to increase readability and to preserve nuances of the original that are important for my analysis.

7. For a treatment that emphasizes, as I will do, the philological and Romantic antecedents of Wagner's account of language origins, see Reinhart Meyer-Kalkus, "Richard Wagners Theorie der Wort-Tonsprache in 'Oper und Drama' und 'Der Ring der Nibelungen,'" *Athenäum: Jahrbuch für Romantik* 6 (1996): 153–95. On the importance of the language-music inter-action for an understanding of Wagner's musical categories, and for a reflec-tion on the ways in which this understanding was embedded in the musical culture of the mid-nineteenth century, see Thomas Grey, *Wagner's Musical Prose: Texts and Contexts* (Cambridge: Cambridge University Press, 1995).

8. Wagner will of course ultimately reinterpret these categories in terms of Arthur Schopenhauer's philosophy. His first real engagement with Scho-penhauer's *The World as Will and Representation* (*Die Welt als Wille und Vorstellung*), however, occurs in the fall of 1854, after the completion of both *Opera and Drama* and the first full drafts of the *Ring* libretto. On the role of Schopenhauer for Wagner's later work, see Bryan Magee, *The Tristan Chord: Wagner and Philosophy* (London: Penguin Press, 2000).

9. Friedrich Schelling, "Über das Verhältnis des Realen und Idealen in der Natur," in *Von der Weltseele,* 2nd ed. (Hamburg: Perthes, 1806), xxi.

10. See also Wagner, "The Artwork of the Future," trans. William Ashton Ellis, in *The Art-Work of the Future and other Works,* in *Richard Wagner's Prose Works,* 1:69–213, here 1:92: "Language is the condensed element [*das verdichtete Element*] of the voice, and the word is the solidified mass [*die gefestigte Masse*] of the tone"; original German: "Das Kunstwerk der Zuku-nft," in *Sämtliche Schriften und Dichtungen,* 3:42–177, here 3:64. Transla-tion modified.

11. It is not clear whether Wagner knew that the two words actually derive from different roots (*Verdichtung* from *dicten,* "to tighten," related to Indo-European **deuk-,* "to lead"; *Dichtung* by way of the Latin *dictare* from Indo-European **deik-,* "to show"). It is also not clear whether he would have been willing to acknowledge this "actually" even if he had been aware of it, since in his understanding of language origins, similar-sounding sounds are always originally semantically related. See OD, 270/132. What *is* clear, however, is that this pseudoetymological wordplay has an implicit precedent in the schol-arly discourse surrounding the origins of the *Stabreim.* See chapter 3, 103–26.

12. The image, like so many others scattered throughout Wagner's theoretical writings, has a strong Grimmian pedigree. The entry for "blood" [*Blut*] in the Grimms' German dictionary explicitly ties blood to breath—and both to the formative principle of organic growth—across the etymological identity of *blasen* (to blow), *Blut* (blood), and *blühen* (to bloom). Jacob Grimm and Wilhelm Grimm, "Blut," in *Deutsches Wörterbuch* (Leipzig: S. Hirzel, 1854–1960), vol. 2, column 170. Jacob Grimm, like Wagner, sees in the vowels the blood of language, to be contrasted with its consonantal skeleton.

13. This is perhaps Wagner's favorite term for the music-language relationship. See, for example, OD, 280/142: "the rapture-inducing marriage [*Vermählung*] of the inseminating poetic thought [*des zeugenden dichterischen Gedankens*] and the infinite child-bearing capacity of music [*unendlichen Gebärungsvermögen der Musik*]." Gendered metaphors in general are pervasive and extravagantly developed. Even the human ear is sexualized. See Grey, *Wagner's Musical Prose*, 130–80.

14. See OD, 254–71/117–33.

15. See OD, 270/132. The hypothesis of a one-sound, one-sense economy actually does little more than generalize to the realm of poetic praxis one of the necessary presuppositions of a scientific etymology: phonetic identities must correspond consistently to semantic ones if laws about phonetic change are to yield meaningful knowledge about the development of language as a whole. From the perspective of the linguistic establishment, however, such a claim can hold good only for words that retain the first letters of their original roots: the conjecture is that a system in its simplest, most elemental form will necessarily eschew the redundancy of homophony in a way that later, more complicated permutations might not. The "illegitimacy" of Wagner's derivations springs therefore from his tendency to ignore this important qualification, despite evidence that he is aware of the possibility of historical sound shifts. On the role of a homophone-free origin for the development of a rigorous etymology, see Jacob Grimm, *On the Origin of Language*, trans. Raymond A. Wiley (Leiden, Netherlands: Brill Publishers, 1984), 17; original German: "Über den Ursprung der Sprache," in *Kleinere Schriften*, ed. Otfrid Ehrismann, in Jacob Grimm und Wilhelm Grimm, *Werke*, ed. Ludwig Erich Schmitt, vol. 1 (Hildesheim: Olms-Weidmann, 1991), 284; and Franz Bopp, *Vocalismus, oder sprachvergleichende Kritiken über J. Grimm's deutsche Grammatik und Graff's althochdeutschen Sprachschatz, mit Begründung einer neuen Theorie des Ablauts* (Berlin: Nicolaische Buchhandlung, 1836), 18.

16. See OD, 279/140 for a reflection on the interplay between linguistic analysis and etymological synthesis as a peculiarity of consonantal alliteration: "To the word-poet, the revelation of a kinship among his privileged accents [. . .] was only possible through the consonantal *Stabreim* of the linguistic roots. What determined this kinship, however, was merely the particularity of their common consonant; no other consonant could rhyme with it, and therefore the kinship was restricted to one specific family." See OD, 269–70/132–33 for a discussion of the *Stabreim* in its capacity to express

"mixed emotions," later exemplified in the line "Die Liebe bringt Lust und Leid" [Love brings pleasure and pain]; *OD*, 292/152.

17. Grimm, *Deutsche Grammatik, Erster Theil*, 2nd ed., 592. Wagner's critique, here, is an expanded and radicalized version of Grimm's own warning against his fellow linguists' tendency to ignore the vowels: "Etymologists who declare the vowel to be a matter of indifference, as it indeed appears to be in some of the languages of the Orient, and who only hold fast to the skeleton of the consonants, lose in this way more than they gain, since knowledge of vowel offers precisely the surest and richest information regarding the origin and derivation of words"; ibid., 5. On Grimm's attempt to grant the vowels a traceable role in the process of language formation (with the help of the concept of *Ablaut*, or apophony), see Ulrich Wyss, *Die Wilde Philologie: Jacob Grimm und der Historismus* (Munich: C. H. Beck Verlag, 1979), 144–60. Grimm's emphasis on a specifically vocalic capacity to *modify* root meanings, as in the case of verb forms such as "sing, sang, sung," or noun pairings such as *Lachs* and *Luchs*, may in fact have directly inspired Wagner's more expansive musical theory of vocalic modulation.

18. For a reading of the Rhinemaidens' famous "*Weia! Waga! Woge, du Welle* [. . .]" in the opening lines of the *Ring* cycle, as an instance of Wagner's attempt to implement precisely such sound rules—and thus also, as an instance of "sound poetry avant la letter"—see Meyer-Kalkus, "Theorie der Wort-Tonsprache," 183–88.

19. With regard to the relationship between Wagner's account of the *Stabreim* rules and his theory of musical modulation, Grey says, "I should stress, incidentally, that even though the issue of *Stabreim* dominates the exposition of the 'poetic-musical period' idea in *Opera and Drama*, I consider it a red herring with respect to any pragmatic interpretation of this piece of text—which is not to say, of course, that Wagner's *Stabreim* doesn't perform other, analyzable roles in the setting of Wagner's dramatic poetry in general"; Grey, *Wagner's Musical Prose*, 204. Having explicitly discounted the role of the *Stabreim* rules, he can then maintain that "the passage in *Opera and Drama* amounts in large part to a reformulation of familiar prescriptions for the harmonizing of textual and musical structures"; ibid. It is, however, in Wagner's *re*formulation of these "familiar prescriptions"—a reformulation, I am arguing here, which transforms the traditional understanding of the language-music interaction precisely by phrasing it *in terms of the Stabreim*—that the historical particularity of Wagner's theoretical position must be sought.

20. Wagner, "Artwork of the Future," in *Richard Wagner's Prose Works*, 1:115 / "Das Kunstwerk der Zukunft," in *Sämtliche Schriften und Dichtungen*, 3:86.

21. See *OD*, 292/152–53: "If we take, for instance, an alliterative verse with completely homogeneous emotional content, such as 'die Liebe giebt Lust zum Leben' [Love gives pleasure to life], then [. . .] the musician here finds no natural inducement to step beyond the already selected key [. . .]. If, on the other hand, we take a verse of mixed emotion, such as 'die Liebe bringt Lust und Leid' [Love brings pleasure and pain], where the *Stabreim* combines two

opposing emotions, the musician will feel called upon to transition away from the original key, which corresponds to the first emotion, toward a different one, which corresponds to the second as it relates to the first."

22. For a study that takes seriously the implications of an inseminating sense for an analysis of Wagner's musical works, see Frank Glass, *The Fertilizing Seed: Wagner's Concept of the Poetic Intent* (Ann Arbor, MI: UMI Research Press, 1983).

23. See chapter 1, 41–42, for a discussion of the Schellingian account in its relationship to the traditional etymology of "rhythm." See chapter 2, 80–82, for a discussion of the Saussurean version.

24. See, for instance, the following image, in *OD*, 278/140: "But the true poet must now come, who, with the clairvoyant eye of his highest, redemption-seeking poet's need [*Dichternoth*], shall recognize in the dirty beggar the redeeming God, shall take from him his crutches and rags; and shall soar with him, on the wind [*Hauch*] of his desirous longing, upwards into infinite space, upon which the breath of the liberated God pours out infinite delights of the most blissful feeling."

25. Wagner, "Artwork of the Future," in *Richard Wagner's Prose Works*, 1:115–16 / "Das Kunstwerk der Zukunft," in *Sämtliche Schriften und Dichtungen*, 3:86–87.

26. See *OD*, 358/211–12: "The prophetic evolution of artistic expression, in its influence on the expression of life, cannot spring initially from artworks whose linguistic foundations lie in the Italian or French languages; rather, of all the modern operatic languages, only German is capable of being employed to reanimate artistic expression, in the manner we have recognized as necessary, because only German has retained in daily life the accent on the root syllable, whereas the other languages place the accent, according to arbitrary, unnatural conventions, on meaningless inflective syllables [*Beugungssylben*]."

27. Jacob Grimm, *Teutonic Mythology*, trans. James Steven Stallybrass, vol. 1 (London: George Bell and Sons, 1888), 131–32 / Grimm, *Deutsche Mythologie*, 2nd ed. (1844), ed. Helmut Birkhan, in Jacob Grimm und Wilhelm Grimm, *Werke*, ed. Ludwig Erich Schmitt, vol. 26 (Hildesheim: Olms-Weidemann, 2003), 120.

28. Grimm goes on to add *Wunsch* (wish), which is also, as he makes clear, a principle of movement. See *Teutonic Mythology*, 137 / *Deutsche Mythologie*, 126–31. *Wunsch* finds its Wagnerian expression in the emphasis on erotic love, which is for Wagner the motivating motive par excellence— namely, the one that drives the desiring subject out of itself in the direction of the desired object. See *OD*, 291/151: "The motive of love is that which drives the subject out of itself, and compels it to bind itself to another."

29. Jacob Grimm and Wilhelm Grimm, "Bewegen," in *Deutsches Wörterbuch*, vol. 1, column 1768.

30. Jacob Grimm, *Teutonic Mythology*,–33 / Grimm, *Deutsche Mythologie*, 121.

31. Richard Wagner, *Siegfried*, trans. Stewart Spencer, in *Wagner's "Ring of the Nibelung": A Companion* (New York: Thames and Hudson, 1993),

211. original German: Wagner, *Der Ring des Nibelungen,* in *Sämtliche Schriften und Dichtungen,* 6:104 (here with the alliterating consonants in bold).

32. See E. Magee, *Richard Wagner and the Nibelungs,* 124–25 and 216–17. Magee points out here that Simrock contributes the figure of the Wanderer to the interpretation of Wotan's name and essence—a figure that takes on great significance for Wagner in the *Ring.*

33. Simrock refers to the spirit-arousing power (*geisterregende Kraft*) attributed to this "*Ur*-font" (*Urbrunnen*) by other *Edda* poems, and cites Grimm's irritation with early Icelandic interpreters who confuse Odin's mead-motivated eagle flight—for Grimm a figure of ecstatic, poetic inspiration—with a representation of "common drunkenness." Simrock, "Erläuterungen," in Simrock, *Die Edda,* 368–502, here 379–80.

34. Ibid., 382; emphasis mine.

35. According to Simrock, this self-reflexive structure is present also in the third stanza, since the "white son of Bölborr"—from whom Odin claims to have learned the runic songs—is in fact none other than Odin himself. Ibid., 382.

36. Ibid., 381.

37. Richard Wagner, *Götterdämmerung,* trans. Stewart Spencer, in *Wagner's "Ring of the Nibelung": A Companion,* 280–81, translation very slightly modified; original German: *Ring des Nibelungen,* in *Sämtliche Schriften und Dichtungen,* 6:178. The entire trajectory of the *Edda* poem is retold here, in the prologue to the *Götterdämmerung,* by the oldest of the three fates, or Norns:

FIRST NORN
(*rises and attaches during her song one end of a golden rope to a branch of the pine tree*)
For good or ill,
I wind the rope and sing.
At the world-ash
once I wove
when, tall and strong,
a forest of sacred branches
blossomed from its bole;
in its cooling shade
there frothed a spring,
whispering wisdom [*Weisheit raunend*],
its ripples ran:
I sang then sacred sense.
A dauntless god
came to drink at the spring;
one of his eyes
he paid as toll for all time:
from the world-ash
Wotan broke off a branch;
the shaft of a spear

the mighty god cut from its trunk.
In the span of many seasons
the wound consumed the wood;
fallow fell the leaves,
barren, the tree grew rotten:
sadly the well-spring's
drink ran dry;
the sense of my singing
grew troubled.
But if I no longer
weave by the world-ash today,
the fir must serve to fasten the rope:
sing, my sister,
—I cast it to you—
do you know what will become of it?

38. Brünnhilde bestows her runes on Siegfried, who loses them again, and with them his memory (of her and of the origin) by drinking what is thus in effect the anti-*Quelle*. See Wagner, *Götterdämmerung*, in *Wagner's "Ring of the Nibelung": A Companion*, 284–87 and 289–93 / *Ring des Nibelungen*, in *Sämtliche Schriften und Dichtungen*, 6:182–86 and 6:189–94.

39. The world tree ends up as the pile of shards upon which Siegfried and Brünnhilde burn.

40. *Götterdämmerung*, 182/284.

41. Wagner, "The Wibelungen," trans. William Ashton Ellis, in *Pilgrimage to Beethoven and Other Essays*, in *Richard Wagner's Prose Works*, 7:257–98, here 7:275; original German: "Die Wibelungen," in *Sämtliche Schriften und Dichtungen*, 2:115–56, here 2:132.

42. According to Wagner's own interpretation, the theoretical consciousness of *Opera and Drama* (which prepares the way for the unprecedented dynamism of the *Ring*) first emerges from an interruption, or *Hemmung*, of its author's artistic drive. See Wagner, "Music of the Future," trans. William Ashton Ellis, in *Judaism in Music and Other Essays*, 293–346, here 296; original German: "Zukunftsmusik: An einen französichen Freund," in *Sämtliche Schriften und Dichtungen*, 7:87–137, here 7:88–89.

43. Wagner, "Artwork of the Future," in *Richard Wagner's Prose Works*, 1:101 / "Das Kunstwerk der Zukunft," in *Sämtliche Schriften und Dichtungen*, 3:72. On the structure of natural development and the status of art as microcosm, see Richard Wagner, "Music of the Future," in *Judaism in Music*, 296 / "Zukunftsmusik," in *Sämtliche Schriften und Dichtungen*, 7:88–89: "If we may designate nature in broad overview as a developmental trajectory from unconsciousness to consciousness, and if this process displays itself most conspicuously in the human individual, then the observation thereof in the life of the artist is certainly one of the most interesting [activities], because in him and his creations the world displays itself and comes to consciousness."

44. Wagner, "Music of the Future," in *Judaism in Music*, 339 / "Zukunftsmusik," in *Sämtliche Schriften und Dichtungen*, 7:132.

45. Jacob Grimm, *Teutonic Mythology*, 69 / *Deutsche Mythologie*, 60.

CHAPTER 5. PYTHAGORAS IN THE LABORATORY

1. On the role played by Wagnerian ideas for the French Symbolist poets, see Léon Guichard's early and still classic work, *La musique et les lettres en France en temps du wagnérisme* (Paris: Presses Universitaires de France, 1963). More recently, see Joseph Acquisto, *French Symbolist Poetry and the Idea of Music* (Burlington, VT: Ashgate, 2006); Margaret Miner, *Resonant Gaps: Between Baudelaire and Wagner* (Athens: University of Georgia Press, 1995); and Wolfgang Storch and Josef Mackert, eds., *Les Symbolistes et Richard Wagner / Die Symbolisten und Richard Wagner* (Berlin: Edition Hentrich, 1991).

2. Charles Baudelaire, "Richard Wagner and *Tannhäuser* in Paris," in *Baudelaire: Selected Writings on Art and Artists*, trans. P. E. Charvet (Cambridge: Cambridge University Press, 1972), 325–57, here 330; original French: "Richard Wagner et *Tannhäuser* à Paris," in *Oeuvres complètes*, ed. Claude Pichois (Paris: Ballimard, 1975), 779–815, here 784. Baudelaire composed the essay in response to the public outcry precipitated by the Paris premiere of *Tannhäuser*. Wagner's own account of the affair, which caused him to pull the opera after only three performances, can be found in Richard Wagner, "A Report on the Production of 'Tannhäuser' in Paris," in *Judaism in Music*, 347–60; original German: "Bericht über die Aufführung des 'Tannhäuser' in Paris," in *Sämtliche Schriften und Dichtungen*, ed. Richard Sternfeld (Leipzig: Breitkopf & Härtel / C. F. W. Siegel, 1911–16), 7:138–49.

3. Baudelaire, "Richard Wagner and *Tannhäuser*," 331 / "Richard Wagner et *Tannhäuser*," 784. The portion of the poem cited by Baudelaire runs as follows, in my own translation (which aims solely to reproduce, as closely as possible, the semantic content of the original as it pertains to my argument, without making any claims to capture the poetic effect):

Nature is a temple in which the living pillars
Sometimes yield confused words;
Man passes there through forests of symbols
which observe him with knowing eyes.
Like long-held echoes, mingling in the distance
Into one deep and shadowy unity,
Vast as the night and as the light,
The scents, the colors and the sounds respond to one another.

[La nature est un temple où de vivants piliers
Laissent parfois sortir de confuses paroles;
L'homme y passé à travers des forêts de symbols
qui l'observent avec des regards familiers.
Comme de longs échos qui de loin se confondent
Dans une ténébreuse et profonde unité,
Vaste comme la nuit et comme la clarté,
Les parfums, les couleurs et les sons se respondent.]

4. Baudelaire, "Richard Wagner and *Tannhäuser*," 331 / "Richard Wagner et *Tannhäuser*," 784 (translation very slightly modified). For a reading of the essay that places great weight on the counterfactual valence of the "as" (*comme*) in order to emphasize the subversiveness of Baudelaire's profoundly post-Wagnerian, urban aesthetics of decadence, or *Verfall*, see Joseph Aquisto, "Uprooting the Lyric: Baudelaire in Wagner's Forests," *Nineteenth-Century French Studies* 32 (2004): 223–37.

5. Hans von Wolzogen, *Poetische Lautsymbolik: Psychische Wirkungen der Sprachlaute im Stabreime aus R. Wagner's "Ring des Nibelungen"* (Leipzig: Edwin Schloemp, 1876), here 43. Wagner claimed to have studied the treatise "with interest."

6. Ibid., 48.

7. Ibid., 44.

8. Ibid., 3. Wolzogen goes on to compare the phenomenon to that of onomatopoetic words whose etymological origins are *not* sound symbols: "Nonetheless, the effect is authentic and in accordance with nature [*naturgemäss und wahrhaft*], not the product of imagination or of arbitrary interpretation." Ibid., 4.

9. Wolzogen did not himself invent the term. He did, however, first apply it to Wagner's work, and is thus responsible for its unique importance in the history of Wagner reception. See Hans von Wolzogen, "Leitmotive," *Bayreuther Blätter* 20 (1897): 313–30. On Wagner's own terminology, which includes *Grundmotiv, musikalisches Motiv*, and *Grundthema*, see Grey, *Wagner's Musical Prose: Texts and Contexts* (Cambridge: Cambridge University Press, 1995), esp. 319–22.

10. Arthur Rimbaud, "Vowels," in *Rimbaud Complete*, trans. Wyatt Mason (New York: Modern Library/Random House, 2002), 104; original French: "Voyelles," *Oeuvres complètes*, ed. Rolland de Renéville and Jules Mouquet, Bibliothèque de la Pléiade (Paris: Gallimard, 1963), 103. René Ghil, *Traité du verbe avec Avant-dire de Stéphane Mallarmé* (Paris: Chez Giraud, 1886), 13–30.

11. On the role played by "Wagnerism" for Russian modernist art in all its forms, see Rosamund Bartlett, *Wagner in Russia* (Cambridge: Cambridge University Press, 1995); and Bernice Glatzer Rosenthal, "Wagner and Wagnerian Ideas in Russia," in *Wagnerism in European Culture and Politics*, ed. David C. Large and William Weber (Ithaca, NY: Cornell University Press, 1984), 198–245. Bely's verse meditation on the sound-sense bonds of an originarily "Aryan" or Indo-European language, in *Glossalalia: A Poem about Sound* (*Glossalolija: Poema o Zvuke*); Kandinsky's merging of dance, painting, poetry, and music in the context of "color-tone dramas" such as *The Yellow Sound* (*Der gelbe Klang*); Scriabin's plans for a monumental, multimedial artwork to (literally) end all artworks and his quest for a synesthetic *Ur*-language in which to write his operatic texts—all are indebted, with varying degrees of acknowledgement, to Wagner's interpretation of early nineteenth-century language theories.

12. Wilhelm Wundt, *Grundzüge der physiologischen Psychologie*, 1st through 6th eds. (Leipzig: Wilhelm Engelmann, 1874–1911).

13. The relationship between psychophysiological principles and avant-garde artistic practices has received astonishingly little sustained scholarly attention, despite widespread acknowledgment of its historical role. On the significance of Wundt-derived theories within the context of Russian Symbolism, Futurism, and early Formalism, see Gerald Janacek's brief treatment in *Zaum: The Transrational Poetry of Russian Futurism* (San Diego: San Diego State University Press, 1996), esp. 15–21, 46–47. On the importance of Wundtian ideas for early Dadaist experiments, see Tobias Wilke, "Da-da: 'Articulatory Gestures' and the Emergence of Sound Poetry," *Modern Language Notes* 128, no. 3 (2013): 639–68. The most extensive investigation to date is a doctoral thesis; see Peter Michael Mowris, *Nerve Languages: The Critical Response to the Physiological Psychology of Wilhelm Wundt by Dada and Surrealism*, PhD diss., University of Texas at Austin (Ann Arbor, MI: ProQuest/UMI, 2010).

14. Scholarly accounts of Wundt's historical role have seldom failed to assume a Helmholtzian influence on the development of psychophysiology, particularly since such influence can also be demonstrated in the case of Wundt's most important psychophysicalist predecessor, Gustav Fechner. Neither Fechner nor Wundt, however, are commonly supposed to have inherited from Helmholtz anything more substantive than a general predilection for the experimental method, which they are then presumed to have implemented in fundamentally un-Helmholtzian ways. (Helmholtz himself had well-publicized, deep-seated reservations about the whole project of psychic measurement, and reportedly considered Wundt's laboratory work to be insufficiently rigorous.) See Michael Heidelberger, *Nature from Within: Gustav Theodor Fechner and His Psychophysical Worldview* (Pittsburgh: University of Pittsburgh Press, 2004), originally published in German under the title *Die innere Seite der Natur: Gustav Theodor Fechners wissenschaftlich-philosophische Weltauffassung* (Frankfurt am Main: Vittorio Klostermann, 1993); and Joel Michell, *Measurement in Psychology: A Critical History of a Methodological Concept* (Cambridge: Cambridge University Press, 1999). A welcome recent attempt to modify this traditional perspective can be found in Alexandra Hull's *The Psychophysical Ear: Musical Experiments, Experimental Sounds* (Cambridge, MA: MIT Press, 2013), which places Helmholtz's work at the center of its comprehensive study of psychophysical acoustic theories.

15. Hermann von Helmholtz, "On the Physiological Causes of Harmony in Music," trans. A. J. Ellis, ed. David Cahan (Chicago: University of Chicago Press, 1995), 46–75; original German: "Die physiologischen Ursachen der musikalischen Harmonie" (1857), *Populäre wissenschaftliche Vorträge*, vol. 1 (Braunschweig: Friedrich Vieweg, 1865), 57–91. Hereafter cited as *PC*, with the English pagination followed by the German. The only available English translation departs at times quite substantially from the phrasing and rhetorical organization, of the German text. I have modified the translation where necessary to preserve aspects of the original that are important for my argument.

16. Hermann von Helmholtz, *On the Sensations of Tone as a Physiological Basis for the Theory of Music*, trans. A. J. Ellis (New York: Longmans, Green, and Co., 1912); original German: *Die Lehre von den Tonempfindungen als physiologische Grundlage für die Theorie der Musik*, 1st through 6th eds. (Braunschweig: Vieweg, 1863–1913). I have frequently modified the translation to preserve aspects of the original German that are relevant for my argument. The translation is based on the fourth edition from 1877, which was the last one to which Helmholtz himself was able to contribute revisions; the German pagination I provide is for the more widely accessible and essentially identical sixth edition, which was published posthumously in 1913. My citations, however, are taken almost exclusively (the single exception is explicitly noted) from parts of the sixth edition text that were already present verbatim in the original, much less easily accessible, 1863 version. For a detailed reading of Helmholtz's monograph against the backdrop of its natural scientific, music theoretical, and technological context, see Matthias Rieger, *Helmholtz Musicus: Die Objektivierung der Musik im 19: Jahrhundert durch Helmholtz' Lehre von den Tonempfindungen* (Darmstadt: Wissenschaftliche Buchgesellschaft, 2006). For a much broader attempt to treat Helmholtz's innovations in relation to a particularly modern phenomenon of analytic, self-reflexive listening—a perspective that aligns closely with my own—see Benjamin Steege, *Helmholtz and the Modern Listener* (Cambridge: Cambridge University Press, 2012).

17. The locus classicus for this perspective, which sees in the sciences of number, harmony, and grammar the privileged paradigms for a science of the cosmos per se, is certainly Plato's Pythagorean-tinged late dialogue, the *Philebus* (discussed in chapter 1, 34–41).

18. On the prelude to the controversy, beginning with the publication of Newton's *Principia Mathematica*, see John T. Cannon and Sigalia Dostrovsky, *The Evolution of Dynamics: Vibration Theory from 1687 to 1742*, Studies in the History of Mathematics and Physical Sciences, vol. 6 (New York: Springer Verlag, 1981). On the controversy itself, see Gerald Wheeler and William Crummett, "The Vibrating String Controversy," *American Journal of Physics* 55, no. 1 (1987): 33–37. On its consequences for the theory of harmony, see Thomas Street Christensen, *Rameau and Musical Thought in the Enlightenment* (Cambridge: Cambridge University Press, 2004), 153–59.

19. Helmholtz, *On the Sensations of Tone*, 231 / *Tonempfindungen*, 378.

20. Gottfried Wilhelm Leibniz, "The Principles of Nature and Grace, Based on Reason," in *Philosophical Papers and Letters*, trans. and ed. Leroy E. Loemker, vol. 2 (Chicago: University of Chicago, 2012), 1033–43, here 1040–41; original French: "Principes de la nature et de la grâce, fondés en raison" (1714), in *Die philosophischen Schriften von Gottfried Wilhelm Leibniz*, ed. C. J. Gerhardt, vol. 6 (1885; repr., Hildesheim: Olms, 1978), 598–606, here 604: "For everything has been regulated in things, once for all, with as much order and agreement as possible; the supreme wisdom and

goodness cannot act except with perfect harmony. The present is great with
the future; the future could be read in the past; the distant is expressed in
the near. One could learn the beauty of the universe in each soul if one could
unravel all that is rolled up in it, but that develops perceptibly only with time.
[. . .] Each soul knows the infinite, knows everything, but confusedly. Thus
when I walk along the seashore and hear the great noise of the sea, I hear the
separate sounds of each wave but do not distinguish them; our confused perceptions are the result of the impressions made on us by the whole universe.
It is the same with each monad. Only God has a distinct knowledge of everything, for he is the source of everything. It has been very well said that he is
everywhere as a center but that his circumference is nowhere, since everything is immediately present to him without being withdrawn at all from
this center." For a treatment of Leibnizian wave analogies in the context of a
fascinating genealogy of the limit-concept of *Rauschen*, see Rüdiger Campe,
"The *Rauschen* of the Waves: On the Margins of Literature," *SubStance* 61
(1990): 21–38.

21. The Kantian passage, which lists "the boundless ocean set into a rage,"
alongside steep cliffs, volcanoes, hurricanes, tall waterfalls, and thunder-
clouds, as examples of an apparently *unlimited* natural power, can be found
in Immanuel Kant, *Critique of the Power of Judgment*, trans. Paul Guyer
and Eric Matthews, ed. Paul Guyer (Cambridge: Cambridge University Press,
2007), 144; original German: *Die Kritik der Urteilskraft*, in *Gesammelte
Werke*, ed. Königlich Preußische Akademie der Wissenschaften (Berlin: De
Gruyter, 1908), 5:261. On the relationship of Caspar David Friedrich's painting to Kant's theory of the sublime, see, for instance, Brad Prager, "Kant and
Caspar David Friedrich's Frames," *Art History* 25, no. 1 (2002): 68–86.

22. *Corps sonore* (sound body) is the term employed by the Baroque composer and music analyst Jean-Philippe Rameau—the theorist of harmony
with whom Helmholtz perhaps has most in common—for the compound
phenomenon which Helmholtz himself calls *Klang*. Many of Rameau's key
categories reappear in Helmholtz's *Sensation of Tones*, modified to make
room for the results of the latter's empirical experiments and technological
innovations. For a treatment of this relationship that emphasizes the modifications, see Matthias Rieger, *Helmholtz musicus*.

23. See *PC*, 99–100/86: "The greater the difference of the vibration-
lengths [*Schwingungsdauer*], the quicker the beats [*Schwebungen*]. As long
as no more than four to six beats occur in a second, the ear readily distinguishes the individual, alternating amplifications of tone. [. . .] If the beats
become more rapid, it becomes increasingly difficult for the ear to hear them
individually, while a rawness of tone remains."

24. For an exploration of the possibility that Helmholtzian music theory
also influenced Wagner, see Steege, *Helmholtz and the Modern Listener*,
224–34.

25. Helmholtz, *On the Sensations of Tone*, 339 / *Tonempfindungen*,
554. With this claim, Helmholtz reaches the conclusion to which he had
pronounced himself committed at the outset of his music-theoretical discussion: "It will be shown that our modern system acquired the form in which

we currently possess it principally through the influence of an increasingly universal use of harmonic chords [*Zusammenklänge*]. Within this system, for the first time, a complete overview was acquired of all the requirements of the harmonic network [*des Harmoniegewebes*]. Owing to the strict closure and consistency [*festgeschlossenen Konsequenz*] of this system, not only can we allow ourselves many liberties in the use of more imperfect consonances and of dissonances, which the older systems had to avoid, but the consistency of the system actually often requires the insertion of thirds into final cadences, as a mode of distinguishing between major and minor, where they were formerly circumvented." Helmholtz, *On the Sensations of Tone*, 339 / *Tonempfindungen*, 345. The explicit reference to Richard Wagner is introduced only in the fourth edition of 1877, though the logic of music-theoretical development remains essentially the same across all the editions.

26. Helmholtz, *On the Sensations of Tone*, 256 / *Tonempfindungen*, 423–24.

27. See Helmholtz, *On the Sensations of Tone*, 369 / *Tonempfindugen*, 594: "The development of harmony gave rise to a much richer unfolding of musical art than was previously possible, because the far clearer articulation of relationships among tones, by means of chords and chordal sequences, allowed for the use of much more remote relationships, and particularly of modulations into distant keys. In this way the richness of the means of expression was increased, as well as the speed with which melodic and harmonic transitions could now be introduced without destroying the musical unity [*Zusammenhang*]."

28. This formulation appears for the first time in the third edition of *Tonempfindungen*, which includes an expanded version of the final chapter. Helmholtz, *On the Sensations of Tone*, 364 / *Tonempfindung*, 585.

29. Helmholtz, *On the Sensations of Tone*, 362–71 / *Tonempfindungen*, 581–99 (the published translation calls the chapter "Esthetical Relations").

30. See *On the Sensations of Tone*, 250 / *Tonempfindungen*, 413: "Every movement is for us an expression of the forces through which it is produced, and we know instinctively how to judge the driving forces when we observe the action they produce. This holds equally and perhaps even more for the movements produced by the force-displays [*Kraftäußerungen*] of the human will and of the human drives than for the mechanical movements of external nature. In this way the melodic movement of tones can become an expression for the most diverse human frames of mind."

31. The Helmholtzian position here quite deliberately recalls the Kantian bifurcation between a mechanistic natural universe and a teleological human reason. The physicist who contemplates the difference between the order of the ocean waves and the order of musical expression reproduces— albeit with consequences that Kant himself would never have condoned— the famous distinction between cognitive and moral faculties, between pure and practical reason, between natural science and human feeling, which constitutes the true philosophical substance of an encounter with the dynamic sublime.

32. On Fechner's early proximity to nature-philosophy, and particularly to the work of the Schelling disciple Lorenz Oken, see Heidelberger, *Nature from Within*, 28ff.

33. Gustav Theodor Fechner, *Elemente der Psychophysik*, 2 vols. (Leipzig: Breitkopf & Härtel, 1860). Only the first volume, which I do not cite here, has been translated into English as Gustav Theodor Fechner, *Elements of Psychophysics*, vol. 1, trans. Helmut E. Adler, ed. David H. Howes and Edwin G. Boring (New York: Holt, Rinehart and Winston, 1966).

34. Fechner, *Elemente der Psychophysik*, vol. 2. Fechner refers to "the Fourier principle" on page 238, in the context of a discussion devoted to the periodicity of consciousness. Citations of Helmholtz permeate the entire middle section of the volume (pages 238–376), in which Fechner investigates the relationship among oscillatory stimuli, oscillatory sensation, and the ostensibly oscillatory activity of the mind.

35. Fechner, *Elemente der Psychophysik*, 2:454ff.

36. Ibid., 2:458. The passage continues as follows: "There are waves produced by a general cause such as the wind, but upon which, in turn, due to more particular causes, puckers or surface ripples [*Oberwellen*] form, which can be seen as disturbances of an underlying wave [*Unterwelle*] that remains independent of them; meanwhile, the whole wave, as it exists, constitutes the main wave [*Hauptwelle*] or total wave [*Totalwelle*]."

37. Ibid., 2:542.

38. Ibid.

39. The rule takes the algebraic form $\Delta y = C(\Delta\beta)/\beta$, where C represents a constant, Δy the change in sensation, and $\Delta\beta$ the change in stimulus. Fechner himself attributes the discovery of this psychophysical "fundamental formula" to his teacher, the Leipzig physiologist Ernst Weber; both Fechner and Wundt consequently call it "the Weber law." For Fechner's own presentation, which uses the symbols dy, $d\beta$, and K, see Fechner, *Elemente der Psychophysik*, 1:10.

40. Fechner, *Elemente der Psychophysik*, 2:542. The phrase "live, move and exist" (*leben, weben und sind*) is a reference to Acts 17:28: "[I]n him we live and move and have our being, as even some of your own poets have said, 'For we are indeed his offspring.'" English Standard Version Bible (London: Crossway, 2012).

41. Fechner, *Elemente der Psychophysik*, 2:468.

42. Wilhelm Wundt, *Grundzüge der physiologischen Psychologie*, 5th ed. (1902–3), 3:617. The first edition was published as a single volume; the subsequent 2nd, 3rd, and 4th editions were published as two volumes; and the 5th and 6th editions were published as three volumes. I will cite here primarily from the later, more expansive versions. Only the first volume of the fifth edition has been translated. See Wilhelm Wundt, *Principles of Physiological Psychology*, vol. 1, trans. Edward Bradford Titchner (London: Swan Sonnenschein and Co., 1904).

43. Wundt, *Grundzüge*, 4th ed. (1893), 2:481.

44. See, for instance, Wundt, *Grundzüge*, 5th ed. (1902–3), 3:17–18, where Wundt suggests that the most obviously rhythmic functions of the psychophysiological apparatus, such as pulse and respiration, should be

considered merely a special case of the all-encompassing rhythm that governs psychophysiological activity as a whole—and that all apparently *arrhythmic* phenomena should be considered deficient permutations, or rather fragments, of rhythmic ones: "The interactions of excitatory [*erregender*] and inhibitory [*hemmender*] forces, upon which all central functions of the nervous system are likely based, hang everywhere together with the inherent tendency toward an *oscillatory* progression of vital processes, to which the rhythmic body movements also belong. [. . .] In this sense, the movements that lie beyond the realm of the rhythmic functions are *fragments of rhythms,* which periodically come closer and less close to the actual rhythms."

45. See, for instance, Wundt, *Grundzüge,* 4th ed. (1893), 1:562–64, esp. 564: "Self-consciousness, since it has its roots in the constant activity [*Wirksamkeit*] of apperception, ultimately falls back on this activity alone, such that, after the completed development of consciousness [*nach vollendeter Bewusstseinsentwicklung*], the will appears, together with its associated emotions and tendencies [*Strebungen*], as the most authentic and indeed only content of self-consciousness."

46. Wundt, *Grundzüge,* 5th ed. (1902–3), 3:617–18.

47. Wundt, *Grundzüge,* 4th ed. (1893), 2:646.

48. Wundt, *Grundzüge,* 4th ed. (1893), 2:648; emphasis in the original. A somewhat different version of this passage concludes the first edition of the *Grundzüge.* See Wundt, *Grundzüge,* 1st ed. (1874), 863. While both the earlier and later versions of the passage make clear that Wundt understands his doctrine of psychophysiological parallelism to be importantly different from the Leibnizian notion of a preestablished harmony between body and mind—a notion to which he in both cases explicitly alludes—only the later version expressly formulates this difference in terms of a nature-philosophical insistence on the role of teleological development.

49. Wundt, *Grundzüge,* 4th ed. (1893), 2:619

50. Wundt, *Grundzüge,* 4th ed. (1893), 2:612.

51. Wilhelm Wundt, *Probleme der Völkerpsychologie* (Leipzig: Wiegandt, 1911), 40: "The most familiar example of such a natural sound association, which has successfully withstood all the artificial etymologies of the older Indo-German linguistics [*Indogermanistik*], is the concept-pair of father and mother, or, as it tends to appear in countless languages and in our children's idiom, Papa and Mama [. . .]. The more such a relationship finds confirmation in the languages of completely unrelated peoples [*stammesfremden Völkern*], i.e., under conditions that eliminate the possibility of deriving the word pairs from a common foundational language [*Grundsprache*], the less it can be considered a product of coincidence." Wundt goes on to make the resistance to etymological explication one of three criteria for identifying true sound metaphors; ibid., 41.

52. Wundt, *Grundzüge,* 4th ed. (1893), 2:619–20.

53. Wundt, *Grundzüge,* 5th ed. (1902–3), 3:628.

54. For one of the most expansive versions of Wundt's overarching narrative about the emergence and development of language—as the product of an expressive drive that first manifests itself in the "roots" of sound

gestures—see Wundt, *Grundzüge*, 4th ed. (1893), 2:615ff. Most of this narrative is already present in the first edition of the *Grundzüge*, and its placement at the very end of the work, in the first as well as all subsequent editions, hints at its key role in the development of Wundt's more general narrative of mental development. For Wundt's position on the psychophysiological meaning of the Germanic sound shifts, see Wilhelm Wundt, *Völkerpsychologie: Eine Untersuchung der Entwicklungsgesetze von Sprache, Mythus, und Sitte*, 2nd ed., vol. 1, *Die Sprache* (Leipzig: Wilhelm Engelmann, 1904), 484–529, esp. 505. On the priority of modern meter, see Wundt, *Grundzüge*, 5th ed. (1902–3), 3:164–65.

55. Wundt begins his work, in the foreword to the first edition of 1900, with a conciliatory gesture aimed at the Neogrammarians, whose scholarly contributions he professes to have mined for his examples, and whose disciplinary jurisdiction he claims not to call into question: "With regard to the above-mentioned divergence between the standpoints of the psychologist and the historian, I will obviously refrain from passing my own judgments about controversial questions pertaining to the histories of language, myth, and custom, insofar as such questions are of purely historical character. Only where historical inferences connect up with, or even, as it indeed occasionally happens, consist exclusively of psychological hypotheses, do I consider myself permitted to go beyond the role of disinterested observer"; Wundt, *Völkerpsychologie*, 2nd ed. (1904), 1:vii. The foreword to the second edition, written four years later, makes clear that despite this proleptic apologia, Wundt's foray into the realm of language science provoked an immediate and overwhelmingly negative response from professional linguists, particularly with respect to the proffered psychophysiological explanation of phonetic change (*Lautenwandel*): Wundt thanks Karl Brugmann and others for their "corrections and additions" but continues to insist on the legitimacy of his specifically psychological, or rather, psychophysiological, approach to historical linguistic study. Wundt, *Völkerpsychologie*, 2nd ed. (1904), 1:ix–x.

56. Eduard Sievers, *Rhythmisch-melodische Studien* (Heidelberg: Carl Winter, 1912).

57. "Only those people who possess a certain minimum degree of *motoric predisposition* are capable of grasping the oppositions in question immediately and without any effort"; Eduard Sievers, "Ziele und Wege der Schallanalyse," in *Stand und Aufgaben der Sprachwissenschaft: Festschrift für Wilhelm Streitberg* (Heidelberg: Carl Winter, 1924), 65–111, here 66.

58. For Sievers's own presentation of his method, see, for instance, Sievers, "Ziele und Wege der Schallanalyse," or "Über ein neues Hilfsmittel philologischer Kritik," in Sievers, *Rhythmisch-melodische Studien*, 78–111. On Sievers's "ear philology" in the wider context of nineteenth- and twentieth-century theories and practices of voice, see Reinhart Meyer-Kalkus, *Stimme und Sprechkünste im 20. Jahrhundert* (Berlin: Akademie Verlag, 2001), 73–142.

59. Eduard Sievers, *Die Eddalieder: Klanglich untersucht und herausgegeben von Eduard Sievers, Abhandlungen der Philologisch-historischen Klasse der Sächsischen Akademie der Wissenschaften* 37, no. 3 (Leipzig:

B. G. Teubner, 1923). See also "Über ein neues Hilfsmittel," in Sievers, *Rhythmisch-melodische Studien*, 101–9.
60. Sievers, "Ziele und Wege der Schallanalyse," 66.
61. Ghil, *Les dates et les oeuvres: Symbolisme et poésie scientifique* (Paris: G. Crès, 1923), ix–x. Wolzogen, *Poetische Lautsymbolik*, 52.
62. Ghil, *Traité du verbe*, 22.

CHAPTER 6. JAKOBSON'S ZEROS

1. Roman Jakobson, "On the Identification of Phonemic Entities" (1949), in *Selected Writings*, vol. 1, *Phonological Studies* (The Hague: Mouton, 1962), 418–25, here 425.
2. Roman Jakobson, *Novejšaja russkaja poèzija: Nabrosok pervyi: Viktor Xlebnikov* [Newest Russian poetry: First attempt: Viktor Khlebnikov] (1921), in *Selected Writings*, vol. 5, *On Verse, Its Masters and Explorers* (The Hague: Mouton, 1979), 299–354; and *O češskom stiche: Preimuščestvenno v sopostavlenii s russkim* [On Czech verse, primarily in comparison with Russian] (1923), in Jakobson, *Selected Writings*, 5:3–130. *Newest Russian Poetry*, which was intended as the introduction to an edition of Jakobson's friend Khlebnikov's verse, has been partially translated into English as "Modern Russian Poetry: Velimir Khlebnikov," trans. E. J. Brown, in *Major Soviet Writers: Essays in Criticism*, ed. E. J. Brown (Oxford: Oxford University Press, 1973), 58–82. A complete bilingual edition is available in German translation under the title "Die neueste russische Poesie: Erster Entwurf: Viktor Chlebnikov," in *Texte der russischen Formalisten*, vol. 2, ed. W. D. Stempel (Munich: Fink, 1972), 18–135. *On Czech Verse* exists in German translation as *Über den tschechischen Vers: Unter besonderer Berücksichtigung des russischen Verses* (Bremen: K Presse, 1974).
3. The arguments against a psychophysical understanding of verse occasionally take on, in both books, distinctly "proto-phonological" contours. Jakobson himself frequently comments on this fact in later years, attributing his prescient instincts to an intense engagement with Russian Futurist sound poetry, and with the work of students influenced by Saussure and the Russian linguist Jan Baudouin de Courtenay. See, for instance, Roman Jakobson, "Structuralisme et Téléologie" [Structuralism and teleology] (1975), in *Selected Writings*, vol. 7, *Contributions to Comparative Mythology: Studies in Linguistics and Philology, 1972–1982* (The Hague: Mouton, 1985), 125–27, here 125–26.
4. Roman Jakobson, "The Concept of the Sound Law and the Teleological Criterion" (1928), in Jakobson, *Selected Writings*, 1:1–2. This little text was originally published in Czech as the précis of a longer paper delivered in 1927 to members of the Prague Linguistic Circle. As the chronologically earliest of all of Jakobson's writings to gain entry into the volume *Phonological Studies*, it has an interesting transitional status. On the relationship between Jakobson's notion of "teleological criterion" and the theories of language change among Russian Formalists more generally, see Anatoly Liberman, "Roman Jakobson and His Contemporaries on Change in Language

and Literature (The Teleological Criterion)," in *Language, Poetry, and Poetics: The Generation of the 1890s: Jakobson, Trubetzkoy, Majakovskij,* ed. Krystyna Pomorska (Berlin: De Gruyter, 1987), 143–56.

5. See, for instance, Roman Jakobson, "Musikwissenschaft und Linguistik" [Musicology and linguistics] (1932), in *Selected Writings,* vol. 2, *Word and Language* (The Hague: Mouton, 1971), 551–53; "On the Identification of Phonemic Entities," in Jakobson, *Selected Writings,* 1:420 and 1:423ff; or Roman Jakobson and Linda R. Waugh, *The Sound Shape of Language* (1979; repr., Berlin: De Gruyter, 2002), 22.

6. For one of the last and most extensive elaborations of this trajectory, see Jakobson and Waugh, *Sound Shape of Language,* 13–17.

7. Roman Jakobson, "Structuralisme et Téléologie" [Structuralism and teleology], in *Selected Writings,* 7:126.

8. This is presumably also why so many sophisticated readers have ultimately concluded that Jakobson's actual "philosophy" of language—which is to say, his position regarding the ontological "essence" of his object—is relatively traditional, particularly when compared with the apparently more radical, because explicitly *anti*teleological, perspective of Saussure. If one takes Jakobson's favorable references to teleology at face value, it is hard to avoid the impression that he is simply reverting to a premechanist, and thus effectively prescientific, paradigm. See Patrice Maniglier, "L'ontologie du négative: Dans la langue n'y a-t-il vraiment que des différences?" *Methodos* 7, July 2007, http://methodos.revues.org/674; and Jean-Claude Milner, "A Roman Jakobson ou le bonheur de la symétrie," in *Le périple structural: Figures et paradigmes* (Paris: Éditions du Seuil, 2002), 131–40. Jakobson himself confronts the problem of his relationship to systemic history, and thus also to the notion of a teleological rather than a mechanical unfolding, most extensively in a late dialogue with Krystyna Pomorska on the role of time in language and literature. His comments there, while illuminating, do little to answer the question of how, exactly, his own notion of a goal-oriented historical trajectory can be said to differ from the nineteenth-century idea of "progress" he disparages. The dialogue includes Jakobson's account of an early exchange with Nikolai Trubetzkoy—"that linguist and associate whom I admired above all others"—in which the latter declares himself to be in full agreement regarding the nonarbitrary, goal-oriented nature of linguistic development. See Roman Jakobson and Krystyna Pomorska, "Dialogue on Time in Language and Literature," in Roman Jakobson, *Verbal Art, Verbal Sign, Verbal Time,* ed. Krystyna Pomorska and Stephen Rudy (Minneapolis: University of Minnesota Press, 1985), 11–24.

9. The year 1939 is also, not coincidentally, the publication date of Nikolai Trubetzkoy's foundational *Principles of Phonology.* The two thinkers collaborated so intensely throughout the 1920s and 1930s that it is nearly impossible to assign priority to one or the other for the elaboration of basic phonological principles. Jakobson's own accounts from these early years are less thoroughly worked out than Trubetzkoy's, but they already contain—as later notes will make clear—the seeds of significant differences in emphasis and perspective. The points where Jakobson implicitly diverges from

Trubetzkoy's brilliant but—I would argue—system-theoretically less radical account are the ones of particular interest to me here.

10. For an attempt to trace the development of Jakobson's position regarding the sense in which the phonological unit can be said to *exist*—a question he repeatedly explicitly brackets but just as repeatedly obliquely tackles throughout his career—see Ľubomír Ďurovič, "The Ontology of the Phoneme in Early Prague Linguistic Circle," in *Jakobson entre l'est et l'ouest, 1915–1939*, ed. Françoise Gadet and Patrick Sériot (Lausanne: Université de Lausanne, 1997), 69–76.

11. Roman Jakobson, "Zur Struktur des Phonems" [On the structure of the phoneme] (1939), in Jakobson, *Selected Writings*, 1:280–310, here 1:280. Hereafter cited as *SP*.

12. Plato, *Philebus*, 16c–d, in *Complete Works*, 404. The passage is discussed in chapter 1, 35–37.

13. "As Saussure himself already expressly emphasized, what is primarily relevant is not the acoustic content of a phoneme in itself but rather its opposition to the other phoneme-contents. It was therefore from the very beginning not the phoneme in itself but the phonemic opposition, or the phoneme as member of an opposition (*l'opposition et l'opposé*), that became the foundational pillar [*Grundpfeiler*], the primary concept of phonological research."

14. There exists, however, an essential difference between the two interpretive undertakings, since written letters, unlike phonological ones, are not actually *designed* to fulfill this kind of differential function. The elements of a phonetic alphabet always have a positive referent in the form of the phonological sound-segments for which they stand, and, as a result, they tend not to be shaped so as to render the relevant oppositional relationships easy to spot. It is in this sense that the process of learning to read phonetically has to be considered parasitic upon the process of learning to "read" the deeper "writing" of the phonological system. Without the foundations provided by the latter, the former would present an impossible challenge: "One can make accessible to a deaf-mute child the meanings of the written words, just as one otherwise makes accessible to children the meanings of spoken words. We know, however, from the praxis of deaf-mute education that the task in question is all but insoluble"; *SP*, 304.

15. "The dichotomy of distinctive features is, in essence, a logical operation, one of the primary logical operations of a child and—if we pass from ontogeny to phylogeny—of mankind." Jakobson, "Identification of Phonemic Entities," in Jakobson, *Selected Writings*, 1:424.

16. Particularly useful with regard to the question of method, in addition to "Zur Struktur des Phonems," are Roman Jakobson, "Observations sur le classement phonologique des consonnes" [Observations on the phonological classification of consonants] (1939), in Jakobson, *Selected Writings*, 1:272–79; Jakobson, "Identification of Phonemic Entities," in Jakobson, *Selected Writings*, 1:418–25; and Roman Jakobson and J. Lotz, "Notes on the French Phonetic Pattern" (1949), in Jakobson, *Selected Writings*, 1:426–34. Since Jakobson also explicitly borrows much of his account of the minimal pairs

technique in "The Identification of Phonemic Entities" from the American linguist William Freeman Twaddell, whose early group theoretical definition of the phoneme is one of Jakobson's favorite later points of reference (though there is no indication that he knew this work already at the time of the 1939 lectures), I will include elements of Twaddell's approach in my reconstruction as well. See W. Freeman Twaddell, *On Defining the Phoneme, Language Monographs* 16 (Baltimore: Waverly Press, 1935). For Jakobson's references to Twaddell's presentation, see Jakobson, "Identification of Phonemic Entities," in Jakobson, *Selected Writings*, 1:420ff; and Jakobson, "Phonology and Phonetics" (1955), in Jakobson, *Selected Writings*, 1:464–504, here 1:472ff. On the difficulty of precisely reconstructing Jakobson's analytic methods from his published writings, see the following personal recollections by his student, Paul Kiparsky: "Jakobson was certainly the most captivating teacher I ever had. There was always a kind of mystery behind his teaching and writing. He explained what the problems were that interested him and why they were important, he dazzled you with his brilliant solutions to them, but if you wanted to try that kind of work yourself you had to figure out on your own how to do it. In his articles he would hide his tracks the way mathematicians do, presenting his results but not revealing how he discovered them"; Paul Kiparsky, "Roman Jakobson and the Grammar of Poetry," in *A Tribute to Roman Jakobson, 1896–1982*, ed. Paul E. Gray (Berlin: Mouton, 1983), 227–38, here 27–28. The tendency seems to have been shared by some of his closest friends and associates. Nikolai Trubetzkoy's *Principles of Phonology*, for instance, which is so relevant for the articulation of the foundational phonological principles as they appear in "Zur Struktur des Phonems," contains almost no discussion of an actual analytical procedure for the discovery of discrete phonological units.

17. See Jakobson and Waugh, *Sound Shape of Language*, 91.

18. These particular English-language examples of paradigm sets are taken from Twaddell, *On Defining the Phoneme*. For corresponding examples from Jakobson, see, for instance, Roman Jakobson, "Identification of Phonemic Entities," in Jakobson, *Selected Writings*, 1:420: "By studying the possible commutations we obtain, e.g., a French phonemic 'paradigm' /bu/ 'boue' : /mu/ 'mou' : /pu/ 'pou' : /vu/ 'vous' : /du/ 'doux' : /gu/ 'goût', and thus we find out that the phoneme b in /bu/ can be decomposed into five commutable elements: b/m, b/p, b/v, b/d, b/g. In examining the same phoneme in other environments we confirm this scheme. Cf. /bo/ 'beau' : /mo/ 'mot' : /vo/ 'veau' : /po/ 'peau' : /do/ 'dos' : /go/ 'gau', etc."

19. There is some debate as to how, exactly, early German-language phonetic categories such as *Breite* (breadth) and *Enge* (narrowness) correspond to the English-language distinctions currently in widespread use. *Breite* and *Enge*, for instance, have been translated as "low" and "high," but they also bear a clear, if somewhat inconsistent, relationship to the categories of "open" and "closed." For a brief reflection on the problem this poses for the translator, see the translator's note in Nikolai Sergevich Trubetzkoy, *Principles of Phonology*, trans. Christiane A. M. Baltaxe (Berkeley: University of California Press, 1969), 227. I have opted in what follows to hew as closely

as possible to the spatial and gestural connotations of the original German in my translations of Jakobson's technical terminology.

20. Cf. chapter 2, 85–87.

21. On the relationship between this "distinctive feature" model of phonology—to which Jakobson gives expression, in Czech, as early as 1932—and Trubetzkoy's more atomistic understanding of the phoneme itself as basic unit, see Tsutomu Akamatsu, "The Development of Functionalism from the Prague School to the Present," in *History of the Language Sciences: An International Handbook on the Evolution of the Study of Language from the Beginnings to the Present*, ed. Sylvain Auroux, E. F. K. Koerner, Hans-Josef Niederehe, and Kees Versteegh, vol. 2. (Berlin: De Gruyter, 2001), 1768–90, here 1773.

22. Roman Jakobson, "Proposition au Premier Congrès International de Linguistes: Quelles sont les méthodes les mieux appropriées à un exposé complet et pratique de la phonologie d'une langue quelconque?" [Proposition to the First International Congress of Linguists: What are the most appropriate methods for a compete and practical exposition of the phonology of a given language?] (1927), in Jakobson, *Selected Writings*, 1:3. Jakobson continues, throughout his career, to attribute the origin of this all-important idea to Trubetzkoy, whose death in 1939 prevented him from carrying to fruition his investigation of its linguistic consequences. See, for instance, Jakobson and Waugh, *Sound Shape of Language*, 92–94. See also Morris Halle, "On the Origins of the Distinctive Features," in *Roman Jakobson: What He Taught Us*, ed. Morris Halle (Columbus: Slavica, 1983), 77–86. For a critique from the perspective of contemporary phonological problems, see Stephen R. Anderson, "Roman Jakobson and the Theory of Distinctive Features," in *Phonology in the Twentieth Century: Theories of Rules and Theories of Representations* (Chicago: University of Chicago Press, 1985), 116–39.

23. *SP*, 302. Jakobson uses the symbols /y/ and /ü/ interchangeably in this passage. Since, however, the table definitively decides in favor of *ü*, I have taken the liberty of doing the same in the cited lines, so as to ensure the comprehensibility of the relationships expressed.

24. Roman Jakobson, "Identification of Phonemic Entities," in Jakobson, *Selected Writings*, 1:420.

25. "As music imposes upon sound matter a graduated scale, similarly language imposes upon it the dichotomous scale which is simply a corollary of the purely differential role played by phonemic entities"; Jakobson, "Identification of Phonemic Entities," in Jakobson, *Selected Writings*, 1:423.

26. "The phoneticians have ascertained that the emission of consonants presents an infinity of degrees and shades with regard to the participation of voice: the glottis can be closed to a greater or lesser degree; the vibrations of the vocal cords can be of different amplitude; and the phase at which they begin or cease may vary. Thus, the glottis is capable of producing diverse nuances in the matter of consonantal voicing, but only the opposition 'presence vs. absence of voicing' is utilized to differentiate word meanings. Since the sound matter of language is a matter organized and formed to serve as a

semiotic instrument, not only the significative function of the distinctive features but even their phonic essence is a cultural artifact"; Jakobson, "Identification of Phonemic Entities," in Jakobson, *Selected Writings*, 1:422–23.

27. "Every language employs these oppositions, on the one hand singularly, on the other as bundles, for the purpose of differentiating words. For example, in Ottoman Turkish the contrast /o/—/u/ or /u/—/y/ contains one opposition, the contrast /o/—/y/ forms a bundle of two, and the contrast /o/—/i/ a bundle of all three vocalic oppositions. In this way the phoneme proves itself to be a complex unit, namely a bundle of distinctive or, put differently, phonemic qualities"; *SP*, 303.

28. Plato, *Sophist*, 253a, trans. Nicholas P. White, in *Complete Works*, 275.

29. Roman Jakobson, "On the So-Called Vowel Alliteration in Germanic Verse" (1963), in Jakobson, *Selected Writings*, 5:189–96, here 5:189. Hereafter cited as *VA*.

30. The opposition tense/untensed corresponds, in Jakobson's analysis, to the traditional opposition *fortis/lenis*. Sounds that require greater muscular tension are usually, but not necessarily, also voiced.

31. In modern phonetic terminology, this makes the former group *fricatives* and the latter *plosives* or *stops*.

32. Jakob Boehme, *Mysterium Magnum, oder Erklärung über das erste Buch Mosis, von der Offenbarung Göttlichen Worts, durch die drei Principia Göttliches Wesens, auch vom Ursprung der Welt und der Schöpfung* (1623), in *Sämtliche Schriften*, vol. 7 (1730; repr., Stuttgart: Frommann, 1956), 331–32; and Johann Georg Hamann, "Neue Apologie des Buchstabens h," in *Sämtliche Werke*, ed. Josef Nadler, vol. 3 (Vienna: Herder 1951), 89–108, here 98. On the role of the *h* in the eighteenth-century orthographic battles that pitted language rationalists against the advocates of (language) spirit, see Jonathan Sheehan, "Enlightenment Details: Theology, Natural History, and the Letter *H*," *Representations* 61 (1998): 25–56. On the role of the *h* in the history of language philosophy more generally, see Daniel Heller-Roazen, *Echolalias: On the Forgetting of Language* (New York: Zone Books, 2005), 33–44.

33. Jakobson and Lotz, "French Phonemic Pattern" in Jakobson, *Selected Writings*, 1:426–34, here 1:431. For an insightful account of the role played by the linguistic zero both at the origins of phonology and in the later French structuralist/poststructuralist context, see Catharine Diehl, "The Empty Space in Structure: Theories of the Zero from Gauthiot to Deleuze," *Diacritics* 38, no. 3 (2008): 93–119. My account deviates from hers only in my interpretation of Jakobson's contribution, which I think successfully escapes, in its final formulation, the (self-consciously) paradoxical incoherence of the various poststructuralist approaches.

34. Douglas Walker, *French Sound Structure* (Calgary: University of Calgary Press, 2001), 78.

35. Jakobson and Waugh, *Sound Shape of Language*, 154–55. See also the definition provided in Philip Carr, *A Glossary of Phonology* (Edinburgh: Edinburgh University Press, 2008), 156: "*schwa*: The name for a vowel

quality which is produced without lip rounding and with the body of the tongue in the neutral position. Transcribed as [ə], it occurs widely in unstressed syllables in many varieties of English, as in the word 'character'. This vowel alternates with a wide variety of other vowels in many languages. It also alternates with zero in many languages, such as French." For an extended rumination on the "endangered" status of this peculiar phoneme, see Heller-Roazen, *Echolalias*, 27–32.

36. Claude Lévi-Strauss, *Introduction to the Work of Marcell Mauss*, trans. Felicity Baker (London: Routledge, 1987), 63–64; original French "Introduction à l'oeuvre de Marcel Mauss" (1950), in Marcel Mauss, *Sociologie et anthropologie*, 4th ed. (Paris: PUF, 1968), xlix–l.

37. Jacques Derrida, "Structure, Sign, and Play in the Discourse of the Human Sciences," in *Writing and Difference*, trans. Alan Bass (London: Routledge, 1978), 278–94; original French: "La structure, le signe, et le jeu dans le discours des sciences humaines," in *L'écriture et la différence* (Paris: Seuil, 1967), 409–28; Gilles Deleuze, "How Do We Recognize Structuralism?" in *Desert Islands and Other Texts, 1953–1974*, trans. Michael Taormina, ed. David Lapoujade (Los Angeles: Semiotext[e], 2004)), 170–92; original French: "À quoi reconnaît-on le structuralisme?" (1972), in *Le XXe siècle, Histoire de la philosophie: Idées, doctrines*, vol. 8, ed. Françoise Châtelet (Paris: Hachette, 2000), 299–335.

38. The ancient Platonic question about the relationship between delimited structure and its undelimited substrate—together with the late nineteenth-century mathematical question about the relationship between the rational number system and the continuum—is very clearly (and occasionally explicitly) at issue in all three of these texts. It comes as no surprise, therefore, to find Jakobson's zero in dialogue, particularly in the case of Derrida, with a whole host of Greek terms and figures (*dynamis, chora, desmos, psyche, metron, logos*) drawn from the horizon of Plato's *Philebus* and *Timeaus*.

39. Jakobson and Waugh, *Sound Shape of Language*, 155.

40. Roman Jakobson, "Zero Sign," in *Russian and Slavic Grammar: Studies, 1931–1938*, trans. Linda R. Waugh, ed. Linda R. Waugh and Morris Halle (Berlin: De Gruyter, 1984), 151–60, here 153. Translation very slightly modified (emphasis mine); original French: "Signe Zéro" (1939), in Jakobson, *Selected Writings*, 2:211–19, here 2:213. Jakobson presented an oral German paraphrase of this article in Copenhagen, also in 1939, under the title "Das Nullzeichen." See Roman Jakobson, "Das Nullzeichen" (1940), in Jakobson, *Selected Writings*, 2:220–22. With the phrase "as I have tried to demonstrate elsewhere," Jakobson refers the reader to his seminal article on the structure of the Russian verb, which contains in its opening lines the very first attempt to formalize the marked/unmarked relationship in terms of a latent zero: "One of the essential properties of phonological correlations is the fact that the two members of a correlational pair are not equivalent: one member possesses the mark in question, the other does not; the first is designated as marked [*merkmalhaltig*], the other as unmarked [*merkmallos*] [. . .] (see N. Trubetzkoy in TCLP, IV, 97). Morphological correlations may be characterized on the basis of the same definition. [. . .] When

a linguist investigates two morphological categories in mutual opposition, he often starts from the assumption that both categories should be of equal value, and that each of them should possess a positive meaning of its own: Category I should signify A, while Category II should signify B; or at least I should signify A and II the absence or negation of A. In reality, the general meanings of correlative categories are distributed in a different way: if Category I announces the existence of A, then category does not announce the existence of A, i.e., it does not state whether A is present or not. The general meaning of the unmarked Category II, as compared to the marked Category I, is restricted to the lack of 'A-signalization'"; Jakobson, "Structure of the Russian Verb," in *Russian and Slavic Grammar: Studies, 1931–1938*, trans. Brent Vine and Olga T. Yokoyama, ed. Linda R. Waugh and Morris Halle (Berlin: De Gruyter, 1984), 1–14, here 1; original German: "Zur Struktur des russischen Verbums" (1932), in Jakobson, *Selected Writings*, 2:3–15, here 2:3.

41. It would thus be hard to imagine a position better designed to polemically negate the nineteenth-century German Idealist prioritization of *Realopposition*, which in Schelling's nature-philosophy takes the form of a dynamic, universe-propelling clash of forces: the neutral state of total indifference, which is to say of absolute rest, is for Schelling "a nonthing" (*ein Unding*). Schelling, *Von der Weltseele*, in *Historische-kritische Ausgabe* (*HKA*), I:6, 79. On the development of Jakobson's ideas about markedness over the course of his career, see Edwin L. Battistella, "The Development of Markedness in Jakobson's Work," in *The Logic of Markedness* (Oxford: Oxford University Press, 1996), 19–34. On the further development of the concept within the context of generative grammar, see the later chapters in Battistella, *The Logic of Markedness*, as well as Catherine V. Chvany, "The Evolution of the Concept of Markedness from the Prague Circle to Generative Grammar," in *The Selected Essays of Catherine V. Chvany*, ed. Olga T. Yokoyama and Emily Klenin (Columbus: Slavica, 1996), 234–41.

42. See, for instance, the discussion of glides in Jakobson and Waugh, *Sound Shape of Language*, 153–56. Jakobson's own attempt to rethink the philosophical tradition of entity formation in terms of the concept of latency is clearly intimately related to (and presumably influenced by) Nikolai Trubetzkoy's equally loaded terminology of phonological neutralization or *Aufhebung*. For Trubetzkoy, a phonological opposition appears as "neutralized" in places where it ceases to differentiate meaning. The German phonemes /t/ and /d/, for instance, are normally opposed to one another by the absence and presence, respectively, of the distinctive feature "voicedness." The voiced phoneme /d/, however, cannot be realized at the end of German words, which means, according to Trubetzkoy, that the opposition between "voiced" and "unvoiced" is temporarily "neutralized" in this position. The phoneme that appears in such cases is neither /t/ nor /d/—both of which only exist, phonologically speaking, insofar as they remain opposed—but rather the representative of what Trubetzkoy calls an "archiphoneme," which he defines as "the sum total of distinctive features common to both phonemes" [*die Gesamtheit der distinktiven Eigenschaften* [. . .] *die zwei Phonemen gemeinsam*

sind]. Trubetzkoy, *Principles of Phonology* 79 / *Grundzüge der Phonologie*, 71. Since the archiphoneme participates in no binary oppositions, it cannot itself appear at the level of the phonological analysis; it is rather, as its name suggests, a common ground or underlying foundation (*arche*) upon which this analysis turns out to rest, and toward which the analyzable phenomenon of neutralization obliquely directs our attention. From the perspective of the theory of the archiphoneme, in other words, the kinds of zeros Jakobson finds so interesting would be mere epiphenomena (and, indeed, Trubetzkoy does not himself devote any attention to the zero phoneme in Jakobson's sense): the latency of any particular unmarked term in the phonological chain will always turn out to rest, for Trubetzkoy, on the deeper latency of a metaphonologically existing, but phonologically inarticulable, *Vergleichsgrundlage*, or criterion of comparison. Jakobson, I am arguing here, entirely dispenses with this rather traditional philosophical framework and, in the process, makes a much more radical—and phonologically writeable—notion of latency into the foundation for his definition of system. Past accounts of the relationship between these two positions have, I think, tended to elide their differences in a way that privileges Trubetzkoy's significantly less challenging approach. For a particularly powerful example of this elision and its conceptual consequences, see Giorgio Agamben, *The Time that Remains: A Commentary on the Letter to the Romans*, trans. Patricia Dailey (Stanford, CA: Stanford University Press, 2005), 99–104; original Italian *Il tempo che resta: Un commento alle Lettera ai Romani* (Turin, Italy: Bollati Boringhieri, 2000), 94–98. Agamben adopts the Derridean reading of the zero phoneme and inserts this reading—quite illuminatingly—into his examination of the theological and *geschichtsphilosophical* paradigm of the end times. The dynamic of "zeroing" implied by the zero phoneme becomes thereby aligned, or even identified, with the Pauline concept of *katargein* and the Hegelian concept of *Aufhebung*. Such an ahistorical identificatory series can be constructed, I would argue, only because Agamben takes Trubetzkoy's understanding of the zero, rather than Jakobson's, to be the definitive formulation of the concept. For an informative overview of the role played by Trubetzkoy's concept of the "archiphoneme" in Prague school linguistics, and its disappearance after 1939, see Tsutomu Akamatsu, "Development of Functionalism," in Auroux et al., *History of the Language Sciences*, esp. 1773ff.

43. It is presumably no coincidence that the dynamic here described finds graphic shape in one of the standard symbolic representations of the empty set: [ ].

44. Roman Jakobson, "Zero Sign," 158 / "Signe Zéro," in Jakobson, *Selected Writings*, 2:218–19.

45. Though analogies can be formed among "ratios" of tension/laxity—the /h/ : /#/ relation of modern Icelandic, for example, can be set equal to other such relations to yield the correlation t/d = s/ɯ = p/b = f/v = k/g = h/#—no underlying feature or unit unites all the relational terms.

46. Paul Valéry, *Oeuvres I*, ed. Jean Hytier (Paris: Gallimard, 1957), 623; cited in Jakobson and Waugh, *Sound Shape of Language*, 155. With thanks to Beau Madison Mount for his help with the translation.

47. It was the Polish linguist Jerzy Kuryłowicz who first proposed, in 1927, that the newly deciphered Hittite *h* could be best understood as the etymological descendant of Saussure's hypothetical phonemes. For Saussure's original argument, which expands upon and radicalizes the earlier work of the Neogrammarian linguist Karl Brugmann, see *Mémoire sur le système primitif des voyelles dans les langues indo-européennes* (Leipzig: B. G. Teubner, 1879). The terminology of the "sonic coefficient" (*coefficient sonantique*) is introduced on page 8. On the contributions of the Saussurean theory and its connection to the "laryngeal hypothesis" in contemporary historical linguistics—according to which Saussure's resonants are to be phonetically interpreted as glides characterized by the single phonetic feature of laryngeal tension—see Anna Morpurgo Davies, "Saussure and Indo-European Linguistics," in *The Cambridge Companion to Saussure*, ed. Carol Sanders (Cambridge: Cambridge University Press, 2004). The great Indo-Europeanist and one-time Saussure student Antoine Meillet famously referred to Saussure's *Mémoire* as "the most beautiful book of comparative grammar that has ever been written." See Alfred Merlin, "Notice sur la vie et les travaux de M. Antoine Meillet Membre de l'Académie," *Comptes-rendus des séances de l'année* 4 (1952): 572–83, here 580.

48. Jakobson may well have adopted the symbol "#" for his own phonetically indeterminate zeros as an oblique tribute to his predecessor's fantasy of a strictly numerical notation, in which structural relationships among Indo-European phonemes would be recorded without reference to the positive, phonetic essence of their sounds: "[O]ne could, without specifying its phonetic nature, catalogue and represent [such a] phoneme by its number in a table of the Indo-European phonemes"; *Course*, 262 / *CLG*, 496:1, translation modified. Cited by Roman Jakobson, "Typological Studies and their Contribution to Historical Comparative Linguistics" (1958), in Jakobson, *Selected Writings*, 1:523–32, here 1:529.

49. Jakobson, "Les lois phoniques du langage enfantin et leur place dans la phonologie générale" [The phonic laws of child language and their place in general phonology] (1939), in Jakobson, *Selected Writings*, 1:317–27.

50. Roman Jakobson, *Child Language, Aphasia, and Phonological Universals*, trans. Allan R. Keiler (The Hague: Mouton, 1968), 11; original German: *Kindersprache, Aphasie, und Allgemeine Lautgesetze* (1941), in Jakobson, *Selected Writings*, 1:328–401, here 1:328. I have modified the translation where necessary to preserve locutions relevant to my argument. The title chosen for the English translation is particularly unfortunate, from my perspective, since the phrase "phonological universals" utterly obscures the fact that Jakobson is explicitly proposing a new, universal kind of *sound law*. Hereafter cited as *CL*, with the English pagination followed by the original German.

51. Grimm, *Deutsche Grammatik, Erster Theil*, 2nd ed. (1822), 588 (see chapter 1, 54–55).

52. Jakobson introduces the notion of a synchronic "developedness" [*Entfaltetsein*] on the first page of his study, in connection with the diachronic development [*Entfaltung*] that precedes and follows it in the twin forms of child language and aphasia: "For the linguist, who is concerned with the

developedness [*Entfaltetsein*] of language structure, its birth and death [*Absterben*] must also provide much that is instructive"; *CL*, 13/328.

53. Jakobson takes the expression from the Polish linguist Jan Baudouin de Courtenay. See Roman Jakobson, *Six Lectures on Sound and Meaning*, trans. John Mepham (Cambridge, MA: MIT Press, 1981), 34.

54. Roman Jakobson and Morris Halle, "Phonology and Phonetics" (1955), in Jakobson, *Selected Writings*, 1:464–504, here 1:499–500. See also *CL*, 80–82/394–96.

55. Daniel Heller-Roazen takes the site of this originary, undifferentiated "all" as the starting point for his meditation on the forgetting of language. See Heller-Roazen, *Echolalias*, 9–12.

56. See *CL*, 25/338.

57. For some accounts of the reception of these ideas by later linguists, including assessments of their currency and validity, see John A. Hawkins, "Language Universals in Relation to Acquisition and Change: A Tribute to Roman Jakobson," in *New Vistas in Grammar: Invariance and Variation*, ed. Linda R. Waugh and Stephen Rudy (Amsterdam: Benjamins, 1991), 473–93; Martin Atkinson, "Jakobson's Theory of Phonological Development," in *Explanations in the Study of Child Language Development* (Cambridge: Cambridge University Press, 1982), 27–37; and Charles A. Ferguson, "New Directions in Phonological Theory: Language Acquisition and Universals Research," in *Current Issues in Linguistic Theory*, ed. Roger W. Cole (Bloomington: Indiana University Press, 1977), 256–70.

58. "One could expect from the outset that this most simple and maximal contrast is predestined [*berufen*] to open up the distinction between the vocalic and consonantal systems at the threshold of child language, and in fact this expectation is confirmed by experience"; *CL*, 69/375.

59. On Jakobson's career-long interest in the phenomenon of *audition colorée*, see Arby Ted Siraki, "Problems of a Linguistic Problem: On Roman Jakobson's Coloured Vowels," *Neophilologus* 93 (2009): 1–9.

60. Emphasis mine. The consonants Jakobson here calls "closure sounds," or *Verschlusslaute*, are "occlusives" in modern English terminology. For traditional grammarians, however, they were the *mutae*, or "mutes," since their articulation supposedly involved no temporal duration.

61. The notion that such a relationship exists is clearly as fascinating to Jakobson as it is phonologically repellant. His attitude toward those scholars, such as child psychologists Wilhelm and Clara Stern, who have covered this ground before him, is—perhaps in response to this subterranean dynamic—unusually ambivalent. See *CL*73/378.

62. The citation is taken from Clara Stern and William Stern, *Die Kindersprache: Eine Psychologische und Sprachtheoretische Untersuchung* (Leipzig: Barth, 1928), 355ff.

63. "The stage of actual language formation begins, as Wundt already correctly realized, as soon as the sound utterances 'involve the definite intention of naming'"; *CL*, 25/338.

64. "The preference for black and red at a stage of child development in which bright colors are not yet differentiated brings to mind the originary

contrast between the bilabial stop and the *a*. One may compare, in addition, the 'agnossia for colours other than red, black and white'"; *CL*, 84/399.

65. Jacob Grimm, *Altdeutsche Wälder*, 14; emphasis mine. On this passage in relation to Grimm's own analysis of the vowels, see Ulrich Wyss, *Die Wilde Philologie: Jacob Grimm und der Historismus* (Munich: C. H. Beck Verlag, 1979), 158–59; on this passage in relation to the methods of the *Altdeutsche Wälder* (and August Wilhelm Schlegel's critique thereof), see Wyss, *Die Wilde Philologie*, 220–29.

66. See Jakobson and Waugh, *Sound Shape of Language*, 159: "One must remember, however, that the relative chronology does not encompass all the constituents of the system. For instance, if the place of semi-vowels or of /h/ was not determined in the ordered rules of the Drafts, it was not for 'lack of appropriate attention', but simply because the data available had not yet permitted the assignment of a definite position in the developmental scale of acquisitions to these items; it is characteristic that even between the three children cited by Gerguson & Farwell there are divergences in this respect."

67. Jakobson, "Retrospect," in Jakobson, *Selected Writings*, 1:631–58, here 1:632. Jakobson wrote a "Retrospect" for each volume of his selected writings.

68. Jakobson, "Retrospect," in Jakobson, *Selected Writings*, 5:572.

69. For a particularly insightful version of this approach, see Boris Gasparov, "Futurism and Phonology: Futurist Roots of Jakobson's Approach to Language," *Cahiers de l'ILSL* 9 (1997): 105–24. For other examples, see Amy Mandelker, "Velimir Chlebnikov and Theories of Phonetic Symbolism in Russian Modernist Poetics," *Die Welt der Slaven* 31 (1986): 20–36; and Angelika Lauhus, "Die imaginäre Phonologie Velimir Chlebnikovs und die neuere Sprachwissenschaft," in *"Tgolí chole Mêstró": Gedenkschrift für Reinhold Olesch*, ed. Renate Lachmann, Angelika Lauhaus, Theodor Lewandowski, and Bodo Zelinsky (Cologne: Böhlau, 1990), 557–80.

70. Velimir Khlebnikov and Aleksei Kruchenykh, "The Letter as Such," in *Collected Works of Velimir Khlebnikov*, vol. 1, *Letters and Theoretical Writings*, trans. Paul Schmidt, ed. Charlotte Douglas (Cambridge, MA: Harvard University Press, 1987), 257–58; original Russian: "Bukva kak takovya," in *Sobranie proizvedenii Velimira Khlebnikova* [Collected Works], ed. N. Stepanov and Y. Tynianov (Leningrad: Izdatelstvo pisatelei v Leningrade, 1928–33), 5:248. On Kruchenykh as the historically first practitioner of pure sound poetry, see, for instance, Gerald Janecek, *Zaum: The Transrational Poetry of Russian Futurism* (San Diego, CA: San Diego State Press, 1996), 53.

71. Velimir Khlebnikov and Aleksei Kruchenykh, "Declaration of the Word as Such" (1913), in *Russian Futurism Through Its Manifestos, 1912–1928*, ed. and trans. Anna Lawton and Herbert Eagle (Ithaca, NY: Cornell University Press, 1988), 68; original Russian: "Slovo kak takovoe," in *Russkii Futurizm: Teoriia, Praktika, Kritika: Vospominaniia*, ed. V. Terekhina and A. Zimenkov (Moscow: Nasledie, 1999), 48.

72. Roman Jakobson, letter to Velimir Khlebnikov, February 1914, in Bengt Jangfeldt, *Jakobson-budetljanin: Sbornik materialov* (Stockholm:

Almquist & Wiksell International, 1992), 154. Cited and translated in Gasp-arov, "Futurism and Phonology," 113.

73. Georg von Gabelentz, "Das lautsymbolische Gefühl," in *Die Sprach-geschichte, ihre Aufgaben, Methoden und bisherigen Ergebnisse* (Leipzig: Weigel, 1891), 219; emphasis mine.

74. Velimir Khlebnikov, "The Warrior of the Kingdom," in *Collected Works*, 1:294 / *Sobranie proizvedenii*, 5:187ff.

75. This seemingly self-contradictory oscillation between archaism and utopism, between the fetishization of the origin and the anticipation of a future goal, has given rise to several scholarly treatments that attempt to dis-tinguish between a backward and a forward-looking Khlebnikov. See, for instance, Aage A. Hansen-Löve, "Velimir Chlebnikov's poetischer Kanni-balismus," *Poetica* 19 (1987): 88–133; and "Kručenych vs. Chlebnikov: Zur Typologie zweier Programme im russischen Futurismus," *Avant Garde* 5/6 (1991): 15–44; Andrew Baruch Wachtel, "Futurist Historians?" in *An Obses-sion with History: Russian Writers Confront the Past* (Stanford, CA: Stan-ford University Press, 1994), 148–76. As I have demonstrated with respect to Wagner, however, this oscillation is in fact a natural outgrowth—or rather, a radicalization—of the nineteenth-century teleological approach to language-historical development.

76. See Velimir Khlebnikov, "Artists of the World! (A Written Language for Planet Earth: A Common System of Hieroglyphs for the People of our Planet)," in *Collected Works*, 1:364–69, here 365–66 / *Sobranie proizvede-nii*, 5:216ff. For the "wavelike motion" of *b*, and for the lists of examples, see "A Checklist: The Alphabet of the Mind," ibid., 314–17/207ff. Other glos-saries can be found in "The Warrior of the Kingdom," ibid., 294–95/187ff.; and "On the Simple Names of Language," ibid., 299–303/203ff.

77. Andrej Bely, *Glossalalia: A Poem about Sound*, trans. Thomas R. Beyer, Jr. (Dornach: Pforte Verlag, 2003), 49–50. For a discussion of the influ-ence of Wundt's psychophysiological theories on European avant-garde poet-ics in general, including Bely's Symbolist production, see chapter 5, 184–86.

78. Velimir Khlebnikov, "Our Fundamentals," *Collected Works*, 1:376–91, here 384; original Russian: *Tvoreniia* [Works], ed. M. Ia. Poliakov, V. P. Grigoriev, and A. E. Parnis (Moscow: Sovetsky Pisatel, 1986), 624ff.

79. Khlebnikov, "Our Fundamentals," *Collected Works*, 1:385.

80. Ibid., 380–81. Neologisms with negative valence are, of course, also pos-sible. *Tsvety* (flowers), for instance, becomes *mvety* (fleurs du mal); ibid., 382.

81. Ibid.; emphasis mine.

82. This is a version of the more general observation that has often been used to distinguish Khlebnikov's poetic practice from Kruchenykh's: Khleb-nikov, as Jakobson remarks already in 1921, nearly always provides the reader with the tools required for following the poet's progression toward *zaum*. See Jakobson, "Modern Russian Poetry," 58–82 / *Novejšaja russkaja poèzija*, in Jakobson, *Selected Writings*, 5:299–354. See also Hansen-Löve, "Kručenych vs. Chlebnikov," 26–36.

83. Khlebnikov, "Our Fundamentals," in *Collected Works*, 1:377. For the forest, see 376. On the role of imaginary numbers in Khlebnikov's poetry,

and of mathematical tropes more generally, see Anke Niederbudde, *Mathematische Konzeptionen in der russischen Moderne: Florenskij, Chlebnikov, Charms* (Munich: Sagner, 2006).

84. Khlebnikov, "Oleg and Kazamir," *Collected Works*, 1:296–98, here 297 / *Sobranie proizvedenii*, 5:191.

85. Khlebnikov, "From the Notebooks," *Collected Works*, 1:400–10, here 403 / *Sobranie proizvedenii*, 5:265–76.

86. Khlebnikov, "Teacher and Student," *Collected Works*, 1:277–87, here 280–81 / *Tvoreniia*, 584ff. For Khlebnikov's different units of measurement—including "an infantry man's march step or a heartbeat (which are identical in time), a single vibration of the A string, and the vibration of the lowest sound in the alphabet, *u*"—of which Z can be a function, see Khlebnikov, "Our Fundamentals," 386–91.

87. Khlebnikov himself makes a modest first attempt at such a poetry in the year of Jakobson's manuscript (others will follow):

> I here offer the first experiments in *zaum* language as the language of the future (with one reservation, that vowels in what follows are incidental and serve the purposes of euphony):
> Instead of saying:
> The Hunnic and Gothic hordes, having united and gathered themselves about Attila, full of warlike enthusiasm, progressed further together, but having been met and repulsed by Aetius, the protector of Rome, they scattered into numerous bands and settled and remained peacefully on their own lands, having poured out into and filled up the emptiness of the steppes.
> Could we not say instead:
> SHa + So (Hunnic and Gothic hordes), Ve Attila, Cha Po, So Do, but Bo + Zo Aetius, KHo of Rome, So Mo Ve + Ka So. Lo SHa of the steppes + Cha.
> And that is what the first *zaum* story played upon the strings of the alphabet sounds like.

See Khlebnikov, "Artists of the World!," 368 / *Sobranie proizvedenii*, 5:216ff.

88. Roman Jakobson, "Closing Statement: Linguistics and Poetics" (1960), in Jakobson, *Selected Writings*, vol. 3, *Poetry of Grammar and Grammar of Poetry* (The Hague: Mouton, 1981), 18–51, here 29.

89. Ibid., 27.

90. Such a writing differs drastically from the writing of language science, since the linguist relates to equivalences in a manner diametrically opposed to the principle of all poetic production: the linguistic metalanguage seeks merely to (re)articulate, in the form of a clean notation, the terms of a given linguistic material; its "set," or orientation, to use Jakobson's Formalist vocabulary, is toward the language code it conceives as object, the particularities of which it wants (only) to represent. The metalinguistic message ("message," for Jakobson, being an event of actualized speech, and thus roughly equivalent to the Saussurean *parole*) thus takes shape as an equation that *names* the code in an unavoidably sequential assertion of "this" is "that." Where the poet, whose focus is on the development of the poetic line—and thus on the shape of the message itself—employs equivalence to build rhythmic sequences of the form x . . . x . . . x, the linguist uses sequence

to express equivalence, in statements of the form x=x. The linguist defines as a way of mirroring the language code; the poet *unfurls* this code in the very process of poeticizing. See Jakobson, "Linguistics and Poetics," in Jakobson, *Selected Writings*, 3:27: "It may be objected that metalanguage also makes sequential use of the equivalent units when combining synonymic expressions into an equational sentence: A = A ('Mare is the female of the horse.'). Poetry and metalanguage, however, are in diametrical opposition to each other: in metalanguage the sequence is used to build an equation, whereas in poetry the equation is used to build a sequence."

91. Ibid., 42.

92. "Thus the distribution of word stresses among the downbeats within the line, the split into strong and weak downbeats, creates a regressive undulatory curve superimposed upon the wavy alternation of downbeats and upbeats. [. . .] The Russian binary meters reveal a stratified arrangement of three undulatory curves [. . .]"; ibid., 33. Jakobson also refers favorably, if cautiously, to Sievers's categorical distinction between *Autorenleser* and *Selbstleser*; ibid., 38.

93. Yuri Tynianov, *The Problem of Verse Language*, ed. and trans. Michael Sosa and Brent Harvey (Ann Arbor: Ardis, 1981), 33; original Russian: *Problema stikhotvornogo iazyka* (Leningrad: Academia,1924; repr., Moscow: KomKniga, 2007), 9. On the question of Tynianov's own vexed relationship to the concept of teleology, see Liberman, "Roman Jakobson and His Contemporaries," 150–53. For Jakobson's perception of Tynianov's role in the (re)emergence of a teleological approach to language and literary change, see Jakobson and Pomorska, "Dialogue on Time," 15–19.

94. Tynianov, *Problem of Verse Language*, 33 / *Problema stikhotvornogo iazyka*, 9.

95. Roman Jakobson, "Modern Russian Poetry," 59 / *Novejšaja russkaja poèzija*, in Jakobson, *Selected Writings*, 5:300. Hereafter cited as *NRP*, with the English pagination, where applicable, followed by the Russian from *Selected Writings*. Since the English translation is not complete, it will not always be possible to provide an English reference; where only one page number is given, it refers to the Russian.

96. Tynianov, *Problem of Verse Language*, 33 / *Problema stikhotvornogo iazyka*, 9.

97. Jakobson cites, as his primary example of the misguided emotional-expressionist approach, the manifesto "The Founding and Manifesto of Futurism" ("Fondazione e Manifesto del Futurismo," 1909) by the Italian Futurist Filippo Tommaso Marinetti. See Marinetti, "The Founding and Manifesto of Futurism," in *Modernism: An Anthology of Sources and Documents*, ed. Vassiliki Kolocotroni, Jane Goldman, and Olga Taxidou (Chicago: University of Chicago Press, 1998), 249–53; original Italian: "Fondazione e Manifesto del Futurismo," in *Teoria e invenzione futurista*, ed. Luciano De Maria (Milan: Mondadori, 1968), 7–14. The choice of Marinetti is significant. Five years before the publication of *Newest Russian Poetry*, in 1914, Marinetti had travelled to Moscow and Saint Petersburg to proselytize for his particular brand of Futurism (and definitively assert himself as

founder of the movement). The visit was divisive in the extreme: most of the key figures of Russian Futurism refused to accept Marinetti's claims to priority; Khlebnikov distributed a hostile leaflet that mocked the event while agitating for a purely Russian, anti-European approach to poetry; and even Jakobson himself is said to have lost his temper. For a first-person account of the visit, see Benedikt Livshit, *The One and a Half-Eyed Archer*, trans. John Bowlt (Newtonvill, MA: Oriental Research Partners, 1977), 181–213; for a scholarly treatment, see Charlotte Douglas, "The New Russian Art and Italian Futurism," *Art Journal* 34, no. 3 (1975): 224–39.

98. See chapter 2, 90–97.

99. These less radical colleagues may or may not include Tynianov, whose approach to the concept of teleology is complex enough to demand an independent investigation. Tynianov's extraordinary analysis of what he calls "equivalents"—with which he has in mind, essentially, systemic structures that have been divested of any substantive relationship to a systemic unity principle—recalls the Jakobsonian notion of an emptied analogical correspondence in more ways than just its name. See Tynianov, *Problem of Verse Language*, 42–47 / *Problema stikhotvornogo iazyka*, 24–32.

100. In the absence of existing terminology for what will later become known as the phonological/phonetic distinction, Jakobson introduces the term "euphonic" to refer to inner-linguistic—and thus potentially meaningful—sound play.

101. "The assault on rhyme by poets from different schools and periods can evidently be explained to a considerable degree by the fact that rhyme constitutes an impoverishment in comparison with the 'artistic selection of vowels or consonants within the speech flow itself' [. . .]. The enrichment of rhyme could be realized in a moment when attention became more intensely fixed on the euphonic structure of the verse, in a moment when the structures that were previously present, albeit concealed (latent), finally get pulled into the bright light of consciousness"; *NRP*, 349.

102. German mathematicians, toward the end of the nineteenth century, had arrived for the first time at a rigorous definition of the *logoi alogoi*—and by extension, of the numerical realm capacious enough to contain them—through a procedure that required them to move from the inside out. They began with the "function" or dynamic of collection that generates the natural, counting numbers ($1, 2, 3, \ldots$), and generalized it to arrive at the relational dynamic that generates the domain of the rationals ($1/2 = 2/4$, $1/3 = 2/6, \ldots$). From here, they generalized once again, in a gesture that effectively *emptied* the rational analogy of its unifying principle, to obtain the domain of all reals. One of the central discoveries of modern set theory involved the counterintuitive *impossibility* of proceeding otherwise, which is to say, of beginning with the not-yet-enumerated continuum. Only the actual activity of passing, via a continual process of functional emptying, from the naturals to the rationals to the reals, can give access to the analyzable "space" of an *articulated* irrationality. It is of course no accident that the most famous and powerful account of this process, Richard Dedekind's *Continuity and Irrational Numbers* (*Stetigkeit und irrationale Zahlen*, 1872), envisions the

irrationals as contentless "cuts" or incisions—and thus also as a kind of negative writing—within the content-filled domain of rational number-things.

103. See *NRP*, 73–79/330–43. The terminology of "poetic etymology" is introduced on *NRP*, 76/334.

104. Khlebnikov, "From the Notebooks," *Collected Works*, 1:400 / *Sobranie proizvedenii*, 5:265–76.

105. Jakobson, "Linguistics and Poetics," in Jakobson, *Selected Writings*, 3:42; emphasis mine.

106. Goethe, *Faust I* and *II*, ed. and trans. Stuart Atkins, in *Goethe's Collected Works*, vol. 2 (Princeton, NJ: Princeton University Press, 1984), 305; original German: *Faust: Der Tragödie zweiter Teil* in *Werke*, partition I, vol. 15, 337. This stanza, and in particular the line Jakobson cites, operates as a kind of motto not only for Wagner himself but also for several of the most prominent Russian Symbolists, including Andrej Bely. Jakobson's subtle yet radical reframing thereof may thus also be read as a submerged anti-Symbolist gesture. For the most significant Wagnerian usage—in a context that aligns Faust's salvation, via the pull of eternal femininity, with the eventual salvation of the German *Volk* via the pull of German *Geist*—see Richard Wagner, "Beethoven," trans. William Ashton Ellis, in *Richard Wagner's Prose Works*, 5:57–126, here 5:124–26; original German: "Beethoven," in *Sämtliche Schriften und Dichtungen*, ed. Richard Sternfeld (Leipzig: Breitkopf & Härtel / C. F. W. Siegel, 1911–16), 9:61–127, here 9:125–27. On the role of Goethe for the Russian Symbolists, see Michael Wachtel, *Russian Symbolism and the Literary Tradition: Goethe, Novalis, and the Poetics of Vyacheslav Ivanov* (Madison: University of Wisconsin Press, 1994).

107. It should not be forgotten, however, that Jakobson's monograph was originally intended to appear *together* with a collection of Khlebnikov's verse, which clearly complicates the relationship between primary object and secondary interpretation as conventionally conceived. An interpretation whose appearance coincides with that of its object, both spatially and temporally, can only retroactively be cleanly severed from the material it quite literally helps to constitute.

108. A characteristically elegant version of this reading, with respect to the zero phoneme, can be found in Agamben, *Time that Remains*, 99–104/ *Il tempo che resta*, 94–98. Agamben interprets the theological backdrop of nineteenth century *Geschichtsphilosophie* less as a fundamental component of the self-understanding of Hegel's own intellectual epoch than as the radically new discovery of twentieth-century modernists and postmodernists (including Agamben himself). He therefore misses, I think, the extent to which the most sophisticated modernist rethinkings, like Jakobson's, actually pervert rather than simply "explicate" this model.

AFTERWORD

1. John Locke, *An Essay Concerning Human Understanding* (1690), ed. Peter Nidditch (Oxford: Oxford University Press, 1975), 408.

2. Despite the relative obviousness of this claim, the profound heterogeneity among various historical versions of the arbitrariness concept has seldom been emphasized by scholars seeking to trace the arc of the tradition. Hans Aarsleff's self-consciously polemical refusal, for instance, to acknowledge the wide gap separating Locke's understanding of the arbitrary from Saussure's, detracts substantially from the value of his otherwise groundbreaking study *From Locke to Saussure: Essays on the Study of Language and Intellectual History* (Minneapolis: University of Minnesota Press, 1982). A similar criticism with respect to the positions of Plato, Aristotle, Varro, Chomsky, and Jakobson could be aimed at John E. Joseph's fascinating and ambitious treatment of the problem in *Limiting the Arbitrary: Linguistic Naturalism and its Opposites in Plato's "Cratylus" and Modern Theories of Language* (Amsterdam: Benjamins, 2000).

3. The earliest exposition of Lewis's approach to the question of what it means for a community to speak a language can be found in the final chapter of his first book, *Convention: A Philosophical Study* (Cambridge, MA: Harvard University Press, 1969). This approach appears again, revised, in Lewis, "Languages and Language," in *Minnesota Studies in the Philosophy of Science*, vol. 7, ed. Keith Gunderson (Minneapolis: University of Minnesota Press, 1975), 3–35. Other relevant contributions include Lewis, "Meaning without Use: Reply to Hawthorne," *Australasian Journal of Philosophy* 70 (1992): 106–10; Lewis, "Humean Supervenience Debugged," *Mind* 103 (1994): 473–90; and Lewis, "Reduction of Mind," in *A Companion to Philosophy of Mind*, ed. Samuel Guttenplan (Oxford: Blackwell Publishers, 1994), 412–31.

4. This particular example stems from a discussion about the relationship between linguistic knowledge and linguistic convention—aimed in large part directly against Lewis's views thereof—which appears in chapter 2 of Chomsky's *Rules and Representations* (New York City: Columbia University Press, 1980), 47–88. Other versions of this antiempiricist, nativist position appear throughout Chomsky's oeuvre.

5. Chomsky is certainly not the only critic of Lewis's conventionalist approach to have made this point. For a version of the language creativity objection as framed by a philosopher rather than a linguist, see John Hawthorne, "A Note on 'Languages and Language,'" *Australasian Journal of Philosophy* 68 (1990): 116–18. Lewis responded directly to this objection in "Meaning without Use: Reply to Hawthorne," which was followed, in turn, by Hawthorne, "Meaning and Evidence: A Reply to Lewis," *Australasian Journal of Philosophy* 71 (1993): 206–11. See also Stephen Schiffer, "Two Perspectives on Knowledge of Language," *Philosophical Issues* 16 (2006): 275–87.

6. For comprehensive statements of the various early Chomskean positions, see Chomsky, *Cartesian Linguistics: A Chapter in the History of Rational Thought* (New York: Harper & Row, 1966); *Language and Mind* (New York: Harcourt Brace, 1968); and *Reflections on Language* (New York: Pantheon Books, 1975). For the first exposition of X-bar theory, see Noam Chomsky, "Remarks on Nominalization," in *Reading in English*

Transformational Grammar, ed. R. Jacobs and P. Rosenbaum (Waltham, MA: Ginn, 1970), 184–221.

7. For Chomsky's more recent views, see Chomsky, *The Minimalist Program* (Cambridge, MA: MIT Press, 1995); Chomsky, "Minimalist Inquiries: The Framework" (1998), reprinted in *Step by Step: Essays on Minimalist Syntax in Honor of Howard Lasnik*, ed. Roger Martin, David Michaels, and Juan Uriagereka (Cambridge, MA: MIT Press, 2000), 89–155; and Chomsky, "Beyond Explanatory Adequacy," in *Structures and Beyond*, ed. Adriana Belletti (Oxford: Oxford University Press, 2004), 104–31. For an early exposition of the notion of bare phrase structure, see Chomsky, "Bare Phrase Structure" (1994), reprinted in *Government and Binding Theory and the Minimalist Program*, ed. Gert Webelhuth (Oxford: Blackwell Publishing, 1995), 383–439. A useful, if somewhat superficial, overview of a wide variety of topics important to Chomsky's current understanding of language can be found in Chomsky, *The Science of Language: Interviews with James McGilvray* (Cambridge: Cambridge University Press, 2012).

8. Chomsky makes this claim several times over the course of the conversations included in *The Science of Language* (see, for instance, 14, 43, and 52).

Aarsleff, Hans. *From Locke to Saussure: Essays on the Study of Language and Intellectual History*. Minneapolis: University of Minnesota Press, 1982.

Acquisto, Joseph. *French Symbolist Poetry and the Idea of Music*. Burlington, VT: Ashgate, 2006.

———. "Uprooting the Lyric: Baudelaire in Wagner's Forests." *Nineteenth-Century French Studies* 32 (2004): 223–37.

Adelung, Johann Christian. *Grammatisch-kritisches Wörterbuch der hochdeutschen Mundart, mit beständiger Vergleichung der übrigen Mundarten, besonders aber der Oberdeutschen*. Leipzig: Johann Gottlob Immanuel Breitkopf, 1774–86.

———. *Umständliches Lehrgebäude der deutschen Sprache zur Erläuterung der deutschen Sprachlehre für Schulen*. Leipzig: Johann Gottlob Immanuel Breitkopf, 1782.

Adler, Jeremy. "The Aesthetics of Magnetism: Science, Philosophy and Poetry in the Dialogue between Goethe and Schelling." In *The Third Culture: Literature and Science*, edited by Elinor S. Shaffer, 66–102. New York: De Gruyter, 1997.

Agamben, Giorgio. *Il tempo che resta: Un commento alle Lettera ai Romani*. Turin: Bollati Boringhieri, 2000.

———. *The Time That Remains: A Commentary on the Letter to the Romans*. Translated by Patricia Dailey. Stanford: Stanford University Press, 2005.

Akamatsu, Tsutomu. "The Development of Functionalism from the Prague School to the Present." In vol. 2 of *History of the Language Sciences: An International Handbook on the Evolution of the Study of Language from the Beginnings to the Present*, edited by E. F. K. Koerner, Sylvain Auroux, Hans-Josef Niederehe, and Kees Versteegh, 1768–90. Berlin: De Gruyter, 2001.

Allen, Fredric. "Ueber den Ursprung des homerischen Versmasses." *Zeitschrift für vergleichende Sprachforschung auf dem Gebiete des Deutschen, Griechischen und Lateinischen* 24 (1879): 556–92.

Anderson, Stephen R. "Roman Jakobson and the Theory of Distinctive Features." In *Phonology in the Twentieth Century: Theories of Rules and Theories of Representations*, 116–39. Chicago: University of Chicago Press, 1985.

Andrews, Edna. *Markedness Theory: The Union of Asymmetry and Semiosis in Language*. Durham, NC: Duke University Press, 1990.

Antonsen, E. H. "Rasmus Rask and Jacob Grimm: Their Relationship in the Investigation of Germanic Vocalism." *Scandinavian Studies* 34 (1962): 183–94.

Aristotle. *Complete Works of Aristotle: The Revised Oxford Translation.* Edited by J. Barnes. 2 vols. Princeton, NJ: Princeton University Press, 1983.

Arrivé, Michel. "Saussure: Le temps et la symbolization." In *Sprachtheorie und Theorie der Sprachwissenschaft: Geschichte und Perspektiven: Festschrift für Rudolf Engler zum 60. Geburtstag*, edited by Ricarda Liver, Iwar Werlen, and Peter Wunderli, 37–47. Tübingen: Narr, 1990.

Asmuth, Christoph. "'Das Schweben ist der Quell aller Realität': Platner, Fichte, Schlegel, Novalis und die produktive Einbildungskraft." In *System and Context: Early Romantic and Early Idealistic Constellations / System und Kontext: Frühromantische und frühidealistische Konstellationen*, edited by Rolf Ahlers, 349–74. New York: New Athenaeum / Toronto: Neues Athenaeum, 2004.

———. *Transformation: Das Platonbild bei Fichte, Schelling, Hegel, Schleiermacher und Schopenhauer und das Legitimationsproblem der Philosophiegeschichte*. Göttingen: Vandenhoeck & Ruprecht, 2006.

Atkinson, Martin. "Jakobson's Theory of Phonological Development." In *Explanations in the Study of Child Language Development*, 27–37. Cambridge: Cambridge University Press, 1982.

Attridge, Derek. "Language as History / History as Language: Saussure and the Romance of Etymology." In *Post-Structuralism and the Question of History*, edited by Geoffrey Bennington Derek Attridge, and Robert Young, 183–211. Cambridge: Cambridge University Press, 1987.

Bach, Thomas. *Biologie und Philosophie bei C. F. Kielmeyer und F. W. J. Schelling*. Stuttgart: Frommann-Holzboog, 2001.

Badir, Sémir. "Ontologie et phénoménologie dans la pensée de Saussure." In *Cahier de L'Herne Ferdinand de Saussure*, edited by Simon Bouquet, 108–20. Paris: L'Herne, 2003.

Barney, Rachel. *Names and Nature in Plato's Cratylus*. New York: Routledge, 2001.

Bartlett, Rosamund. *Wagner in Russia*. Cambridge: Cambridge University Press, 1995.

Bartsch, Karl. *Der saturnische Vers und die altdeutsche Langzeile: Ein Beitrag zur vergleichenden Metrik*. Leipzig: G. Teubner, 1867.

Battistella, Edwin L. *The Logic of Markedness*. Oxford: Oxford University Press, 1996.

Baudelaire, Charles. "Richard Wagner and Tannhäuser in Paris." Translated by P. E. Charvet. In *Baudelaire: Selected Writings on Art and Artists*, 325–57. Cambridge: Cambridge University Press, 1972.

———. "Richard Wagner et Tannhäuser à Paris." In *Oeuvres Complètes*, edited by Claude Pichois, 779–815. Paris: Ballimard, 1975.

Beiser, Frederick. *German Idealism: The Struggle against Subjectivism, 1781–1801*: Cambridge, MA: Harvard University Press, 2002.

———. *The Romantic Imperative: The Concept of Early German Romanticism*. Cambridge, MA: Harvard University Press, 2003.

Bely, Andrej. *Glossalalia: A Poem about Sound*. Translated by Thomas R. Beyer. Dornach: Pforte Verlag, 2003.

Bennett, Jonathan. "Leibniz's Two Realms." In *Leibniz: Nature and Freedom*, edited by Donald Rutherford and J. A. Cover, 135–55. Oxford: Oxford University Press, 2005.

Benveniste, Émile. "La notion de 'rythme' dans son expression linguistique." In *Problèmes de linguistique générale*, 327–35. Paris: Gallimard, 1966.

———. "Lettres de Ferdinand de Saussure à Antoine Meillet." *Cahiers Ferdinand de Saussure* 21 (1964): 93–135.

Bernhardi, August Ferdinand. *Anfangsgründe der Sprachwissenschaft*. Berlin: Heinrich Frölich, 1805.

Bersier, Gabrielle. "Visualizing Carl Friedrich Kielmeyer's Organic Forces: Goethe's Morphology on the Threshhold of Evolution." *Monatshefte* 97, no. 1 (2005): 18–32.

Bierbach, Christine. *Sprache als "Fait Social": Die linguistische Theorie F. de Saussures und ihr Verhältnis zu den positivistischen Sozialwissenschaften*. Tübingen: Niemeyer, 1978.

Blumenbach, Johann Friedrich. *An Essay on Generation*. Translated by A. Crichton. London: T. Cadell, Strand, 1792.

———. *Handbuch der Naturgeschichte*. 12th ed. Göttingen: Dieterichsche Buchhandlung, 1830.

———. *Über den Bildungstrieb und das Zeugungsgeschäft*. Göttingen: Johann Christian Dieterich, 1781.

Boehme, Jakob. *Mysterium Magnum, oder Erklärung über das erste Buch Mosis, von der Offenbarung Göttlichen Worts, durch die drei Principia*

Göttliches Wesens, auch vom Ursprung der Welt und der Schöpfung.
1730. Vol. 7 of *Sämtliche Schriften,* edited by Will-Erich Peuckert. 11
vols. Stuttgart: Frommann, 1956.

Bopp, Franz. *Analytical Comparison of the Sanskrit, Greek, Latin and Teutonic Languages, Showing the Original Identity of Their Grammatical
Structure.* Amsterdam: Benjamins, 1974.

———. *A Comparative Grammar of the Sanskrit, Zend, Greek, Latin,
Lithuanian, Gothic, German, and Slavonic Languages.* Translated by
Edward B. Eastwick. 2 vols. London: Williams and Norgate, 1885.

———. *Über das Conjugationssystem der Sanskritsprache in Vergleichung
mit jenem der griechischen, lateinischen, persischen und germanischen
Sprache.* Foundations of Indo-European Comparative Philology 1, edited
by Roy Harris. London: Routledge, 1999.

———. *Vergleichende grammatik des sanskrit, send, armenischen,
griechischen, lateinischen, litauischen, altslavischen, gothischen und
deutschen.* 3 vols. Berlin: F. Dümmler, 1857–61.

———. *Vocalismus, oder sprachvergleichende Kritiken über J. Grimm's
deutsche Grammatik und Graff's althochdeutschen Sprachschatz, mit
Begründung einer neuen Theorie des Ablauts.* Berlin: Nicolaische Buchhandlung, 1836.

Bouquet, Simon. *Introduction à la lecture de Saussure.* Paris: Payot &
Rivages, 1997.

———. "La linguistique générale de Ferdinand de Saussure: Textes et retour
aux textes." *Historiographia Linguistica* 27 (2000): 265–77.

Breitbach, Olaf. *Goethes Metamorphosenlehre.* Munich: Fink, 2006.

Brugman, Karl. *Zum heutigen Stand der Sprachwissenschaft.* Strassbourg:
K. J. Trübner, 1885.

Byrne, Christopher. "Aristotle on Physical Necessity and the Limits of Teleological Explanation." *Apeiron* 35 (2002): 20–46.

Campe, Rüdiger. "The Rauschen of the Waves: On the Margins of Literature." *SubStance* 61 (1990): 21–38.

Cannon, Garland. "Jones's 'Sprung from Some Common Source': 1786–
1986." In *Sprung from Some Common Source: Investigations into the
Prehistory of Languages,* edited by Sydney M. Lamb and E. Douglas
Mitchell, 23–47. Stanford: Stanford University Press, 1991.

Cannon, John T., and Sigalia Dostrovsky. *The Evolution of Dynamics:
Vibration Theory from 1687 to 1742.* Studies in the History of Mathematics and Physical Sciences 6. New York: Springer Verlag, 1981.

Carr, Philip. *A Glossary of Phonology.* Edinburgh: Edinburgh University
Press, 2008.

Chomsky, Noam. "Bare Phrase Structure." In *Government and Binding*

Theory and the Minimalist Program, edited by Gert Webelhuth, 383–439. Oxford: Blackwell 1995.

———. "Beyond Explanatory Adequacy." In *Structures and Beyond*, edited by Adriana Belletti, 104–31. Oxford: Oxford University Press, 2004.

———. *Cartesian Linguistics: A Chapter in the History of Rational Thought*. New York: Harper & Row, 1966.

———. *Language and Mind*. New York: Harcourt Brace, 1968.

———. "Minimalist Inquiries: The Framework." In *Step by Step: Essays on Minimalist Syntax in Honor of Howard Lasnik*, edited by David Michaels Roger Martin and Juan Uriagereka, 89–155. Cambridge, MA: MIT Press, 2000.

———. *The Minimalist Program*. Cambridge, MA: MIT Press, 1995.

———. *Reflections on Language*. New York: Pantheon Books, 1975.

———. "Remarks on Nominalization." In *Reading in English Transformational Grammar*, edited by R. Jacobs and P. Rosenbaum, 184–221. Waltham, MA: Ginn, 1970.

———. *Rules and Representations*. New York: Columbia University Press, 1980.

———. *The Science of Language: Interviews with James Mcgilvray*. Cambridge: Cambridge University Press, 2012.

Christensen, Thomas Street. *Rameau and Musical Thought in the Enlightenment*. Cambridge: Cambridge University Press, 2004.

Chvany, Catherine V. "The Evolution of the Concept of Markedness from the Prague Circle to Generative Grammar." In *The Selected Essays of Catherine V. Chvany*, edited by Olga T. Yokoyama and Emily Klenin, 234–41. Columbus, OH: Slavica, 1996.

Clarke, D. E., ed. and trans. *The Havamal, with Selections from Other Poems of the Edda, Illustrating the Wisdom of the North in Heathen Times*. Cambridge: Cambridge University Press, 1923.

Collinge, N. E. "The Introduction of the Historical Principle into the Study of Languages: Grimm." In vol. 2 of *History of the Language Sciences: An International Handbook on the Evolution of the Study of Language from the Beginnings to the Present*, edited by E. F. K. Koerner Sylvain Auroux, Hans-Josef Niederehe, and Kees Versteegh, 1210–22. Berlin: De Gruyter, 2001.

Culler, Jonathan. *Saussure*. Ithaca, NY: Cornell University Press, 1976.

Delbrück, Berthold. *Einleitung in das Sprachstudium: Ein Beitrag zur Geschichte und Methodik der vergleichenden Sprachforschung*. Leipzig: Breitkopf & Härtel, 1893.

———. *Introduction to the Study of Language: A Critical Survey of the*

History and Methods of Comparative Philology of the Indo-Europeans Languages. Translated by Eva Channing. Philadelphia: Benjamins, 1974.

Deleuze, Gilles. "À quoi reconnaît-on le structuralisme?" In *Le xxe siècle, histoire de la philosophie: Idées, doctrines,* edited by Françoise Châtelet, 299–335. Paris: Hachette, 2000.

———. *Différence et repetition.* Paris: Presses Universitaires de France, 1968.

———. "How Do We Recognize Structuralism?" Translated by Michael Taormina. In *Desert Islands and Other Texts, 1953–1974,* edited by David Lapoujade, 170–92. Los Angeles: Semiotext[e], 2004.

———. *The Logic of Sense.* Translated by Mark Lester and Charles Stivale. New York: Columbia University Press, 1990.

———. *Logique du sens.* Paris: Minuit, 1969.

———. *Mille plateaux.* Paris: Éditions de Minuit, 1980.

Derrida, Jacques. *De la grammatologie.* Paris: Minuit, 1967.

———. *Of Grammatology.* Translated by Gayatri Chakravorty Spivak. Baltimore: Johns Hopkins University Press, 1997.

———. "La structure, le signe et le jeu dans le discours des sciences humaines." In *L'écriture et la difference,* 409–28. Paris: Seuil, 1967.

———. "Structure, Sign, and Play in the Discourse of the Human Sciences." Translated by Alan Bass. In *Writing and Difference,* 278–94. London: Routledge, 1978.

Diehl, Catharine. "The Empty Space in Structure: Theories of the Zero from Gauthiot to Deleuze." *Diacritics* 38, no. 3 (2008): 93–119.

Dosse, François. *Histoire du structuralisme.* 2 vols. Paris: La Découverte, 1991–92.

Doublas, Charlotte. "The New Russian Art and Italian Futurism." *Art Journal* 34, no. 3 (1975): 224–39.

Ďurovič, Ľubomír. "The Ontology of the Phoneme in Early Prague Linguistic Circle." In *Jakobson entre l'est et l'ouest 1915–1939,* edited by Françoise Gadet and Patrick Sériot, 69–76. Lausanne: Université de Lausanne, 1997.

Eggers, Michael. "Von Pflanzen und Engeln: Friedrich Schlegels Sprachdenken im Kontext der frühen Biologie." In *Die Lesbarkeit der Romantik: Material, Medium, Diskurs,* edited by Erich Kleinschmidt, 159–83. Berlin: De Gruyter, 2009.

Engler, Rudolf. "À propos de la réflexion phonologique de F. de Saussure." *Historiographia Linguistica* 30 (2003): 99–128.

———. "Die Zeichentheorie F. de Saussures und die Semantik im 20. Jahrhundert." In vol. 2 of *History of the Language Sciences: An International Handbook on the Evolution of the Study of Language from the*

Beginnings to the Present, edited by E. F. K. Koerner Sylvain Auroux, Hans-Josef Niederehe, and Kees Versteegh, 2130–52. Berlin: De Gruyter, 2001.

———. "La géographie linguistique." In *Histoire des idées linguistiques*, edited by Sylvain Auroux, 239–52. Brussels: Mardaga, 2000.

———. "Heureux qui comme Ulysse a fait un beau voyage. . . ." *Cahiers Ferdinand de Saussure* 45 (1991): 151–65.

———. "Ni par nature ni par intention." In *Recherches de linguistique: Hommages à Maurice Leroy*, edited by André Coupez and Francine Mawet Jean Bingen, 74–81. Brussels: Éditions de l'Université de Bruxelles, 1980.

———. "Sémiologies Saussuriennes." *Cahiers Ferdinand de Saussure* 29 (1974–75): 45–73.

Ettmüller, Ludwig, ed. and trans. *Die Lieder der Edda von den Nibelungen, stabreimende Verdeutschung nebst Erläuterungen*. Zürich: Orell, Füssli & Co., 1837.

Fechner, Gustav Theodor. *Elemente der Psychophysik*. 2 vols. Leipzig: Breitkopf & Härtel, 1860.

———. *Elements of Psychophysics*. Vol. 1. Translated by Helmut E. Adler. New York: Holt, Rinehart and Winston, 1966.

Fehr, Johannes. "'La vie sémiologique de la langue': Equisse d'une lecture de notes manuscrites de Saussure." *Languages* 107 (1992): 73–82.

Ferguson, Charles A. "New Directions in Phonological Theory: Language Acquisition and Universals Research." In *Current Issues in Linguistic Theory*, edited by Roger W. Cole, 256–70. Bloomington: Indiana University Press, 1977.

Foley, John Miles. *Homer's Traditional Art*. University Park: Pennsylvania State University Press, 1999.

Fontanelle, Bernard Le Bovier de. *Lettres Galantes*. Vol. 1 of *Oeuvres*. Paris: Libraires Associés, 1766.

Formigari, Lia, and Tullio de Mauro, eds. *Leibniz, Humboldt, and the Origins of Comparativism*. Amsterdam: Benjamins, 1990.

Förster, Eckart. "Die Bedeutung von §§76, 77 der Kritik der Urteilskraft für die Entwicklung der nachkantischen Philosophie [Teil 1]." *Zeitschrift für philosophische Forschung* 56 (2002): 185–86.

Foucault, Michel. *Les mots et les choses: Une archéologie des sciences humaines*. Paris: Gallimard, 1966.

———. *The Order of Things: An Archeology of the Human Sciences*. Translated by Alan Sheridan. New York: Vintage, 1973.

Frank, Manfred. *Unendliche Annäherung: Die Anfänge der philosophischen Frühromantik*. Frankfurt: Suhrkamp, 1997.

Franz, Michael. *Schellings Tübinger Platon-Studien*. Göttingen: Vanderhoeck & Ruprecht, 1996.

Gabelentz, Georg von. *Die Sprachgeschichte, Ihre Aufgaben, Methoden und bisherigen Ergebnisse*. Leipzig: Weigel, 1891.

Gandon, Francis. "Le dernier Saussure: Double articulation, anagrammes, brahmanisme." *Semiotica* 133 (2001): 69–78.

Garber, Daniel. *Descartes' Metaphysical Physics*. Chicago: University of Chicago Press, 1992.

Gasparov, Boris. *Beyond Pure Reason: Ferdinand de Saussure's Philosophy of Language and Its Early Romantic Antecedents*. New York: Columbia University Press, 2013.

———. "Futurism and Phonology: Futurist Roots of Jakobson's Approach to Language." *Cahiers de l'ILSL* 9 (1997): 105–24.

Genette, Gérard. *Mimologics*. Translated by Thäis E. Morgan. Lincoln: University of Nebraska Press, 1995.

———. *Mimologiques: Voyage en Cratylie*. Paris: Editions du Seuil, 1976.

Geoghegan, Bernard Dionysius. "From Information Theory to French Theory: Jakobson, Lévi-Strauss and the Cybernetic Apparatus." *Critical Inquiry* 38, no. 1 (2011): 96–126.

Ghil, René. *Les dates et les oeuvres: Symbolisme et poésie scientifique*. Paris: G. Crès, 1923.

———. *Traité du verbe avec Avant-Dire de Stéphane Mallarmé*. Paris: Chez Giraud, 1886.

Glass, Frank. *The Fertilizing Seed: Wagner's Concept of the Poetic Intent*. Ann Arbor, MI: UMI Research Press, 1983.

Gloyna, Tanja. *Kosmos und System: Schellings Weg in die Philosophie*. Stuttgart: Frommann-Holzboog, 2002.

Goethe, Johann Wolfgang von. *Faust I & II*. Translated by Stuart Atkins. Vol. 2 of *Goethe's Collected Works*, edited by Stuart Atkins. Princeton, NJ: Princeton University Press, 1984.

———. *Werke: Weimarer Ausgabe*. 133 vols. Weimar: Hermann Böhlau, 1887–1919.

Grant, Iain Hamilton. "F. W. J. Schelling, 'On the World Soul,' Translation and Introduction." In *Collapse*. Vol. 6, *Geo/Philosophy*, edited by R. Mackay, 58–95. Bristol: University of West of England, 2010.

———. *Philosophies of Nature after Schelling*. London: Continuum, 2006.

Green, André. "Linguistique de la parole et psychisme non conscient." In *Cahier De L'Herne Ferdinand de Saussure*, edited by Simon Bouquet, 272–84. Paris: L'Herne, 2003.

Grey, Thomas. *Wagner's Musical Prose: Texts and Contexts.* Cambridge: Cambridge University Press, 1995.

Grimm, Jacob. *Deutsche Grammatik, Erster Theil.* 1st ed. Göttingen: Dieterichsche Buchhandlung, 1819.

———. *Deutsche Grammatik, Erster Theil.* 2nd ed. Göttingen: Dieterichsche Buchhandlung, 1822.

———. *Deutsche Grammatik, Erster Theil.* 3rd ed. Göttingen: Dieterichsche Buchhandlung, 1840.

———. *Deutsche Grammatik, Vierter Theil.* 2nd ed. Göttingen: Dieterichsche Buchhandlung, 1837.

———. *Deutsche Grammatik, Zweiter Theil.* 1826. 1st ed. Berlin: Dümmlers Verlag, 1878.

———. *On the Origin of Language.* Translated by Raymond A. Wiley. Leiden: Brill Publishers, 1984.

———. *Teutonic Mythology.* Translated by James Steven Stallybrass. 4 vols. London: Goerge Bell and Sons, 1888.

Grimm, Jacob, and Wilhelm Grimm. *Deutsches Wörterbuch.* Leipzig: S. Hirzel, 1854–1960.

———. *Werke: Forschungsausgabe.* Edited by Ludwig Erich Schmitt. Hildesheim: Olms-Weidmann, 1985ff.

Grimm, Wilhelm. *Über deutsche Runen.* Berlin: Akademie Verlag, 1987.

———. *Zur Geschichte des Reims.* Berlin: Akademie der Wissenschaften, 1852.

Guichard, Léon. *La musique et les lettres en France en temps du wagnérisme.* Paris: Presses Universitaires de France, 1963.

Hacking, Ian. *Historical Ontology.* Cambridge, MA: Harvard University Press, 2002.

———. "Night Thoughts on Philology." In *Michel Foucault: Critical Assessments,* edited by Barry Smart, 266–77. London: Routledge, 1994.

Halle, Morris. "On the Origins of the Distinctive Features." In *Roman Jakobson: What He Taught Us,* edited by Morris Halle, 77–86. Columbus, OH: Slavica, 1983.

Hamann, Johann Georg. "Neue Apologie des Buchstabens H." In vol. 3 of *Sämtliche Werke,* edited by Josef Nadler, 89–108. Vienna: Herder, 1951.

Hansen-Löve, Aage A. "Kručenych vs. Chlebnikov: Zur Typologie zweier Programme im Russischen Futurismus." *Avant Garde* 5/6 (1991): 15–44.

———. "Velimir Chlebnikov's Poetischer Kannibalismus." *Poetica* 19 (1987): 88–133.

Hardenberg, Friedrich von. *Novalis' Schriften: Historische-kritische Ausgabe.*

Edited by Richard Samuel, Hans-Joachim Mähl, and Gerhard Schulz. 6 vols. Stuttgart: W. Kohlhammer, 1960–2006.

———. *Philosophical Writings*. Translated by Margaret Mahony Stoljar. Albany, NY: SUNY Press, 1997.

Harris, Roy, ed. *Foundations of Indo-European Comparative Philology, 1800–1850*. 13 vols. New York: Routledge, 1999.

Hawkins, John A. "Language Universals in Relation to Acquisition and Change: A Tribute to Roman Jakobson." In *New Vistas in Grammar: Invariance and Variation*, edited by Linda R. Waugh and Stephen Rudy, 473–93. Amsterdam: Benjamins, 1991.

Hawthorne, John. "A Note on 'Languages and Language.'" *Australasian Journal of Philosophy* 68 (1990): 116–18.

———. "Meaning and Evidence: A Reply to Lewis." *Australasian Journal of Philosophy* 71 (1993): 206–11.

Heidelberger, Michael. *Die innere Seite der Natur: Gustav Theodor Fechners wissenschaftlich-philosophische Weltauffassung*. Frankfurt am Main: Vittorio Klostermann, 1993.

———. *Nature from Within: Gustav Theodor Fechner and His Psychophysical Worldview*. Pittsburg: University of Pittsburgh Press, 2004.

Heller-Roazen, Daniel. *Echolalias: On the Forgetting of Language*. New York: Zone Books, 2005.

Helmholtz, Hermann von. *Die Lehre von den Tonempfindungen als physiologische Grundlage für die Theorie der Musik*. Braunschweig: Vieweg, 1863 (1st ed.)–1913 (6th ed.).

———. *On the Sensations of Tone as a Physiological Basis for the Theory of Music*. Translated by A. J. Ellis. New York: Longmans, Green, and Co., 1912.

———. *Populäre wissenschaftliche Vorträge*. 3 vols. Braunschweig: Vieweg, 1865.

———. *Science and Culture: Popular and Philosophical Essays*. 1885. Translated by A. J. Ellis. Edited by David Cahan. Chicago: University of Chicago Press, 1995.

———. *Selected Writings of Hermann von Helmholtz*. Translated by Russell Kahl. Middletown, CT: Wesleyan University Press, 1971.

———. *Vorträge und Reden*. Braunschweig: Vieweg, 1903.

Henrich, Dieter. "Der Weg des spekulativen Idealismus: Ein Resumé und eine Aufgabe." In *Jakob Zwillings Nachlaß: Eine Rekonstruktion*, edited by Dieter Henrich and Christoph. Hegel-Studien Beiheft 28. Bonn: Jamme, 1986.

Hess, Volker. "Das Ende der 'Historia Naturalis'?" In *Philosophie des Organischen in der Goethezeit: Studien zu Werk und Wirkung des*

Naturforschers Carl Friedrich Kielmeyer (1765–1844), edited by Kai Torsten Kanz, 153–73. Stuttgart: Franz Steiner, 1994.

Hesse, Mary. "Aristotle's Logic of Analogy." *Philosophical Quarterly* 15, no. 61 (1965): 328–40.

Horky, Phillip Sydney. *Plato and Pythagoreanism*. Oxford: Oxford University Press, 2013.

Gennaro, Rocco, and Charles Huenemann, eds. *New Essays on the Rationalists*. Oxford: Oxford University Press, 1999.

Huffman, Carl A. *Archytas of Tarentum: Pythagorean, Philosopher, and Mathematician King*. Cambridge: Cambridge University Press, 2005.

———. "The Philolaic Method: The Pythagoreanism behind the Philebus." In *Essays in Ancient Greek Philosophy*. Vol. 6, *Before Plato*, edited by A. Preus, 67–85. Albany, NY: SUNY Press, 2001.

Hull, Alexandra. *The Psychophysical Ear: Musical Experiments, Experimental Sounds*. Cambridge, MA: MIT Press, 2013.

Humboldt, Wilhelm von. "On the Comparative Study of Language and Its Relation to the Different Periods of Language Development." Translated by John Wieczorek and Ian Roe. In *Essays on Language*, edited by Theo Harden and Dan Farrelly, 1–22. New York: Peter Lang Publishers, 1997.

———. "Über das vergleichende Sprachstudium in Beziehung auf die verschiedenen Epochen der Sprachentwicklung." In vol. 4 of *Gesammelte Schriften*, edited by Königlich Preussischen Akademie der Wissenschaften, 1–34. Berlin: Behr, 1905.

Huneman, Philippe. "Reflexive Judgment and Wolffian Embryology: Kant's Shift between the First and the Third Critiques." In *Understanding Purpose: Kant and the Philosophy of Biology*, edited by Philippe Huneman, 75–100. Rochester, NY: University of Rochester Press, 2007.

Jäckle, Erwin. "Goethes Morphologie und Schellings Weltseele." *Deutsche Vierteljahrsschrift für Literaturwissenschaft und Geistesgeschichte* 15 (1937): 295–440.

Jäger, Ludwig. *Ferdinand de Saussure: Zur Einführung*. Hamburg: Junius, 2010.

———. "Neurosemiologie." *Cahiers Ferdinand de Saussure* 54 (2001): 239–337.

Jakobson, Roman. *Child Language, Aphasia, and Phonological Universals*. Translated by Allan R. Keiler. The Hague: Mouton, 1968.

———. "Die neueste russische Poesie: Erster Entwurf: Viktor Chlebnikov." In vol. 2 of *Texte der russischen Formalisten*, edited by W. D. Stempel, 18–135. Munich: Fink, 1972.

———. "Modern Russian Poetry: Velimir Khlebnikov." In *Major Soviet*

Writers: Essays in Criticism, translated and edited by E. J. Brown, 58–82. Oxford: Oxford University Press, 1973.

———. *Selected Writings*. 7 vols. The Hague: Mouton, 1962–85.

———. *Six Lectures on Sound and Meaning*. Translated by John Mepham. Cambridge, MA: MIT Press, 1981.

———. "Structure of the Russian Verb." Translated by Brent Vine and Olga T. Yokoyama. In *Russian and Slavic Grammar: Studies, 1931–1938*, edited by Linda R. Waugh and Morris Halle, 1–14. Berlin: De Gruyter, 1984.

———. *Über den tschechischen Vers: Unter besonderer Berücksichtigung des russischen Verses*. Bremen: K-Presse, 1974.

———. "Zero Sign." Translated by Linda R. Waugh. In *Russian and Slavic Grammar: Studies, 1931–1938*, edited by Linda R. Waugh and Morris Halle, 151–60. Berlin: De Gruyter, 1984.

Jakobson, Roman, and Krystyna Pomorska. "Dialogue on Time in Language and Literature." In *Verbal Art, Verbal Sign, Verbal Time*, edited by Krystyna Pomorska and Stephen Rudy, 11–24. Minneapolis: University of Minnesota Press, 1985.

Jakobson, Roman, and Linda R. Waugh. *The Sound Shape of Language*. Berlin: De Gruyter, 2002. First published 1979 by Bloomington: Indiana University Press.

Jameson, Fredric. *The Prison-House of Language: A Critical Account of Structuralist and Russian Formalism*. Princeton, NJ: Princeton University Press, 1972.

Janacek, Gerald. *Zaum: The Transrational Poetry of Russian Futurism*. San Diego: San Diego State University Press, 1996.

Jespersen, Otto. *Language: Its Nature, Development, and Origin*. New York: Henry Holt and Company, 1922.

Johnson, Monte Ransom. *Aristotle on Teleology*. Oxford: Oxford University Press, 2005.

Jones, William. "The Third Anniversary Discourse, on the Hindus, Delivered to the Asiatic Society, 2 February 1786." In *Sir William Jones: Selected Poetical and Prose Works*, edited by Michael Franklin, 355–70. Cardiff: University of Wales Press, 1995.

Jordan, Wilhelm. *Die Nibelunge*. 2 vols. Frankfurt am Main: W. Jordans Selbstverlag, 1869 and 1875.

Joseph, John E. "The Exportation of Structuralist Ideas from Linguistics to Other Fields: An Overview." In vol. 2 of *History of the Language Sciences: An International Handbook on the Evolution of the Study of Language from the Beginnings to the Present*, edited by E. F. K. Koerner Sylvain Auroux, Hans-Josef Niederehe, and Kees Versteegh, 1880–908. Berlin: De Gruyter, 2001.

———. *Limiting the Arbitrary: Linguistic Naturalism and Its Opposites in Plato's Cratylus and Modern Theories of Language*. Amsterdam: Benjamins, 2000.

Kant, Immanuel. *Critique of the Power of Judgment*. Translated by Paul Guyer and Eric Matthews. Cambridge: Cambridge University Press, 2007.

———. *Critique of Pure Reason*. Translated by Paul Guyer and Allen W. Wood. Cambridge: Cambridge University Press, 2007.

———. *Gesammelte Werke: Akademie Ausgabe*. 29 vols. Berlin: De Gruyter, 1900ff.

Keller, Otto. *Der saturnische Vers als rythmisch Erwiesen*. Leipzig: G. Freytag, 1881.

Khlebnikov, Velimir. *Collected Works of Velimir Khlebnikov*. Edited by Charlotte Douglas. 3 vols. Cambridge, MA: Harvard University Press, 1987.

———. *Sobranie proizvedenii Velimira Khlebnikova*. Edited by N. Stepanov and Iu. Tynianov. 5 vols. Leningrad: Izdatelstvo pisatelei v Leningrade, 1928–33.

———. *Tvoreniia*. Edited by V. P. Grigoriev, M. Ia. Poliakov, and A. E. Parnis. Moscow: Sovetsky Pisatel, 1986.

Khlebnikov, Velimir, and Aleksei Kruchenykh. "Bukva kak takovya." In *Sobranie proizvedenii Velimira Khlebnikova*, edited by N. Stepanov and Iu. Tynianov, 248. Leningrad: Izdatelstvo pisatelei v Leningrade, 1928–33.

———. "Declaration of the Word as Such." Translated by Anna Lawton and Herbert Eagle. In *Russian Futurism through Its Manifestos, 1912–1928*, edited by Anna Lawton and Herbert Eagle, 68. Ithaca, NY: Cornell University Press, 1988.

———. "The Letter as Such." Translated by Paul Schmidt. In *Collected Works of Velimir Khlebnikov*, edited by Charlotte Douglas, 257–58. Cambridge, MA: Harvard University Press, 1987.

———. "Slovo kak takovoe." In *Russkii Futurizm: Teoriia, Praktika, Kritika: Vospominaniia*, edited by V. Terekhina and A. Zimenkov, 48. Moscow: Nasledie, 1999.

Kielmeyer, Carl Friedrich. *Gesammelte Schriften*. Berlin: Keiper, 1938.

Kinser, Samuel. "Saussure's Anagrams: Ideological Work." *Modern Language Notes* 94 (1979): 1105–38.

Kiparsky, Paul. "The Phonological Basis of Sound Change." In *The Handbook of Phonological Theory*, edited by J. A. Goldsmith, 640–70. Oxford: Blackwell, 1995.

———. "Roman Jakobson and the Grammar of Poetry." In *A Tribute to Roman Jakobson, 1896–1982*, edited by Paul E. Gray, 227–38. Berlin: Mouton, 1983.

Koerner, E. F. K. "The Concept of 'Revolution' in Linguistics: Historical, Philosophical, and Methodological Issues." In *Linguistic Historiography: Projects and Prospects*, 85–96. Amsterdam: Benjamins, 1999.

———. *Ferdinand de Saussure: Origin and Development of His Linguistic Thought in Western Studies of Language*. Braunschweig: Vieweg, 1973.

———. *Professing Linguistic Historiography*. Philadelphia: Benjamins North America, 1995.

Kolk, Rainer. *Berlin oder Leipzig? Eine Studie zur sozialen Organisation der Germanistik im "Nibelungenstreit."* Tübingen: Niemeyer, 1990.

Kristeva, Julia. "Pour une sémiologie des paragrammes." In *Semeiotike: Recherches pour une sémanalyse*, 113–46. Paris: Seuil, 1969.

Kühnel, Jürgen B. *Untersuchungen zum germanischen Stabreimvers*. Göppingen: Kümmerle Verlag, 1978.

Lachmann, Karl. *Über althochdeutsche Prosodie und Verskunst mit Beiträgen von Jacob Grimm*. Tübingen: Niemeyer, 1990.

Lauhus, Angelika. "Die imaginäre Phonologie Velimir Chlebnikovs und die neuere Sprachwissenschaft." In *"Tgolí Chole Mêstró": Gedenkschrift für Reinhold Olesch*, edited by Angelika Lauhaus Renate Lachmann, Theodor Lewandowski, and Bodo Zelinsky, 557–80. Cologne: Böhlau, 1990.

Lehmann, Winfred. *Historical Linguistics: An Introduction*. London: Routledge, 1992.

Leibniz, Gottfried Wilhelm. "Principes de la nature et de la grâce, fondés en raison." In vol. 6 of *Die philosophischen Schriften von Gottfried Wilhelm Leibniz*, edited by C. J. Gerhardt, 598–606. Hildesheim: Olms, 1978.

———. "The Principles of Nature and Grace, Based on Reason." Translated by Leroy E. Loemker. In vol. 2 of *Philosophical Papers and Letters*, edited by Leroy E. Loemker, 1033–43. Chicago: University of Chicago Press, 2012.

Lenoir, Timothy. "Kant, Blumenbach, and Vital Materialism in German Biology." *Isis* 71 (1980): 77–108.

———. *The Strategy of Life: Teleology and Mechanics in Nineteenth-Century German Biology*. Chicago: University of Chicago Press, 1982.

Lentricchia, Frank. *After the New Criticism*. Chicago: Chicago University Press, 1980.

Leroux, Jean. "An Epistemological Assessment of the Neogrammarian Movement." In *History of Linguistics 2005: Selected Papers from the Tenth International Conference on the History of the Language Sciences*, edited by Douglas Kibbee, 262–73. Amsterdam: Benjamins, 2007.

Leunissen, Mariska. *Explanation and Teleology in Aristotle's Science of Nature*. Cambridge: Cambridge University Press, 2010.

Lévi-Strauss, Claude. "Introduction à l'œuvre de Marcel Mauss." In *Marcel Mauss: Sociologie et qnthropologie*, vii–lii. Paris: PUF, 1968.

———. *Introduction to the Work of Marcell Mauss*. Translated by Felicity Baker. London: Routledge, 1987.

Lewis, David. *Convention: A Philosophical Study*. Cambridge, MA: Harvard University Press, 1969.

———. "Humean Supervenience Debugged." *Mind* 103 (1994): 473–90.

———. "Languages and Language." In vol. 7 of *Minnesota Studies in the Philosophy of Science*, edited by Keith Gunderson, 3–35. Minneapolis: University of Minnesota Press, 1975.

———. "Meaning without Use: Reply to Hawthorne." *Australasian Journal of Philosophy* 70 (1992): 106–10.

———. "Reduction of Mind." In *A Companion to Philosophy of Mind*, edited by Samuel Guttenplan, 412–31. Oxford: Blackwell Publishers, 1994.

Liberman, Anatoly. "Roman Jakobson and His Contemporaries on Change in Language and Literature (the Teleological Criterion)." In *Language, Poetry, and Poetics: The Generation of the 1890s: Jakobson, Trubetzkoy, Majakovskij*, edited by Krystyna Pomorska, 143–56. Berlin: De Gruyter, 1987.

Liliencron, Rochus von. "Zur Runenlehre." In *Zur Runenlehre*, edited by Rochus von Liliencron and Karl Müllenhoff, 1–25. Halle: Schwetschke & Sohn, 1852.

Livshit, Benedikt. *The One and a Half-Eyed Archer*. Translated by John Bowlt. Newtonville, MA: Oriental Research Partners, 1977.

Locke, John. *An Essay concerning Human Understanding*. 1690. Oxford: Oxford University Press, 1975.

Lotringer, Sylvère. "Flagrant délire (introduction)." *Semiotext(e)* 1 (1974): 7–24.

———. "Le 'complexe' de Saussure." *Semiotext(e)* 1 (1974): 90–112.

Magee, Bryan. *The Tristan Chord: Wagner and Philosophy*. London: Penguin Press, 2000.

Magee, Elizabeth. *Richard Wagner and the Nibelungs*. Oxford: Clarendon Press, 1990.

Man, Paul de. "Hypogram and Inscription." In *The Resistance to Theory*, 27–53. Minneapolis: University of Minnesota Press, 1986.

Mandelker, Amy. "Velimir Chlebnikov and Theories of Phonetic Symbolism in Russian Modernist Poetics." *Die Welt der Slaven* 31 (1986): 20–36.

Maniglier, Patrice. *La vie énigmatique des signes: Saussure et la naissance du structuralisme*. Paris: Éditions Léo Scheer, 2006.

———. "Les choses du langage: de Saussure au structuralisme." *Figures de la psychanalyse* 12 (2005): 27–44.

———. "L'ontologie du negative: Dans la langue n'y a-t-il vraiment que des différences?" *Methodos* 7 (2007). Available at http://methodos.revues.org/674.

Marinetti, Filippo Tommaso. "Fondazione e Manifesto del Futurismo." In *Teoria e invenzione futurista*, edited by Luciano De Maria, 7–14. Milan: Mondadori, 1968.

———. "The Founding and Manifesto of Futurism." In *Modernism: An Anthology of Sources and Documents*, edited by Jane Goldman Vassiliki Kolocotroni and Olga Taxidou, 249–53. Chicago: University of Chicago Press, 1998.

Matthews, Bruce. *Schelling's Organic Form of Philosophy: Life as the Schema of Freedom*. Albany, NY: SUNY Press, 2012.

Menninghaus, Winfried. *Unendliche Verdopplung: Die frühromantische Grundlegung der Kunsttheorie im Begriff absoluter Selbstreflexion*. Frankfurt am Main: Suhrkamp, 1987.

Mensch, Jennifer. *Kant's Organicism: Epigenesis and the Development of Critical Philosophy*. Chicago: University of Chicago Press, 2013.

Merlin, Alfred. "Notice sur la vie et les travaux de M. Antoine Meillet Membre de l'Académie." *Comptes-rendus des séances de l'année* 4 (1952): 572–83.

Meyer-Kalkus, Reinhart. "Richard Wagners Theorie der Wort-Tonsprache in 'Oper Und Drama' und 'Der Ring der Nibelungen.'" *Athenäum: Jahrbuch für Romantik* 6 (1996): 153–95.

———. *Stimme und Sprechkünste im 20. Jahrhundert*. Berlin: Akademie Verlag, 2001.

Michell, Joel. *Measurement in Psychology: A Critical History of a Methodological Concept*. Cambridge: Cambridge University Press, 1999.

Milner, Jean-Claude. *Introduction à une science du langage*. Paris: Éditions du Seuil, 1973.

———. *L'amour de la langue*. Paris: Éditions du Seuil, 1978.

———. *Le périple structural: Figures et paradigmes*. Paris: Éditions du Seuil, 2002.

Milner, Jean-Claude, Ann Banfield, and Daniel Heller-Roazen. "Interview with Jean-Claude Milner." *S: Journal of the Jan Van Eyck Circle for Lacanian Ideology Critique* 3 (2010): 4–21.

Miner, Margaret. *Resonant Gaps: Between Baudelaire and Wagner*. Athens: University of Georgia Press, 1995.

Mojsisch, Burkhard, and Orrin Summerell, eds. *Platonismus im deutschen*

Idealismus: Die platonische Tradition in der klassischen deutschen Philosophie. Munich: K. G. Saur, 2003.

Morpurgo Davies, Anna. "'Organic' and 'Organism' in Franz Bopp." In *Biological Metaphor and Cladistic Classification: An Interdisciplinary Perspective*, edited by Henry M. Hoenigswald and Linda F. Weiner, 81–107. Philadelphia: University of Pennsylvania Press, 1987.

———. "Saussure and Indo-European Linguistics." In *The Cambridge Companion to Saussure*, edited by Carol Sanders, 9–29. Cambridge: Cambridge University Press, 2004.

Mowris, Peter Michael. "Nerve Languages: The Critical Response to the Physiological Psychology of Wilhelm Wundt by Dada and Surrealism." PhD diss., University of Texas at Austin, 2010.

Müller-Sievers, Helmut. *Self-Generation: Biology, Philosophy, and Literature around 1800*. Stanford: Stanford University Press, 1997.

Nassar, Dalia. *The Romantic Absolute: Being and Knowing in Early German Romantic Philosophy*. Chicago: University of Chicago Press, 2014.

Niederbudde, Anke. *Mathematische Konzeptionen in der russischen Moderne: Florenskij, Chlebnikov, Charms*. Munich: Sagner, 2006.

Noordegraaf, Jan. "Dutch Philologists and General Linguistic Theory: Anglo-Dutch Relations in the Eighteenth Century." In *Linguists and Their Diversions: A Festschrift for R. H. Robins on His 75th Birthday*, edited by Vivian A. Law and Werner Hüllen, 211–43. Münster: Nodus Publikationen, 1996.

Osthoff, Hermann, and Karl Brugman. "Preface to *Morphological Investigations in the Sphere of the Indo-European Languages*." In *A Reader in Nineteenth Century Historical Indo-European Linguistics*, translated and edited by Winfred P. Lehmann, 198–209. Bloomington: Indiana University Press, 1967.

———. "Vorwort." *Morphologische Untersuchungen auf dem Gebiete der indogermanischen Sprachen* 1 (1878): iii–xx.

Panzer, Friedrich. "Richard Wagner und Fouqué." *Jahrbuch des freien deutschen Hochstifts* (1907): 157–94.

Paul, Hermann. *Principles of the History of Language*. Translated by Herbert Augustus Strong. London: Longmans, Green and Co., 1891.

———. *Prinzipien der Sprachgeschichte*. 3rd ed. Halle: Max Niemeyer, 1898.

Plato. *Complete Works*. Edited by John M. Cooper and D. S. Hutchison. Indianapolis, IN: Hackett, 1997.

Pokorny, Julius. *Indogermanisches etymologisches Wörterbuch*. 11th ed. Berlin: A. Francke AG, 1957.

Pott, August Friedrich. *Etymologische Forschungen auf dem Gebiete der Indo-Germanischen Sprachen mit besonderem Bezug auf die*

Lautumwandlung im Sanskrit, Griechischen, Lateinischen, Littauischen und Gothischen. Foundations of Indo-European Comparative Philology 12, edited by Roy Harris. London: Routledge, 1999.

Prager, Brad. "Kant and Caspar David Friedrich's Frames." *Art History* 25, no. 1 (2002): 68–86.

Prier, Raymond Adolph. *Thauma Idesthai: The Phenomenology of Sight and Appearance in Archaic Greek.* Tallahassee: Florida State University Press, 1989.

Rask, Rasmus. "On Etymology in General." Translated by Niels Ege. In *Investigation of the Origin of the Old Norse or Icelandic Language*, 11–53. Copenhagen: The Linguistic Circle of Copenhagen, 1993.

———. *Von der Etymologie überhaupt: Eine Einleitung in die Sprachvergleichung.* Translated by Uwe Petersen. Tübingen: Gunter Narr Verlag, 1992.

Rey, Jean-Michel. "Saussure Avec Freud." *Critique* 29, no. 309 (1973): 136–67.

Richards, Robert J. "Kant and Blumenbach on the Bildungstrieb: A Historical Misunderstanding." *Studies in the History and Philosophy of Biological and Biomedical Science* 31, no. 1 (2000): 11–32.

———. *The Romantic Conception of Life: Science and Philosophy in the Age of Goethe.* Chicago: University of Chicago Press, 2002.

Rieger, Max. "Die alt- und angelsächsische Verskunst." *Zeitschrift für deutsche Philologie* 7 (1876): 1–64.

Rieger, Matthias. *Helmholtz Musicus: Die Objektivierung der Musik im 19. Jahrhundert durch Helmholtz' Lehre von den Tonempfindungen.* Darmstadt: Wissenschaftliche Buchgesellschaft, 2006.

Rimbaud, Arthur. "Vowels." Translated by Wyatt Mason. In *Rimbaud Complete*, 104. New York: Modern Library/Random House, 2002.

———. "Voyelles." In *Oeuvres Complètes*, edited by Rolland de Renéville and Jules Mouquet, 103. Paris: Gallimard, 1963.

Roe, Shirley T. *Matter, Life, and Generation: 18th Century Embryology and the Haller-Wolff Debate.* Cambridge: Cambridge University Press, 1981.

Rosenthal, Bernice Glatzer. "Wagner and Wagnerian Ideas in Russia." In *Wagnerism in European Culture and Politics*, edited by David C. Large and William Weber, 198–245. Ithaca, NY: Cornell University Press, 1984.

Rousseau, Jean. "La genèse de la grammaire comparée." In vol. 2 of *History of the Language Sciences: An International Handbook on the Evolution of the Study of Language from the Beginnings to the Present*, edited by E. F. K. Koerner Sylvain Auroux, Hans-Josef Niederehe, and Kees Versteegh, 1197–210. Berlin: De Gruyter, 2001.

Sallis, John. "Secluded Nature: The Point of Schelling's Reinscription of the Timaeus." *Pli* 8 (1999): 71–85.

Sandkaulen-Bock, Birgit. *Ausgang vom Unbedingten: Über den Anfang in der Philosophie Schellings.* Göttingen: Vandenhoeck and Ruprecht, 1990.

Saussure, Ferdinand de. *Cours de linguistique générale.* Édition critique. Edited by Rudolf Engler. 2 vols. Wiesbaden: Otto Harrassowitz, 1967 and 1974.

———. *Cours de linguistique générale: Édition critique.* Edited by Tullio de Mauro. Paris: Payot, 1972.

———. *Course in General Linguistics.* Translated by Roy Harris. Chicago: Open Court, 1983.

———. *Écrits de linguistique générale.* Edited by Simon Bouquet and Rudolf Engler. Paris: Gallimard, 2002.

———. "Légendes et récits d'europe du nord: De Sigfrid à Tristan." Selected and edited by Béatrice Turpin. In *Ferdinand de Saussure: Cahiers de L'Herne,* edited by Simon Bouquet, 351–429. Paris: Éditions de l'Herne, 2003

———. *Le leggende germaniche.* Edited by Anna Marinetti and Meli Marcello. Este: Zielo, 1986.

———. *Linguistik und Semiologie: Notizen aus dem Nachlaß: Texte, Briefe, und Dokumente.* Translated by Johannes Fehr. Frankfurt am Main: Suhrkamp, 1997.

———. *Mémoire sur le système primitif des voyelles dans les langues indo-européennes.* Leipzig: B. G. Teubner, 1879.

———. *Saussure's First Course of Lectures on General Linguistics (1907): From the Notebooks of Albert Riedlinger.* Translated by Eisuke Komatsu and George Wolf. Oxford: Pergamon Press, 1996.

———. *Saussure's Second Course of Lectures on General Linguistics (1908–9): From the Notebooks of Albert Riedlinger and Charles Patois.* Translated by Eisuke Komatsu and George Wolf. Oxford: Pergamon Press, 1997.

———. *Saussure's Third Course of Lectures on General Linguistics (1910–11): From the Notebooks of Emile Constantin.* Translated by Eisuke Komatsu and Roy Harris. Oxford Pergamon Press, 1993.

———. *Writings in General Linguistics.* Translated by Carol Sanders and Matthew Pires. Oxford: Oxford University Press, 2006.

Schelling, Friedrich. *First Outline of a System of the Philosophy of Nature.* Translated by Keith R. Peterson. Albany, NY: SUNY Press, 2004.

———. *Ideas for a Philosophy of Nature.* Translated by Errol E. Harris and Peter Heath. Cambridge: Cambridge University Press, 1988.

———. *Sämtliche Werke.* Edited by K. F. A. Schelling. 14 vols. Stuttgart: Cotta'scher Verlag, 1858.

———. *Timaeus.* Stuttgart: Frommann-Holzboog, 1994.

———. *Von der Weltseele.* 2nd ed. Hamburg: Perthes, 1806.

———. *Werke: Historisch-kritische Ausgabe.* Edited by Wilhelm G. Jacobs Hans Michael Baumgartner, Jörg Jantzen, and Hermann Krings. Stuttgart: Friedrich Frommann Verlag, 1976ff.

Schiffer, Stephen. "Two Perspectives on Knowledge of Language." *Philosophical Issues* 16 (2006): 275–87.

Schlegel, August Wilhelm. *Sämtliche Werke.* Edited by Eduard Böcking. 16 vols. Hildesheim: Olms, 1971.

Schlegel, Friedrich. *On the Language and Wisdom of the Indians.* In *A Reader in Nineteenth-Century Historical Indo-European Linguistics,* translated and edited by Winfred P. Lehmann. Bloomington: Indiana University Press, 1967.

———. *Über die Sprache und Weisheit der Indier: Ein Beitrag zur Begründung der Alterthumskunde.* Amsterdam: John Benjamins, 1977.

Schleicher, August. *Compendium der vergleichenden Grammatik der indogermanischen Sprache.* Weimar: Böhlau, 1861.

———. *A Compendium of the Comparative Grammar of the Indo-European, Sanskrit, Greek, and Latin Languages.* Translated by Herbert Bendall. Cambridge: Cambridge University Press, 2014.

———. "The Darwinian Theory and the Science of Language." Translated by Alexander V. W. Bikkers. In *Linguistics and Evolutionary Theory: Three Essays,* edited by E. F. K. Koerner, 13–69. Philadelphia: Benjamins, 1983.

———. *Die darwinische Theorie und die Sprachwissenschaft: Offenes Sendschreiben an Herrn Dr. Ernst Häckel.* Weimar: Hermann Böhlau, 1873.

———. "Fabel in indogermanischer ursprache." *Beiträge zur vergleichenden Sprachforschung auf dem Gebiete der arischen, celtischen und slawischen Sprachen* 5 (1868): 206–8.

———. "On the Significance of Language for the Natural History of Man." Translated by J. Peter Maher. In *Linguistics and Evolutionary Theory: Three Essays,* edited by E. F. K. Koerner, 75–82. Philadelphia: Benjamins, 1983.

———. *Über die Bedeutung der Sprache für die Naturgeschichte des Menschen.* Weimar: Hermann Böhlau, 1865.

Schmidt, Wolf Gerhard. "Der ungenannte Quellentext. Zur Wirkung von Friedrich de la Motte Fouqués 'Held des Nordens' auf Richard Wagners 'Ring'-Tetralogie." *Jahrbuch der Fouqué-Gesellschaft Berlin-Brandenburg* (2000): 7–42.

———. *Friedrich de la Motte Fouqués Nibelungen-Trilogie "Der Held Des Nordens": Studien zu Stoff, Struktur, und Rezeption.* St. Ingbert: Röhrig, 2000.

Schrift, Alan. "Foucault and Derrida on Nietzsche and the End(s) of Man." In *Michel Foucault: Critical Assesments*, edited by Barry Smart, 277–92. London: Routledge, 1994.

Sedley, David. *Plato's Cratylus.* Cambridge: Cambridge University Press, 2003.

Sheehan, Jonathan. "Enlightenment Details: Theology, Natural History, and the Letter H." *Representations* 61 (1998): 25–56.

Shepheard, David. "Saussure et la loi poétique." In *Présence de Saussure: Actes du colloque international de Genéve (21–23 Mars 1988)*, edited by René Amacker and Rudolf Engler, 235–46. Geneva: Droz, 1990.

Sievers, Eduard. *Altgermanische Metrik.* Halle: Niemeyer, 1893.

———. *Die Eddalieder: Klanglich untersucht und herausgegeben von Eduard Sievers.* Abhandlungen der philologisch-historischen Klasse der sächsischen Akademie der Wissenschaften. Leipzig: B. G. Teubner, 1923.

———. "Foundations of Phonetics (selections)." In *A Reader in Nineteenth-Century Historical Indo-European Linguistics*, translated and edited by Winfred P. Lehmann, 257–66. Bloomington: Indiana University Press, 1967.

———. *Grundzüge der Phonetik.* 3rd ed. Leipzig: Breitkopf & Härtel, 1885.

———. *Rhythmisch-melodische Studien.* Heidelberg: Carl Winter, 1912.

———. "Ziele und Wege der Schallanalyse." In *Stand und Aufgaben der Sprachwissenschaft: Festschrift für Wilhelm Streitberg*, 65–111. Heidelberg: Carl Winter, 1924.

Simrock, Karl, ed. and trans. *Die Edda, die Ältere und Jüngere, nebst den mythischen Erzählungen der Skalda.* Stuttgart: Cotta'scher Verlag, 1882.

———. *Die Nibelungenstrophe und ihr Ursprung: Beitrag zur deutschen Metrik.* Bonn: Eduard Weber, 1858.

Siraki, Arby Ted. "Problems of a Linguistic Problem: On Roman Jakobson's Coloured Vowels." *Neophilologus* 93 (2009): 1–9.

Smith, Justin E. H. *Divine Machines: Leibniz and the Sciences of Life.* Princeton, NJ: Princeton University Press, 2011.

Spinoza, Baruch. *Abrégé de grammaire hébraique.* Paris: Librairie Philosophique J. Vrin, 1968.

Starobinski, Jean. *Les mots sous les mots: Les anagrammes de Ferdinand de Saussure.* Paris: Éditions Gallimard, 1971.

———. *Words upon Words: The Anagrams of Ferdinand de Saussure.* Translated by Olivia Emmet. New Haven, CT: Yale University Press, 1979.

Steege, Benjamin. *Helmholtz and the Modern Listener*. Cambridge: Cambridge University Press, 2012.

Stengel, Edmund. "Romanische Verslehre." In vol. 2 of *Grundriss der romanischen Philologie*, edited by Gustav Gröber, 1–96. Strassbourg: Karl Trübner, 1902.

Stern, Clara, and William Stern. *Die Kindersprache: Eine psychologische und sprachtheoretische Untersuchung*. Leipzig: Barth, 1928.

Storch, Wolfgang, and Josef Mackert, eds. *Les Symbolistes et Richard Wagner / Die Symbolisten und Richard Wagner*. Berlin: Edition Hentrich, 1991.

Sulzer, Johann Georg. *Kurzer Begriff aller Wissenschaften und anderer Theile der Gelehrsamkeit, worin jeder nach seinem Inhalt, Nuzen, und Vollkommenheit kürzlich beschrieben wird*. 2nd ed., revised and expanded. Leipzig: Johann Christian Langenheim, 1759.

Tacitus. *Germania*. Translated by J. B. Rives. Oxford: Clarendon Press, 1999.

Tieck, Ludwig. *Minnelieder aus dem schwäbischen Zeitalter*. Hildesheim: Olms, 1966.

Trabant, Jürgen. "Signe et articulation: La solution humboldtienne d'un mystère saussurien." *Cahiers Ferdinand de Saussure* 54 (2001): 269–88.

Trop, Gabriel. "Poetry and Morphology: Goethe's 'Parabase' and the Intensification of the Morphological Gaze." *Monatshefte* 105, no. 3 (2013): 389–406.

Trubetzkoy, Nikolai. *Grundzüge der Phonologie*. Travaux du cercle linguistique de Prague 7. 7th ed. Göttingen: Vandenhoeck & Ruprecht, 1989.

———. *Principles of Phonology*. Translated by Christiane A. M. Baltaxe. Berkeley: University of California Press, 1969.

Turpin, Béatrice. "Légendes—Mythes—Histoire: La circulation des signes." In *Ferdinand de Saussure: Cahiers de L'Herne*, edited by Simon Bouquet, 307–16. Paris: Éditions de l'Herne, 2003.

Twaddell, W. Freeman. *On Defining the Phoneme*. Baltimore, MD: Waverly Press, 1935.

Tynianov, Yuri. *Problema stikhotvornogo iazyka*. Leningrad: Academia, 1924. Repr., Moscow: KomKniga, 2007.

———. *The Problem of Verse Language*. Translated by Michael Sosa and Brent Harvey. Ann Arbor, MI: Ardis, 1981.

Usener, Hermann. *Altgriechischer Versbau: Ein Versuch vergleichender Metrik*. Osnabrück: Otto Zeller, 1965.

———. *Vorträge und Aufsätze*. Leipzig: B. G. Teubner, 1907.

Vachek, Josef. "Prolegomena to the History of the Prague School of

Linguistics." Translated by Z. Kirschner. In vol. 4 of *Prague Linguistic Circle Papers*, 3–51. Philadelphia: Benjamins, 2002.

Valéry, Paul. *Oeuvres I*. Edited by Jean Hytier. Paris: Gallimard, 1957.

Veit, Walter. "Goethe's Fantasies about the Orient." *Eighteenth-Century Life* 26, no. 3 (2002): 164–80.

Verburg, Pieter A. "The Background to the Linguistic Conceptions of Bopp." *Lingua* 2 (1950): 438–68.

Vetter, Ferdinand. *Zum Muspilli und zur germanischen Alliterationspoesie: Metrisches—Kritisches—Dogmatisches*. Vienna: Carl Gerold's Sohn, 1872.

Wachtel, Andrew Baruch. "Futurist Historians?" In *An Obsession with History: Russian Writers Confront the Past*, 148–76. Stanford: Stanford University Press, 1994.

Wachtel, Michael. *Russian Symbolism and the Literary Tradition: Goethe, Novalis, and the Poetics of Vyacheslav Ivanov*. Madison: University of Wisconsin Press, 1994.

Wagner, Richard. *Judaism in Music and Other Essays*. 1894. Translated by William Ashton Ellis. Omaha: University of Nebraska Press, 1994.

———. *Richard Wagner's Prose Works*. 1893–99. Translated by William Ashton Ellis. 8 vols. Omaha: University of Nebraska, 1995.

———. *Sämtliche Schriften und Dichtungen*. Edited by Richard Sternfeld. 16 vols. Leipzig: Breitkopf & Härtel / C. F. W. Siegel, 1911–16.

———. *Wagner's "Ring of the Nibelung": A Companion*. Translated by Stewart Spencer. New York: Thames and Hudson, 1993.

Walker, Douglas. *French Sound Structure*. Calgary, AB: University of Calgary Press, 2001.

Walle, Jürgen van de. "Roman Jakobson, Cybernetics, and Information Theory." *Folia Linguistica Historica* 29, no. 1 (2008): 87–123.

Weatherby, Leif. *Transplanting the Metaphysical Organ: German Romanticism between Leibniz and Marx*. New York: Fordham University Press, 2016.

Westphal, Rudolf. *Allgemeine Metrik der indogermanischen und semitischen Völker auf Grundlage der vergleichenden Sprachwissenschaft*. Berlin: S. Calvary & Co., 1892.

———. *Philosophisch-historische Grammatik der deutschen Sprache*. Jena: Mauke, 1869.

———. "Zur vergleichenden Metrik der indogermanischen Völker." *Zeitschrift für vergleichende Sprachforschung auf dem Gebiete des Deutschen, Griechischen und Lateinischen* 9 (1860): 437–58.

Wheeler, Gerald, and William Crummett. "The Vibrating String Controversy." *American Journal of Physics* 55, no. 1 (1987): 33–37.

Wiessner, Hermann. *Der Stabreimvers in Richard Wagners "Ring Des Nibelungen."* Berlin: Matthiesen Verlag, 1924.

Wilczek, Markus. *Das Artikulierte und das Inartikulierte: Eine Archäologie strukturalistischen Denkens.* Berlin: De Gruyter, 2012.

Wilke, Tobias. "Da-Da: 'Articulatory Gestures' and the Emergence of Sound Poetry." *Modern Language Notes* 128, no. 3 (2013): 639–68.

Willer, Stefan. "Haki Kraki: Über romantische Etymologie." In *Romantische Wissenspoetik: Die Künste und die Wissenschaften um 1800*, edited by Gabrielle Brandstetter and Gerhard Neumann, 393–412. Würzburg: Königshausen & Neumann, 2004.

Wolzogen, Hans von. "Leitmotive." *Bayreuther Blätter* 20 (1897): 313–30.

———. *Poetische Lautsymbolik: Psychische Wirkungen der Sprachlaute im Stabreime aus R. Wagner's "Ring Des Nibelungen."* Leipzig: Edwin Schloemp, 1876.

Wunderli, Peter. "Ferdinand de Saussure: 1er Cahier à lire préliminairement: Ein Basistext seiner Anagrammstudien." *Zeitschrift für französische Sprache und Literatur* 82 (1972): 193–216.

———. *Ferdinand de Saussure und die Anagramme:* Tübingen: Niemeyer, 1972.

Wundt, Wilhelm. *Grundzüge der physiologischen Psychologie:* Leipzig: Wilhelm Engelmann, 1874 (1st ed.)–1911 (6th ed.).

———. *Principles of Physiological Psychology.* Vol. 1. Translated by Edward Bradford Titchner. London: Swan Sonnenschein and Co., 1904.

———. *Probleme der Völkerpsychologie.* Leipzig: Wiegandt, 1911.

———. *Völkerpsychologie: Eine Untersuchung der Entwicklungsgesetze von Sprache, Mythus, und Sitte.* 2nd ed. 3 vols. Leipzig: Wilhelm Engelmann, 1904.

Wyss, Ulrich. *Die wilde Philologie: Jacob Grimm und der Historismus.* Munich: C. H. Beck Verlag, 1979.

Printed and bound by CPI Group (UK) Ltd, Croydon, CR0 4YY

13/04/2025

14656497-0001